AF553870

Annals of the Israel Physical Society Volume 2.

STATISTICAL PHYSICS –

"STATPHYS 13"

Proceedings of the 13th IUPAP Conference

on Statistical Physics, held at the Technion –

Israel Institute of Technology, August 24 – 30, 1977

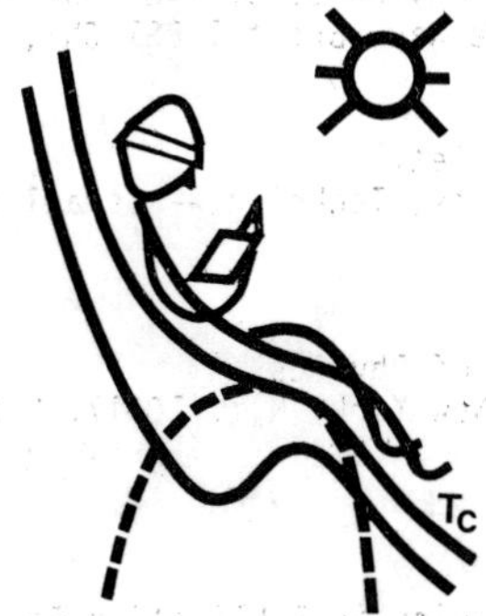

Edited on behalf of the Israel Physical Society by

Dario Cabib, Charles G. Kuper and Ilan Riess,

Technion – Israel Institute of Technology, Haifa

Executive Editor : Chanoch Weil

Part 1: Invited Lectures

Published by
Adam Hilger, Bristol
and
The Israel Physical Society,
in association with
The American Institute of Physics, New York

Published Jointly by

Adam Hilger Ltd (a company owned by The Institute of Physics)
Techno House, Redcliffe Way, Bristol BS1 6NX, England

and

The Israel Physical Society
c/o Department of Physics, Technion -- Israel Institute of Technology
Haifa, Israel

in association with

The American Institute of Physics
335 East 45th Street, New York, N.Y. 10017, U.S.A.

British Library Cataloguing in Publication Data

IUPAP Conference on Statistical Physics, 13th,
Israel Institute of Technology, 1977
Statistical Physics, 'Statphys 13'
—(Israel Physical Society. Annals; Vol.2; 0309-8710).
1. Statistical Mechanics — Congresses
I. Title II. Cabib, Dario III. Kuper, Charles Goethe
IV. Riess, Ilan V. Weil, Chanoch VI. Series
530.1'3 QC 174.7
ISBN 0-85274-356-4
ISSN 0309-8710
CODEN: AIPSDK

Printed by AYALON OFFSET LTD, ISRAEL

C O N T E N T S

PART 2

FOREWORD

The Israel Physical Society provides a forum for interaction between Israeli physicists. Besides organizing an annual two-day general conference, it co-sponsors many international conferences held in Israel. It publishes the "Annals" from time to time, as a vehicle for rapid publication of the Proceedings of international conferences.

We hope that this series will succeed in rationalizing the publication of conference proceedings; the former practice of publishing them ad hoc made it difficult for libraries and researchers to keep a complete and easily accessible record.

On behalf of the Society, we wish the series every possible success.

C.G. Kuper, President — I. P. S.
J. Felsteiner, Secretary

PREFACE

Amongst the recurrent Category A international conferences sponsored by the IUPAP, the Statistical Physics Series has an honourable place. And we, at the Technion, in Haifa, were deeply honoured that the IUPAP Commission for Thermo-dynamics and Statistical Physics saw fit to entrust us with the 13th Conference in this series. The conference was held from 24th to 30th August 1977.

In these Proceedings we present the complete scientific record; the invited lectures form the content of Part 1 and the contributed papers constitute Part 2. This is a departure from tradition – at both the 11th and 12th Conferences, only the invited lectures were included in the Proceedings. We believe – and hope – that the extra cost of publishing the complete proceedings is justified by their increased utility.

In addition to the financial support from the formal sponsors, the Conference was helped by some friends of the Technion, who are personally acknowledged at the end of Part 2. We wish to thank these contributors warmly for their generosity. We owe a special vote of thanks to Mr. Alexander Goldberg, former President of the Technion, for his enthusiastic (and highly successful) efforts in enlisting their help.

We wish to record our gratitude to the Local Organizing Committee, to the International Scientific Program Committee, and to the panel of referees for their work and enthusiasm, and our particular thanks to Mrs. Zippora Saphir, the Organizing Secretary and to Ms. Miriam Bishop, the Editorial Secretary.

The Editorial Committee

Lars Onsager — An Appreciation

by

Cyril Domb
Kings College, London, England
Bar-Ilan University, Israel

The first International Statistical Physics Conference under the auspices of IUPAP was held at Florence in 1949. Onsager was present, as he was at many of the subsequent Statphys Conferences. He rarely gave formal papers - his contribution was usually confined to discussion remarks. At the Florence Conference he made two such remarks. After one lecture he went to the blackboard and stated that Bruria Kaufman and he had calculated the long range order of the two-dimensional Ising model; it was $(1-K^2)^{1/8}$ where K is the modulus of the elliptic integral in the expression for the free energy. He gave no indication of how this extremely difficult calculation had been performed. A second remark dealt with the properties of liquid helium. No one understood it properly, but when Feynman's theory of liquid helium appeared a few years later, it became clear that this was what Onsager had in mind.

No one would challenge that Onsager was the most distinguished participant at these Conferences. When he was awarded a Nobel Prize in 1968 we could legitimately regard him as our Nobel Prize winner. After all, the award was for establishing a completely new branch of thermodynamics, that of irreversible processes, and his arguments were rooted in statistical physics. Of comparable distinction was his work on the two-dimensional Ising model, which inaugurated a new era in the theory of phase transitions and enabled many of us to start up in research.

Any Conference session in which Onsager was present was a little more exciting. He was rather like a quiescent volcano - for long periods he would remain silent and say nothing, but occasionally he would erupt, and whenever he erupted something of interest and significance would emerge.

By present-day standards Onsager's publications were neither lengthy nor numerous. His Nobel Prize was awarded for his discovering of "Reciprocal relations in irreversible processes" and the citation noted that the publication (two papers in Physical Review containing 22 and 15 pages respectively) was one of the smallest ever to be awarded a Nobel Prize. Onsager's total publication list runs to 60-odd papers, a number which could surely be matched by many of us here. But the quality of his publications is superb, and he did not put his name to anything trivial or unfinished. Every Onsager publication was an event. Like his distinguished predecessor Willard Gibbs whose chair he occupied for 28 years at Yale, there were no loose ends in his work. Anyone who has read Gibbs's papers will confirm that every different way of looking at a problem is explored. A similar completeness is to be found in Onsager's papers.

It is fair to describe Onsager as a 20th century universalist. Besides his work on irreversible thermodynamics and the Ising model there are at least four fields in which his contributions ensure him a place of distinction, the theory of electrolytes, liquid helium, metals, and ice - in fact many of his ideas are already incorporated in standard texts. Also he spent a lot of time and thought on problems of turbulence in fluids - but his publications are confined to Conference discussions. It was the general view of those near to him that the unpublished contents of his drawers would have sufficed to establish the reputation of quite a few scientists with less exacting standards than his own.

Onsager always described himself as a chemist, but it is clear that most of the problems he tackled were also in the realm of physics, and anyone seeing how he manipulated operators, Lie groups, elliptic functions and elliptic modular functions could be excused for thinking that he was trained as a mathematician. (He once confided to Elliott Montroll that as a student he worked through all the problems in Whittaker and Watson's "Modern Analysis".)

His knowledge of different aspects of science was immense. I remember travelling with Onsager and Sir Francis Simon from Oxford to Cambridge in 1951 when he was spending a sabbatical year at Cambridge. The discussion turned to minerals and rocks, and then Onsager made a comment about fossils. We both looked a little blank, and Onsager said "But its in all the standard books on palaentology." A scientist is expected to be familiar with everything in standard texts!

Many people were surprised at his wide "earthy" knowledge in science. During the above visit to Oxford Simon arranged for him to be shown around the experimental laboratories. The experimentalists described their apparatus to him, mentioning incidentally the details of how stray radiation was eliminated, good thermal contact maintained, etc. They did not expect a theorist to comment on these points. But they were quite wrong. There was scarcely any feature on which Onsager did not have something pertinent to say.

Even his famous "Principle of microscopic reversibility" had its origin in an earthy problem in organic chemistry. As Onsager himself reported in his Nobel Prize address "An unusual problem in chemical kinetics attracted my attention at the same time. C.N. Ruber was studying the muta-rotation of various sugars by several precise methods; optical rotation, refractive index (interferometer) and volume changes (dilatometer). He discovered that there were (at least) three modifications of galactose, and the possibility that any one of these might transform into either of the others gave rise to a little problem in mathematics. In analysing it I assumed, as any sensible chemist would, that in the state of equilibrium the reaction 1→2 would occur just as often as 2→1 etc., even though this is not a necessary condition for equilibrium which might be maintained by a cyclic process. A new hypothesis of "microscopic reversibility" was needed. But once this had been postulated, the "principle of least dissipation" introduced by Helmholtz in his theory of galvanic diffusion cells and by Kelvin in his theory of thermoelectricity could be deduced; also the Kelvin reciprocal relations in thermoelectricity and conduction. His classic papers tied all these things together and established the new field of irreversible thermodynamics. It is relevant whenever several irreversible processes take place simultaneously; the mundane example given by Prof. Claesson of the Royal Swedish Academy of Sciences was a cold lump of sugar dissolving in a hot cup of tea, which involves conduction of heat and diffusion.

Onsager was a very modest man; in fact it was because of his great humility that he was a poor public lecturer. He always assumed that the audience had the same capacity as himself for understanding a subject. During the same visit to Cambridge in 1951 he was asked to talk to the Kapitza Club on his recent work with Bruria Kaufman on the Ising model. The Kapitza Club (named after its founder) was a forum for meeting of experimentalists and theorists at which important developments in theory or experiment were presented. Onsager was warned that experimentalists would represent a substantial fraction of the audience and he should therefore be careful not to use sophisticated theoretical concepts. Nevertheless he plunged immediately into the mathematics of spinors. After a short while the thought occurred to him that some of the experimentalists might not be familiar with spinors. He turned to the audience and asked with his characteristic smile (almost a grin) "Is there anyone here who does not know what a spinor is?" One or two brave spirits put up their hands. Onsager said "Right, I'll tell you - spinors are matrices isomorphic with the rotation group in three dimensions" and went on assuming that this description was enough for any experimentalist!

In spontaneous discussion he came over better and was often entertaining as well as informative. After a long and sophisticated lecture on a new topic Onsager would get up and say "I think Dr. X has made a good start on this problem, and if he takes his approximation to the second stage he should find the following ... But the real difficulties will arise at the third stage when terms of a new kind come in". This is what I meant in my opening remarks when I talked of a "quiescent volcano". He also had an endearing habit of treating a very distinguished audience like a class of school children. At a Royal Society discussion on some new chemical development he would go to the blackboard, and turn to ask with his characteristic smile "Does anyone here know the best substance to use to illustrate A, B, C? Well it is sulphur, because ...".

In private discussion it was much easier to make contact with Onsager provided you were courageous enough to persist in questioning when you did not understand. He would then drop the level one stage at a time until the gap could be bridged. And he had many collaborators and students whom he started on their research career all of whom testify to his generosity and the inspiration of working with him on scientific problems.

In my opening remarks I mentioned the calculation of the

long-range order for two dimensional Ising model no details of which were ever published. The same result was derived by Yang and published in 1951. It was only 26 years later in the Battelle Conference in Gstaad 1969 that we learned how he had gotten the answer and why he had not published the details. In computing the long range order he had been led to a general consideration of Toeplitz matrices, but he did not know how "to fill out the holes in the mathematics, the epsilons and deltas" and by the time he had achieved this to his satisfaction he found that "the mathematicians got there first".

Onsager was given honorary doctorates by many Universities and received awards and honours from National Academies. But he always remained simple, unspoilt and approachable. It is nice to be able to report that he remained healthy, vigorous and clear thinking right up to his death.

I should like to end with a talmudic quotation which I hope will be appropriate. The Talmud states that

"תלמידי חכמים אין להם מנוחה לא בעולם הזה ולא בעולם הבא"

("Great scholars have no rest within this world or the next world.") This may seem a little harsh on first hearing but perhaps the meaning is as follows: "They cannot rest in this world because they are continuously occupied with creative activity and thought; but their creativity does not cease when they die and pass into the next world since other people continue to study their writings and develop their ideas. In this sense we can feel assured that Onsager will have little rest during this Conference."

CHAPTER 1

THE BOLTZMANN AWARD LECTURE
NON-LINEAR RELAXATION AND FLUCTUATION

Ryogo Kubo

University of Tokyo, Tokyo, Japan

THE BOLTZMANN AWARD: RYOGO KUBO

Citation read by H.B. Callen,
Chairman, IUPAP Commission for Thermodynamics and Statistical Physics

On behalf of the IUPAP Commission for Thermodynamics and Statistical Mechanics, I hereby award the second Boltzmann Medal to Professor Ryogo Kubo, of the University of Tokyo. The award consists of a gold medal, and is given every three years in recognition of outstanding contributions to statistical physics.

Professor Kubo is honoured for his contributions to the theory of non-equilibrium ststistical mechanics, and to the theory of fluctuation phenomena. He is cited particularly for his work in the establishment of the basic relations between transport coefficients and equilibrium time correlation functions: relations with which his name is generally associated.

Professor Kubo obtained both his Batchelor's and Doctor's degrees at the University of Tokyo, serves on its faculty of physics, and has been Dean of its Faculty of Science. He has been active in IUPAP as Chairman of the Commission on Magnetism (1963-66), Secretary of the Commission on Thermodynamics and Statistical Physics (1963-66), and as a Vice-President of IUPAP (1972-). He has been President of the Physical Society of Japan, and is an honorary Foreign Member of the American Academy of Arts and Sciences and of the National Academy of Sciences. He has been the recipient of numerous awards and prizes, including the Order of Culture (Japan), the Nishina Memorial Award, the Matsunaga Prize, the Imperial Prize of the Japanese Academy, and the Fujiwara Prize.

Haifa, August 24, 1977.

ABSTRACT

This reviews a recent work of our group on fluctuation of macrovariables governed by non-linear laws of relaxation. Under a certain condition the stochastic distribution function of a macrosystem is shown to have an extensive property, which breaks only when the system passes through a critical state where bifurcation may occur. Relaxation from an unstable equilibrium is accompanied by anomalous enhancement of fluctuation. Path integral formalism is also discussed.

1. INTRODUCTION

When we make observation of a system, we may take one of the three basic ways, namely observation of response, relaxation, or fluctuation. For a system near equilibrium, a well established theorem tells us that these three kinds of observation are internally related. This is manifested in the socalled fluctuation-dissipation theorem. The history of the theorem goes back to Einstein who showed that the mobility and the diffusion constant of a Brownian particle are related with each other. The Nyquist theorem was a great generalization of the same idea to general Brownian motion. It states that the power of thermal fluctuation of a macrovariable is determined by the corresponding impedance. The Onsager principle, which Onsager postulated for his proof of the reciprocity of kinetic coefficients, states that average regression of fluctuation of a thermal quantity is governed by the corresponding law of macroscopic relaxation. All of these are now unified in the fluctuation-dissipation theorem, which is the central theme of the linear response theory and says that a response function is simply related to a correlation function of relevant physical variables.

We ask ourselves how and to what extent the same sort of general relationship could be established for systems not necessarily near equilibrium. The system may be in a steady state with a flow or in a non-steady state subject to a variable external condition. Still we may exert disturbance on such a system, or change the external condition and observe how the system responds to it. Or we may observe fluctuation. There we expect some general relationship to exist between different sorts of observation. The question is too broad to be answered in a unique way, but something may be said with a considerable degree of generality.

We consider a large system consisting of a large number of constituent elements. The states of the system are characterized by a set of macrovariables each of which is extensive in nature. These macrovariables show some fluctuation around their average values. The question is the relationship existing between the evolution laws of the average motion and fluctuation. There has been a fairly large amount of work on this subject. Among others here we quote a classic paper of van Kampen[1] to which we owe much of our work, although there is some difference of the viewpoints between his work and ours. This paper is a rather brief review of the work developed in our group in the last few years.

2. EXTENSIVITY OF DISTRIBUTION FUNCTIONS IN NON-EQUILIBRIUM STATES

A set of extensive macrovariables will be denoted by X and the corresponding densities by x, namely

$$x = X/\Omega = \varepsilon X \qquad \varepsilon = \Omega^{-1} \tag{1}$$

Ω being the size of the system. In a given macrostate of the system, the variables may realize various values and so are considered as stochastic. The distribution function at time t is denoted by $P(X,t)$ or by $P(x,t)$. If there exists the thermodynamic limit

$$\Omega^{-1}\log P(x,t) = \phi_\varepsilon(x,t), \qquad \lim \phi_\varepsilon(x,t) = \phi_0(x,t) \tag{2}$$

then the process of X(t) is said to possess the extensive property and we may write the asymptotic form

$$P(x,t) \sim C \exp \Omega\phi_0(x,t) \tag{3}$$

for a sufficiently large size of the system. This is equivalent to the extensivity of the generating function of P, namely

$$Q(\lambda, t) = \int e^{\lambda X} P(X,t)dX = \exp \Omega\psi_\varepsilon(\lambda,t), \quad \lim \psi_\varepsilon(\lambda,t) = \psi_0(\lambda,t) \tag{4}$$

This was conjectured by Kubo[2] as an Ansatz, an extension of the central limit theorem to a non-equilibrium macrosystem. On the basis of this Ansatz, Kubo reformulated the $\varepsilon^{\frac{1}{2}}$ expansion method of van Kampen[1]. The extensivity of P(X,t) stems from certain extensivity of the macrosystem and the macrovariables in question. Thus it can be proved under some appropriate conditions as summarized as follows.

1) A Markoffian process with extensive transition probabilities. If the macrovariables X are considered to make a Markoffian process, the Chapman-Kolmogorov equation (master equation) may be written as

$$\frac{\partial P(X,t)}{\partial t} = -\int W(X\to X+\Delta)\, P(X,t)\, d\Delta + \int W(X-\Delta\to X)\, P(X-\Delta,t)\, d\Delta \tag{5}$$

The transition probability W is said to be extensive if it holds that

$$W(X\to X+\Delta,\, t) = \Omega\, w(x,\Delta,t) \tag{6}$$

Defining the Hamiltonian by

$$H(p,x,t) = \int d\Delta\,(1 - e^{-p\Delta})\, w(x,\Delta,t) = \sum_{n=1}^{\infty} \frac{(-)^{n-1}}{n!}\, p^n\, c_n(x,t) \tag{7}$$

where

$$c_n(x,t) = \int d\Delta\cdot\Delta^n w(x,\Delta,t), \tag{8}$$

we write Eq.(5) as

$$\varepsilon\frac{\partial P(x,t)}{\partial t} = -H(\varepsilon\frac{\partial}{\partial x},\, x,\, t)\, P(x,t) \tag{9}$$

which resembles a Schroedinger equation. The above expressions assume a single variable, but generalization to multi-variable cases is straightforward. Kubo et al[3] showed that an expansion of ψ_ε , Eq.(4), in powers of ε gives a consistent solution of Eq.(9). In other words, if P(x,t) is extensive at an initial time, then it remains so at later times, provided that the convergence of the expansion is guaranteed all the way through. Thus the extensivity propagates. If the system passes through a critical condition on its way, the convergence may break and bifurcation may occur. A more rigorous proof of existence of the limits, (2) and (4), has been given by Suzuki[4]. The limiting function, $\phi_0(x,t)$ or $\psi_0(\lambda,t)$, determines the asymptotic behavior of the macrovariables. They satisfy the equations,

$$\frac{\partial \phi_o}{\partial t} + H(\frac{\partial \phi_o}{\partial x}, x,t) = 0, \qquad \frac{\partial \psi_o}{\partial t} + H(-\lambda, \frac{\partial \psi_o}{\partial \lambda}, t) = 0 \tag{10}$$

2) Dynamical systems with extensive properties. Consider a large dynamical system occupying a large domain Ω in space. The dynamical Hamiltonian and the macrovariable under observation are assumed to have the form,

$$\mathcal{H}(t) = \int dr \ h(r,t), \qquad X = \int dr \ x(r) \tag{11}$$

$\mathcal{H}$ and X are extensive in the sense that they are expressed as integrals of local densities. This will be true if the system consists of particles interacting with sufficiently short-ranged forces. The statistical operators $\rho(t)$ of the dynamical system is governed by a quantal or a classical Liouville equation with the Hamiltonian (11). The generating function Q, Eq.(4), is defined by

$$Q(\lambda , t) = \mathrm{Tr}\ \rho(t)\ e^{\lambda X} \tag{12}$$

and the distribution function P(X,t) by an inverse transformation of Q. Suzuki[5] gave a proof of the extensivity of Q and P, namely the existence of the thermodynamic limits by assuming an initial condition that log $\rho(0)$ is extensive in the sense of Eq.(11) and assuming that all of these local density operators satisfy appropriate conditions of boundedness. Suzuki's proof also applies to kinetic models, in which the dynamics is represented by stochastic evolution like in a Glauber model.

3. DETERMINISTIC MOTION AND FLUCTUATION

In the limit of a large system, the most probable value of x maximizes P(x,t),

$$\phi_0(x,t) = \text{maximum for } x = y(t) \tag{13}$$

and describes the deterministic motion. If ϕ_0 is analytic around $x = y(t)$, it can be expanded in terms of $\Delta x(t) = x - y(t)$. This means that the fluctuation around the deterministic motion is essentially gaussian or normal and is characterized by the variance,

$$\langle\Delta x_j(t)\Delta x_k(t)\rangle = \varepsilon\sigma_{jk}(t) \quad \text{with} \quad \sigma^{-1}_{jk}(t) = \left.\frac{-\partial^2\phi_0}{\partial x_j \partial x_k}\right]_{x=y(t)} \tag{14}$$

If we know the function $\phi_0(x,t)$, or the law of its evolution, we should be able to answer the question raised in INTRODUCTION. Generally it is not easy to derive such a law of evolution from a very microscopic basis, but it may be done. If there exists a distinct separation of time scales of macrovariables and microvariables, we can choose a relatively few number of macrovariables and regard them stochastic and Markoffian. In such cases we may use Eq.(10) to derive evolution equations for the cumulants of macrovariables. In particular we get

$$\dot{y}(t) = c_1(y), \qquad \dot{\sigma}(t) = 2c_1'(y)\sigma + c_2(y) \tag{15}$$

for a single variable case, and

$$\dot{y}_j(t) = c_{1j}(y), \qquad \sigma_{jk}(t) = \sum_{\ell}\left\{\frac{\partial c_{1j}}{\partial y_\ell}\sigma_{\ell k} + \sigma_{j\ell}\frac{c_{1k}}{\partial y_\ell}\right\} + c_{2jk}(y) \tag{16}$$

for a multivariable case. Here c_{1j} and c_{2jk} are the first and the second moments of the transition probability defined by Eq.(8) generalized to a multivariable case.

Equation (15) and (16) were obtained first by van Kampen and were rederived by Kubo from the extensivity Ansatz[2,3]. The correlation of fluctuation is given by

$$\langle\Delta x_j(\tau)\ \Delta x_k(t)\rangle = \sum_{\ell}\sigma_{j\ell}(\tau)\frac{\partial y_k(t)}{\partial y_\ell(\tau)} \tag{17}$$

Thus fluctuations are governed by the deterministic motion.

When the system is in a stationary environment, the transition probability is independent of t and so are c_1 and c_2. Then the equilibrium points are determined by $c_1(y_e) = 0$. If there is a unique stable equilibrium, the system relaxes toward it if initially it is off equilibrium. The ideal Brownian motion is the case where

$$c_1 = -\gamma y \qquad \text{and } c_2 = \text{const.}, \tag{18}$$

for which we have

$$y(t) = y_0 e^{-\gamma t}, \quad \sigma(t) = \sigma_0 + (\sigma_e - \sigma_0)e^{-2\gamma t}, \quad \sigma_e = 2\gamma/c_2 \tag{19}$$

When the relaxation is non-linear, the non-linearity generally causes some unusual behavior of fluctuation. Figure 1 illustrates such an example, for which we assume

$$c_1 = 2 - y - y^3, \quad c_2 = \text{const.} = 1 \tag{20}$$

Starting from an initial state $y(0) = -1.6$ and $\sigma(0)$ as indicated on each curve, the variance shows rather complicated change as the system approaches the equilibirum, $y_e = 1$ with $\sigma_e = 0.125$.

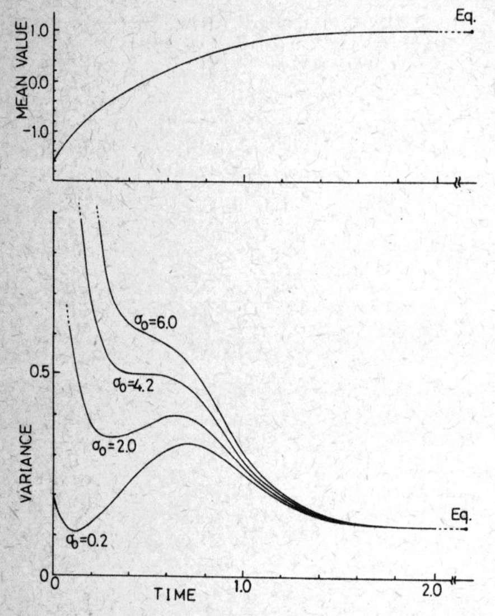

Fig. 1. Fluctuation accompanying a non-linear relaxation.

Fig. 2. Anomalous enhancement of fluctuation in transition from an unstable to a stable point.

The effect of non-linearity is much more emphasized in anomalous enhancement of fluctuation in a transient process from an unstable to a stable equilibrium[6]. Near an unstable point, c'_1 is positive so that σ increases exponentially. As the system relaxes to the final stable point, fluctuation must attain the equilibrium values. In between, fluctuation grows anomalously large. This enhanced fluctuation is a general phenomenon, and had been observed in lasers[6]. Calculations have been made for superradiance[7] and for a kinetic Ising model[8]. Figure 2 shows a recent observation in transient process of excitation of oscillation in a Wien bridge oscillater[9]. The average amplitude of oscillation follows the equation,

$$\dot{y}(t) = ay - by^3, \quad a, b > 0.$$

As the amplitude grows to attain the final equilibrium value, the variance of fluctuation experiences a very large peak. The observation fits nicely to calculations using Eq.(15) with a constant c_2 and assuming a small but finite initial value of y.

If the system is initially at an exact point of unstable equilibrium, Eq.(15) loses its power to describe the transient process. This means that the treatment based on the extensivity fails under the circumstances. In fact there may occur bifurcation of the distribution function, for instance into two peaks if two stable points are available. This problem has been studied in great details by Suzuki[10]. The whole process can then be divided into three regimes. In the initial regime, as one sees from the second equation of (15), the variance grows linearly by diffusion and then exponentially as $e^{2\gamma t}$, where $\gamma = c'_1(0)$. For large t's a scaled time $\tau = e^{2\gamma t}$ is convenient. In this scaled regime, evolution of the distribution function is dominated by drift, namely

$$\frac{\partial P(x,t)}{\partial t} = -\frac{\partial}{\partial x} c_1(x)P(x,t) \tag{21}$$

and is broadened over a finite domain of x. Then P(x,t) starts to build up to attain the final distribution around the stable points. Suzuki devised a connection scheme of the three regimes of evolution which has been applied to the problem of lasers and super-radiance.

4. PATH INTEGRAL

Equation (9) is transformed into an equation of the form,

$$\frac{\partial \psi_\varepsilon(\lambda, t)}{\partial t} = F[\psi_\varepsilon(\lambda, t)]$$

for ψ_ε defined by Eq. (4). For a short time interval t, we write the increment of ψ_ε as

$$\Delta\psi_\varepsilon = \psi_\varepsilon(\lambda, t + \Delta t) - \psi_\varepsilon(\lambda, t) = \Delta t\, F[\psi_\varepsilon(\lambda, t)]$$

Taking $\psi_\varepsilon(\lambda,t) = \lambda x'$, we find easily

$$\Delta\psi_\varepsilon = -\Delta t\, H(-\lambda,x',t)$$

The transition probability of the Markoffian process over the time interval Δt is obtained by the inverse Laplace transform,

$$P(x',t \to x,t+\Delta t) = \frac{1}{2\pi i}\int_{-p-i\infty}^{-p+i\infty} d\lambda \exp[\frac{-\Delta t}{\varepsilon}\{H(-\lambda,x',t) + \lambda\frac{x-x'}{\Delta t}\}] \tag{22}$$

For a given Δt, ε is made so small that $\Delta t/\varepsilon$ is large enough to allow the application of the steepest descent evaluation. Thus we put $-\lambda = p + iq$ choosing p by the condition of the cole, namely

$$\frac{\partial H(p,x,t)}{\partial p} = \frac{x - x'}{\Delta t} \sim \dot{x} \tag{23}$$

and integrate over q noticing that

$$\partial^2 H/\partial p^2 = -\int e^{-p\Delta} w(x,\Delta,t)\, \Delta^2 d\Delta < 0.$$

By summing up all possible paths from x_0 to x over a time interval (t_0, t), we write the Markoffian transition probability in a path integral form,

$$P(x_0 t_0 \to x,t) = \int \mathcal{D}[x(s)] \exp \frac{1}{\varepsilon}\int_{t_0}^{t} L[x(s),\dot{x}(s),s]\, ds \tag{24}$$

where the Lagrangian L is defined by

$$L = -H + p\dot{x}, \qquad \dot{x} = \frac{\partial H}{\partial p} \tag{25}$$

This is somewhat different from ordinary path integrals. If $H(p,x,t)$ is only quadratic in p, there is no need to appeal to the steepest descent method. This is the case if the Markoffian process (5) is a Fokker-Planck type. For a general class of Markoffian process, however, we have to go beyond of such a limitation. By considering large systems, we have taken the advantage of asymptotic evaluation. The measure of paths, denoted by $\mathcal{D}[x(s)]$ in Eq.(24), is more complex than usual, but this is not important for asymptotic evaluation of the probability.

The path integral (24) indicates the extensive character. The integral can be approximated by taking only the path with the greatest probability. Then the action integral on the exponent is nothing but $\phi_0(x,t)$ satisfying Eq.(11). Thus the problem can also be formulated in the Hamiltonian formalism which is particularly useful when the process is no longer normal[3,11].

The path integral representation is deeply connected with the Langevin equation, or the stochastic equation,

$$\dot{x} = g(x,t), \quad \text{or,} \quad \dot{X} = G(X,t) \tag{26}$$

where g or G is considered as a stochastic process with an infinitely short correlation time and with the extensivity character. Thus we may assume

$$\underset{t<t_j<t+\Delta t}{\int\cdots\int} \langle G(X,t_1)G(X,t_2)\ldots G(X,t_n)\rangle_c dt_1 dt_2 \ldots = \Omega c_n(x,t)\Delta t \tag{27}$$

for the n-th cumulant of G. The integral vanishes if the intervals of integration do not coincide. Then the characteristic functional of G can be written as

$$\langle \exp \int_{t_1}^{t} \lambda(t')G(t')dt' \rangle = \exp\{-\frac{1}{\varepsilon}\int_{t_0}^{t} H[-\lambda(t'),x'.t']\, dt'\} \tag{28}$$

where the Hamiltonian H is that introduced by Eq.(7). For a small time interval Δt we recover Eq.(22) by taking the integral in Eq.(28) as $\Delta t H(-\lambda(t'),t')$ and interpreting Eq.(26) as $x - x' = g(x',t)\Delta t$. Connecting such segments over the interval (t_0,t), we have the path integral (24). Here we follow Itoh's interpretation of the equation (26) because it is the simplest[12].

The above can be generalized to a field of macrovariables, which we denote by u(r,t). The entire system occupying a volume Ω is divided into blocks each having a volume Ω_0 which is still large enough to allow asymptotic evaluation. We write the Langevin equation,

$$\dot{u}(r,t) = g(r,t), \quad \text{or} \quad \dot{U}(r,t) = G(r,t) \tag{29}$$

where $U=\Omega_0 u$ and $G=\Omega_0 g$. In order to derive a stochastic process of u(r,t) as a local Markoffian process, we assume the random field g or G to be infinitely short-correlated in space and time, namely

$$\underset{r<r_j<r+\Delta r,\ t<t_j<t+\Delta t}{\int\cdots\cdots\int} \langle G(r_1,t_1)\ldots\ldots G(r_n,t_n)\rangle_c \; dr_1 dt_1\ldots\ldots dr_n dt_n = \Omega_0 c_n[u(r,t),t]\Delta t \tag{30}$$

so that

$$\langle \exp[\iint \lambda(r,t)g(r,t)drdt\rangle = \exp[-\Omega_0 \sum_r \int H([-\lambda(r,t),u(r,t),t]dt] \tag{31}$$

where the sum is made over the blocks and the Hamiltonian density is defined by

$$H(p(r,t),u(r,t),t) = \sum_{n=1}^{\infty} \frac{(-)^{n-1}}{n!} p(r,t)^n \, c_n[u(r,t),t] \tag{32}$$

By repeating the previous procedure for each block and by writing the sum as an integral we get the path integral

$$P[u(r,t_0) \to u(r,t)] = \int \mathcal{D}[u(r,s)] \exp \iint L(u(r,s),\dot{u}(r,s),s] drds \qquad (33)$$

with the Lagrangian density defined by

$$L[u,\dot{u},t] = - H[p,u,t) + p\,\dot{u}, \qquad \dot{u}(r,t) = \delta H/\delta u \qquad (34)$$

A time dependent GL equation with a white noise corresponds to the choice

$$\dot{c}_1 = -Au - Bu^3 + D\nabla^2 , \quad L = \frac{[\dot{u} - c_1]^2}{2\,c_2} , \quad H = pc_1[u] - \frac{1}{2}p^2 c_2[u] \qquad (35)$$

The path integral has been used for renormalization calculations of dynamic critical indices[13]. This formalism seems useful for similar or more general problems.

The concept of the probability of paths is related to a generalization of entropy for non-equilibrium processes as was shown many years ago by Onsager and Mashlup[14] for the case of linear relaxation. As we have seen, this can be generalized to nonlinear cases.

REFERENCES

1. N.G. van Kampen, Can. J. Phys. 49 551 (1961). *Fundamental Problems in Statistical Mechanics*, E. G. D. Cohen ed., North Holland Amsterdam (1962) p.173.
2. R. Kubo, *Synergetics*, H. Haken ed., Teubner Stuttgart (1973) p.28.
3. R. Kubo, K. Matsuo and K. Kitahara, J. Stat. Phys. 9 51 (1973).
4. M. Suzuki, Prog. Theor. Phys. 55 383 (1976).
5. M. Suzuki, Prog. Theor. Phys. 53 1657 (1975). J. Stat. Phys. 14 129 (1976).
6. F. T. Arecchi, V. Degiorgio and B. Querzola, Phys. Rev. Letters 19 1168 (1967). F. Arecchi and V. Degiorgio, Phys. Rev. A3 1108 (1971).
7. R. Bonifacio, P. Schwendimann and F. Haake, Phys. Rev. A4 854 (1971). Y. Takahashi and F. Shibata, J. Stat. Phys. 14 67 (1976). M. Suzuki, Physica 84A 48 (1976).
8. Y. Saito and R. Kubo, J. Stat. Phys. 15 233 (1976).
9. S. Kabashima, M. Itsumi, T. Kawakubo and T. Nagashima, J. Phys. Soc. Japan 39 1183 (1975).
10. M. Suzuki, Prog. Theor. Phys. 57 380 (1977). J. Stat. Phys. 16 11 (1977).
11. K. Kitahara, Doctor Thesis, Faculté des Sciences, Université Libre de Bruxelles, 1974. K. Matsuo, Doctor Thesis, Department of Physics, University of Tokyo 1975.
12. See for instance, R. E. Mortensen, J. Stat. Phys. 1 27 (1967).
13. H. Yahata, Prog. Theor. Phys. 52 871 (1974). C. De Dominicis, Lett. Nuovo Cimento 12 567 (1975). C. De Dominicis, El Brezin and J. Zinn-Justin, Phys. Rev. B12 4945 (1975).
14. L. Onsager and S. Machlup, Phys. Rev. 91 1505 (1953).

CHAPTER 2

REVIEW ON MULTICRITICAL PHASE DIAGRAMS

Amnon Aharony

Department of Physics & Astronomy, Tel Aviv University, Israel

ABSTRACT

Present theoretical knowledge on phase diagrams which exhibit tricritical, bicritical, tetracritical points etc. is reviewed, from the viewpoint of structural displacive phase transitions. Special emphasis is put on recent results concerning the transformation of a first order into a second order transition, on Potts model transitions, and on other effects induced by anisotropic stress. Complicated phase diagrams, which may arise for lower than cubic symmetries, are also mentioned.

1. INTRODUCTION

I was asked by the organizing committee to give an introduction to, and a review of, the ideas of renormalization group, fixed points, multicritical points, complex phase diagrams, etc. etc. Clearly, this is not easy given time and space constraints. Moreover, many of these topics have been recently reviewed in books, papers and talks.[1-6] Instead of repeating these, I therefore prefer to introduce all these ideas through an explicit example, i.e. that of structural displavice phase transitions which occur in perovskite crystals. Most of the above mentioned presentations dwelled on magnetic phase diagrams, especially those of antiferromagnets in a magnetic field.[5] I believe that insufficient attention was devoted to the situation with regards to structural transitions,[7] which have several attractive advantages. First, anything one observes in the magnetic case can also be observed in the structural case, and, in most cases was observed there first.[8,9] Second, the Ginzburg-Landau-Wilson effective Hamiltonian, which one has to <u>assume</u>[1] in the magnetic case, is explicitly <u>calculable</u> in the structural case, using the unharmonic lattice theory.[10] Third, the assumptions concerning <u>cubic symmetry</u> of the Hamiltonian follow directly from the symmetry of the lattice, whereas in the magnetic case they involve additional assumptions, which may not be satisfied, and in the absence of which the phase diagrams may be more complicated (see below, Sec. 4).

In the following I shall show, that appropriate experiments, using mainly anisotropic stress, yield phase diagrams which exhibit tricritical, bicritical, tetracritical and other multicritical points. More explicitly, details will be given of the fixed point structure of the cubic problem, and of the phase diagrams generated by various anisotropic stresses.

2. THE CUBIC PROBLEM

We start with the usual Ginzburg-Landau-Wilson effective Hamiltonian for cubic systems,[10-15]

$$\beta\mathcal{H} = \int d^dx \,\{\tfrac{1}{2}[r_o|\vec{S}|^2 + (\vec{\nabla}\vec{S})^2] + u_o|\vec{S}|^4 + v_o\sum_{\alpha=1}^{n} S_\alpha^4 + \dots\} \quad , \tag{1}$$

where $\vec{S}$ is a continuous n-component local order parameter, and r_o is linear in the temperature T. For the structural displacive phase transitions in the perovskites, $n = 3$ and $\vec{S}$ represents the local pseudocubic axial vector of the soft optic mode which lies at the [111] corner of the Brillouin zone.[11-13] We shall later introduce realistic complications into this Hamiltonian.

The sign of the cubic anisotropy parameter v_o determines whether the spins tend to align along a cubic axis ($v_o < 0$, tetragonal distortion) or along a diagonal ($v_o > 0$, trigonal distortion). Renormalization group study near $d = 4$ shows,[14] that the critical behavior of the cubic system may be characterized by one of four fixed points, as shown in Fig. 1.[15] The isotropic "Heisenberg"-like fixed point, $v^* = 0$, is stable

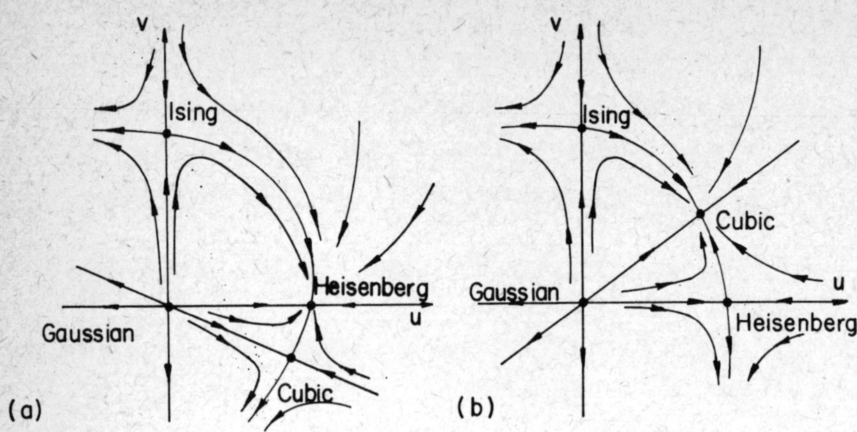

Fig. 1. Schematic flow diagrams and fixed points for the cubic problem. (a) $n < n_c$, (b) $n > n_c$.

for small values of n, $n < n_c(d)$, and the "cubic" one is stable for $n > n_c(d)$. The value of $n_c(d)$ is quite problematic; it has expansions about $n_c(4) = 4$[15,16] and about $n_c(2) = 2$,[17,18] but these are difficult to interpolate for $n_c(3)$. Padé approximants seem to indicate that $n_c(3) > 3$, in agreement with approximate recursion relations at d = 3.[19] More recent approximate calculations[20] give $n_c(3) \simeq 2.3$. Thus, the stable fixed point is presently not finally identified.

The transition is <u>second order</u>, with divergent correlations, as long as the "Hamiltonian flow", under the renormalization group iterations, leads to the stable fixed point. If one starts with a Hamiltonian which is not in the "domain of attraction" of this fixed point, one encounters a "<u>runaway</u>" (See Fig. 1), and the stable fixed point is <u>inaccessible</u>. A direct integration of the renormalization group recursion relations, to yield the free energy, then shows that under such circumstances the transition becomes <u>first order</u>.[21-25] For $n < n_c$, the "cubic" and the "Ising" fixed points thus represent <u>tricritical</u> behavior,[26] separating the first and second order regions. If there existed an externally variable field, which would vary the ratio v_o/u_o e.g. through the value v_c/u_c ($n < n_c$) where c denotes "cubic", then a tricritical phase diagram would result. An example of such a field is the concentration of imurities with a different strength of cubic interaction v.[27] We shall return to another example below.

The critical properties at each of the four fixed points have been reviewed in detail elsewhere,[15] and we shall not dwell upon them further here.

Up to this point, we discussed only the effects of cubic symmetry due to quartic terms in the order parameters. Just as important are the quadratic terms, of the form[22,23,28,29]

$$\beta \mathcal{H}_1 = -\tfrac{1}{2} f \int d^d x \sum_{\alpha=1}^{n} (\partial S_\alpha / \partial x_\alpha)^2 \qquad (2)$$

Note that the coupling between the order parameter and the space coordinates enforces the identity $n = d$. This may be relaxed[28] by introducing an artificial anisotropy, to yield $n \leq d$. The parameter f generates the parameter v_o even if the latter is initially absent. Moreover, if f is sufficiently close to unity, it "pushes" v to the first order regions.[23] Even if f is smaller than its "tricritical" value, it decays to zero very slowly,[28,29] and thus one observes effective exponents, which depend on f, in any practical temperature range.[28] Explicit expressions for these have recently been given in Refs. 22, 23.

3. COUPLING TO EXTERNAL UNIFORM STRESS, THROUGH ELASTIC DEGREES OF FREEDOM

The Hamiltonian (1) is appropraite only for a rigid lattice, with all the elastic degrees of freedom frozen. In practice, we must supplement it by contributions from the strain degrees of freedom $\{\varepsilon_{\alpha\beta}\}$ and from their coupling with the order parameter,[8,11-13]

$$\beta \mathcal{H}_{int} = -\int d^d x \Big[B_o \sum_\alpha \varepsilon_{\alpha\alpha} \vec{S}^2 + B_1 \sum_\alpha \varepsilon_{\alpha\alpha} (n S_\alpha^2 - \vec{S}^2) + B_2 \sum_{\alpha<\beta} \varepsilon_{\alpha\beta} S_\alpha S_\beta \Big] . \qquad (3)$$

The variables $\varepsilon_{\alpha\beta}$ involve both uniform strains $\{\varepsilon^o_{\alpha\beta}\}$, which couple linearly to the external unifrom stresses $\{\sigma_{\alpha\beta}\}$, and "phonon"-like strains, which have no zero wavevector Fourier component.[30-34] Integration over these latter degrees of freedom yields a complicated effective Hamiltonian, with terms like[30] $(\int d^d x\, \vec{S}^2)^2$, terms with complicated non-analytic coefficients, quartic in the order parameter,[31-34] terms like[33,34] $(\int d^d x\, S_\alpha S_\beta)^2$, etc. The "rigid" fixed points, discussed above, are unstable with respect to some of these new terms, leading to a mechanical first order transition.[31] However, the crossover process is very slow, and this effect will probably be observable only very close to the critical temperature. We therefore ignore these effects here.

Integration over the uniform strain degrees of freedom yields similar quartic terms, which we ignore. In addition, it yields terms like those in Eq. (3), with the stress $\sigma_{\alpha\beta}$ replacing the strain $\varepsilon_{\alpha\beta}$. The first term, B_o, simply represents a shift in the critical temperature of the "isotropic" n-component phase transition.

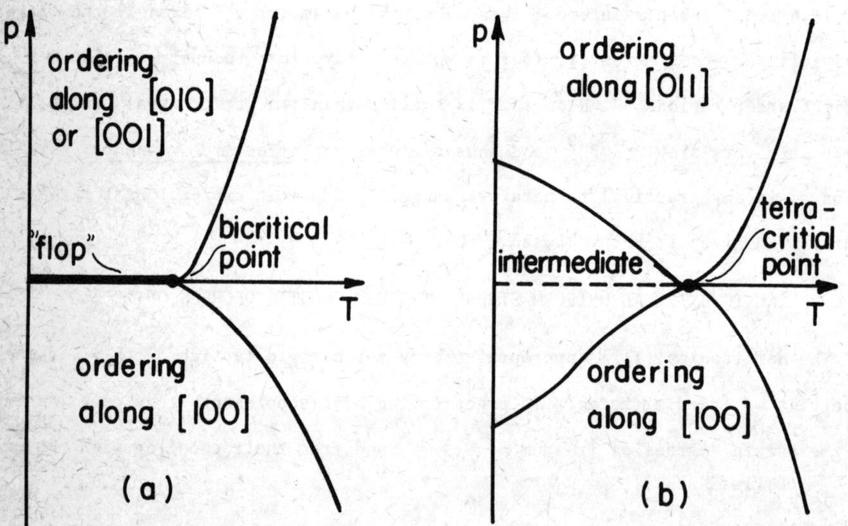

Fig. 2. Phase diagrams for stress p along the 1-axis, for n = 3. (a) Bicritical point and first order "flop" line, $v_c/u_c < v/u \lesssim 0$. (b) Tetracritical point, $v > 0$.

The coefficients B_1 and B_2 represent crossover to lower symmetries.

If the stress is along the 1-axis,[8] $\sigma_{\alpha\beta} = -p\,\delta_{\alpha 1}\,\delta_{\beta 1}$, then B_1 represents uniaxial anisotropy in the 1-direction. The resulting phase diagrams are shown in

Figs. 2 and 3: The transition is second order if the Hamiltonian "flows" to the stable fixed point (Fig. 2), and first order if it does not (Fig. 3).[35] In the second order regime, the phase diagram involves a bicritical point if $v < 0$, and a tetracritical point if $v > 0$.[36] In the first order regime, the order of the transition can be changed, through a tricritical point, for sufficiently large uniaxial stress (Fig. 3.).[35]

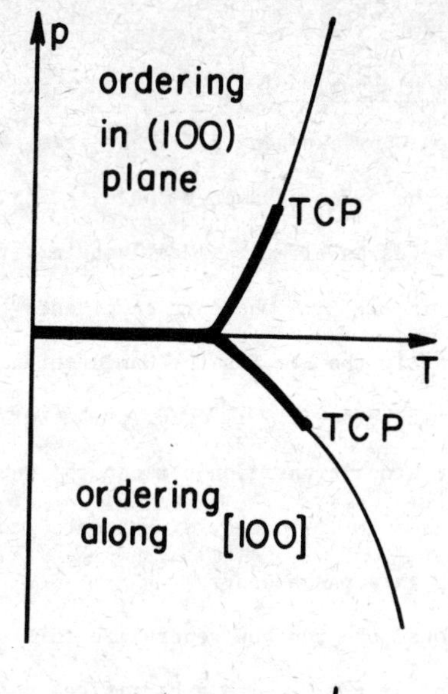

Fig. 3. Phase diagram for stress along the 1-axis, $n = 3$, $v/u < v_c/u_c$. TCP denotes a tricritical point. (Ref. 35).

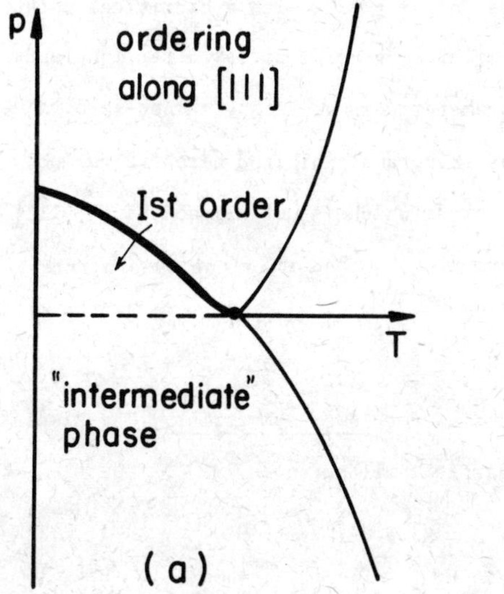

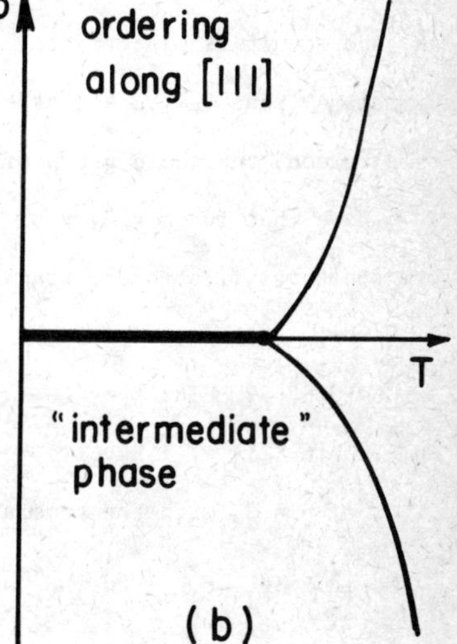

Fig. 4. Phase diagram for stress along the diagonal [111], $n = 3$. (a) $v < 0$, (b) $v > 0$.

The analysis of the pahse diagrams, Figs. 2 and 3, is always based on integation over the non critical order parameter, to obtain an effective Hamiltonian for the critical one.[35,36] A different procedure may lead to wrong results, e.g. differing from Fig. 3.[37]

We next review the case of stress along the diagonal [1,1,...1], $\sigma_{\alpha\beta} \equiv -p/n$.[9,36] The important variable now is B_2, which has a crossover exponent different from that of B_1 at the cubic fixed point.[38] A rotation of the order parameter space, so that one component, $\vec{S}_1$, points along the diagonal, yields an effective Hamiltonain with three parts, leading to the phase diagrams shown in Fig. 4: One part represents the Ising-like ordering of $\vec{S}_1$. Another part is exactly the same as the Hamiltonian of the n-state Potts model,[39,40] in terms of the transverse (n - 1) order parameter components.[41] For n = 3, this leads to the first order transition between the two ordered pahses.[9,42,43] The third part couples these two. Indeed, experiments[9] confirm renormalization group predictions concerning this phase diagram.

Having seen two examples of stress directions, one can now generalize this to any direction. For n = 2, the result will resemble Fig. 2a, with a bicritical point.[44] For larger n, the result is expected to be similar: a general stress direction leads to an "intermediate" phase, with all order parameter components having non-zero values. For this reason, the whole group of phase diagrams exhibited here are called "bicritical". Note that for n = 3, v < 0, a stress which is not aligned along [111] may transform the Potts first order transition into a second order one, via a tricritical point.

4. EFFECTS OF LOWER SYMMETRY

The Hamiltonian (1) will have to be modified for lower symmetries. The simplest example is that of n = 2, with the general quartic Hamiltonian

$$\beta \mathcal{H}_{(4)} = \int d^d x \, (u_{11} S_1^4 + 2u_{12} S_1^2 S_2^2 + u_{22} S_2^4) \quad . \tag{4}$$

This differs from (1) by a tetragonal symmetry term, $(S_1^4 - S_2^4)$, proportional to $(u_{11} - u_{22})$. This term is irrelevant, for $n < n_c(d)$, near the isotropic fixed point.[45,46] However, as has been emphasized above,[36] an irrelevant variable may be dangerous,[47] in that it still affects strongly the shape of the pahse diagram. Indeed, if $u_{11} \neq u_{22}$ then the Hamiltonian flow in the three dimensional parameter space $u_{11} - u_{12} - u_{22}$ may cross from the tetracritical region, $u_{12}^2 < u_{11} u_{22}$, to the bicritical one.[48] This is impossible in the cubic case $u_{11} = u_{22}$,[36] but may happen in the case of the uniaxial antiferromagnet in a field.[5] Lyuksyutov *et al*[48] analyze various possible phase diagrams. Some of their diagrams should really look like Figs. 2 and 3 above. In other cases, new diagrams result. For example, if initially $u_{11} \neq u_{22}$, and $u_{12}^2 < u_{11} u_{22}$, but the flow approaches the isotropic fixed point through the bicritical region, then Figs. 2a and 2b combine to give Fig. 5. Larger differences between u_{11} and u_{22} lead to modifications in Fig. 3. A full analysis, to find the detailed scaling crossover properties of such modified phase diagram, is still a project for the future. The absence of these effects in the pure cubic case, relevant for the structural displacive transitions, justifies the belief that the phase diagrams of these are simpler and better understood.

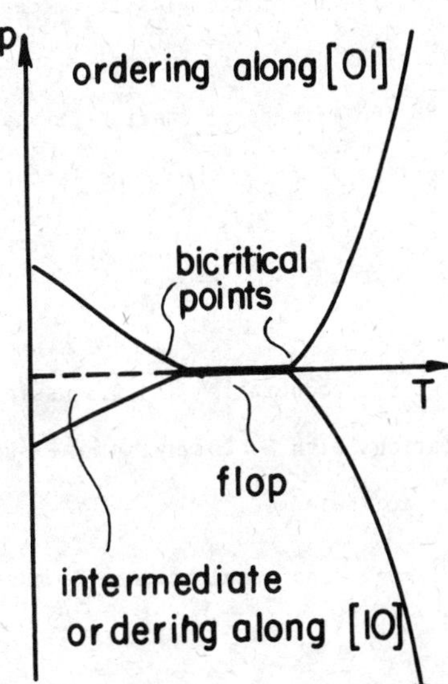

Fig. 5. Possible phase diagram for n = 2, if flow crosses over from "tetracritical" to "bicritical" region. (Ref. 48).

5. FINAL REMARKS

In conclusion, we emphasize again our main points. First, the cubic parameter v was shown to lead to one of three possible phase diagrams, Figs.2 and 3. If v is varied, by an external parameter, so that v = 0, then the shape of the tetracritical phase diagram, Fig. 2b, the temperature dependence of the transverse susceptibility below T_c, etc. will be determined by the hexagonal term[49] $\sum_\alpha S_\alpha^6$. The phase lines bounding the intermediate phase will then approach the tetracritical point much closer to the T-axis $\left[\psi_2 = \frac{1}{2} - 4\varepsilon/(n+8) + \ldots , \text{ see Ref. } 36\right]$. Second, anisotropic stress was shown to lead to many possible interesting phase transition lines, depending on the direction of the stress. In particular, tricritical points separating first and second order regions, for $v/u < v_c/u_c$, and Potts-model-like transitions were emphasized.

Third, the advantages of cubic symmetry, as compared with lower symmetries, was stressed, with regards to the complications the latter may introduce into the phase diagrams.

Finally, we should note that in addition to the effects of the elastic degrees of freedom, discussed in Sec. 3, many other complications may arise in real systems. For example, a whole discussion should be devoted to the effects of configurational disorder,[50] to more complicated microscopic interactions,[15] etc. These are left to other occasions.

ACKNOWLEDGEMENTS

During the past few years, I have enjoyed fruitful cooperation and discussions with A.D. Bruce and K.A. Müller. Helpful communications with E. Domany, M.E. Fisher, T. Nattermann, D.R. Nelson and many others are also acknowledged.

* Supported by a grant from the United-States - Israel Binational Science Foundation (BSF), Jerusalem, Israel.

1. K.G. Wilson and J. Kogut, Phys. Repts. 12C, 75 (1974).
2. S.-k. Ma, "Modern Thorory of Critical Phenomena", (Benjamin, Reading, Mass, 1976).
3. M.E. Fisher, Rev. Mod. Phys. 46, 597 (1974).
4. C. Domb and M.S. Green, eds., "Phase Transitions and Critical Phenomena", Vol. VI (Academic, New York, 1976).
5. M.E. Fisher, A.I.P. Conf. Proc. 24, 273 (1975).
6. A. Aharony, Physica 86-88B, 545 (1977).

7. K.A. Müller and A. Rigamonti, eds., "Local Properties at Phase Transitions" Enrico Fermi Course LIX (North Holland, Amsterdam, 1976).
8. A. Aharony and A.D. Bruce, Phys. Rev. Lett. 33, 427 (1974).
9. A. Aharony, K.A. Müller and W. Berlinger, Phys. Rev. Lett. 38, 33 (1977).
10. A.D. Bruce and R.A. Cowley, J. Phys. C6, 2422 (1973).
11. J.C. Slonczewski and H. Thomas, Phys. Rev. B1, 3599 (1970).
12. J.C. Slonczewski, Phys. Rev. B2, 4646 (1970).
13. K.A. Müller, W. Berlinger and J.C. Slonczewski, Phys. Rev. Lett. 25, 734 (1970).
14. A. Aharony, Phys. Rev. B8, 4270 (1973).
15. A. Aharony, Ref. 4, p. 357.
16. I.J. Ketley and D.J. Wallace, J. Phys. A6, 1667 (1973).
17. R.A. Pelcovitz and D.R. Nelson, Phys. Lett. 57A, 23 (1976).
18. E. Brézin and J. Zinn-Justin, Phys. Rev. B14, 3110 (1976).
19. M.K. Grover, L.P. Kadanoff and F. Wegner, Phys. Rev. B6, 311 (1972).
20. M.C. Yalabik and A. Houghton, Phys. Lett. 61A, 1 (1977).
21. I.F. Lyuksyutov and V.L. Pokrovskii, JETP Lett. 21, 9 (1975).
22. T. Nattermann and S. Trimper, J. Phys. A8, 2000 (1975).
23. T. Nattermann, J. Phys. C9, 3337 (1976).
24. J. Rudnick, unpublished.
25. I.E. Dzyaloshinskii, papers presented at the Amsterdam Magnetism Conference, 1976 and at this conference.
26. A. Aharony, paper presented at the Conference on Critical Phenomena in Multi-component Systems, Athens, Ga. (1974).
27. A. Aharony, Phys. Rev. B12, 1038 (1975).
28. A. Aharony, Phys. Rev. B8, 3349 (1973).
29. A.D. Bruce, J. Phys. C7, 1089 (1974).
30. J. Sak, Phys. Rev. B10, 3957 (1974).
31. D.J. Bergman and B.I. Halperin, Phys. Rev. B 13, 2145 (1976).
32. M.A. de Moura, T.C. Lubensky, Y. Imry and A. Aharony, Phys. Rev. B13, 2176 (1976).
33. K. Murata, preprint (1976).
34. T. Nattermann, preprint (1976).
35. E. Domany, D. Mukamel and M.E. Fisher, Phys. Rev. B15, in press (1977).
36. A.D. Bruce and A. Aharony, Phys. Rev. B11, 478 (1975).
37. I.F. Lyuksyutov, Phys.Lett. 56A, 135 (1976).
38. A. Aharony, Phys. Lett. 59A, 163 (1976).
39. R.K.P. Zia and D.J. Wallace, J. Phys. A 8, 1495 (1975).
40. D.J. Amit, J. Phys. A9, 1441 (1976).
41. See also D. Mukamel, M.E. Fisher and E. Domany, Phys. Rev. Lett. 37, 565 (1976).
42. G.R. Golner, Phys. Rev. B8, 3419 (1973).
43. J. Rudnick, J. Phys. A8, 1125 (1975).
44. D. Mukamel, Phys. Rev. B14, 1303 (1976).
45. E. Brézin, J.C. Le Guillou and J. Zinn-Justin, Phys. Rev. B10, 892 (1974).
46. D.R. Nelson, J.M. Kosterlitz and M.E. Fisher, Phys. Rev. Lett. 33, 813 (1974).
47. M.E. Fisher, in Renormalization Group in Critical Phenomena and Quantum Field Theory: Proceedings of a Conference, J.D. Gunton and M.S. Green, eds (Temple University, 1973).
48. I.F. Lyuksyutov, V.L. Pokrovskii and D.E. Khmel'nitzkii, Zh. Eksp. Teor. Fiz. 69, 1817 (1975) [Sov. Phys. - JETP 42, 923 (1976)] .
49. D.R. Nelson, Phys. Rev. B13, 2222 (1976).
50. A. Aharony, paper presented at the Int. Conf. on Magnetic Alloys and Oxides, Haifa, August, 1977.

CHAPTER 3

THERMODYNAMICS AND THERMODYNAMIC DECODING OF LONG MOLECULES (DNA, RNA, ETC.)

M.Ya. Azbel

Department of Physics & Astronomy, Tel Aviv University, Israel

ABSTRACT

We propose an approach which allows us to obtain the explicit analytical formula for the free energy of long molecules, such as DNA and double-strand replicative RNA, of any nucleotide sequence. In particular, the obtained formula elucidates the fine structure of a differential coiling curve in the vicinity of the helix-coil transition and it makes clear the character of this transition in a general case. We also present a direct method for the stable decoding of the structure of the DNA and RNA nucleotide sequence from thermodynamic data. In the case of a long sequence one obtains the sequence statistical structure with respect to certain segments which form the sequence; in the case of a short enough sequence (several hundreds or thousands of nucleotides) one obtains a detailed sequence structure. The effectiveness of the method is checked for a certain known DNA sequence.

1. INTRODUCTION

It is well-known that DNA and double-strand replicative RNA consist of two nucleotide strands connected by hydrogen bonds. The binding energy depends on the type of the complementary nucleotide pair: adenine-thymine (AT) and guanine-cytosine (GC) in DNA; adenine-uracil (AU) or GC in replicative RNA. Let us denote AT- and AU-pair by "zero", GC pair by "unity" and speak for brevity only about DNA. An increase in temperature results in a separation of strands and in the so-called helix-coil transition. When strands are separated, new degrees of freedom are released (of nucleotides and of strands which may entangle now), so the entropy increases. From the physical viewpoint the transition from bounded strands ("helix" state or phase) to separated ones ("coiled" state or phase) is described[1] by a modified two-component Ising model. An energy H of an arbitrary DNA thermodynamic state (which is formed by consecutive helix and coiled segments) is calculated according to the following rules: I) each helix "zero"-pair adds an energy f_o; II) each helix "unity"-pair adds an energy f_1; III) each completely coiled segment containing L links (i.e. L complementary nucleotide pairs) and bordering on helix segments adds an energy $(V+bkT\ell nL)$, where T is the absolute temperature, k is the Boltzmann constant, $b\sim 1$ is a constant (its value is estimated in Section 3), and V is the "boundary energy" between coiled and helix phases $(V\sim 3000^o)$; IV) a totally coiled segment which includes the first or the last DNA link adds an energy V/2; V) the energy of the completely coiled DNA is zero. Energies f_o, f_1 refer in fact to the difference between the free energies (per link) of completely helical and completely coiled DNA. Therefore, they include the above mentioned entropy, which is related to the released "inner" degrees of freedom of a "coiled" nucleotide pair. Therefore, they depend on temperature[1] (and on the type of solution in which the DNA is dissolved) and they may be of any sign, but always $V>>kT,|f_o|,|f_1|$. The most interesting temperature region of helix-coil transition is $f_o>0>f_1$, and only this region is investigated in our paper.

The specificity of DNA Ising model (besides the long-ranged interaction from Rule III) is related to the specific character of its component sequence. This sequence deals with the genetic information. Therefore, it is neither regular in any sense, nor is it random, and no a priori (and probably a posteriori as well) suppositions can be made about it. Yet until recently all papers on DNA thermodynamics either evaluated numerically the free energy of a certain arbitrarily given sequence, or dealt with some simple kind of sequences, such as homogeneous, periodical, quasiperiodical, or random ones (see References in Reviews 1). But, in 1972 it was suggested[2] that the inequality $V>>kT,|f_o|,|f_1|$ allows us to present in an explicit form the successive approximation for the free energy F of an arbitrary component sequence. Each of the approximations is related to very specific statistical characteristics of the sequence which are determined by the approximation itself. It was proved[2,3] that the leading approximation for F is the ground state energy E=minH (where minimum is taken over all possible thermodynamic states of DNA) and it is related[3] to three certain sequence distribution functions. These functions depend only on the sequence and do not depend on temperature, and each function has four specific variables.

In this paper we present an explicit formula for the free energy of an arbitrary sequence (in the sense that was indicated above) and investigate the character of the helix-coil phase transition which occurs when $b>1$ in Rule III due to the long-range interaction. This formula for F shows that if the DNA specific heat or its optical density (related [1] to a relative number of coiled links), is measured experimentally as a function of three arbitrary independent variables (such as temperature and concentrations C_1, C_2 of two low-molecular weight substances present in the solution together with DNA), then one can obtain the composition of certain segments of the DNA. This information is stable up to a certain (finite but high) accuracy, which is determined by small values of $kT/V, |f_o|/V, |f_1|/V$, and it is very specific for sequences containing few hundreds or thousands of links.

2. GROUND STATE AND DECODING

The DNA ground state is the leading approximation[2,3] for the free energy F and it is the starting point for its evaluation. Suppose $h_t(m)$ is the helix energy of the first m links as shown in Fig. 1, where $h_t(m)=h_t(m-1)+f_o(1-q_m)+f_1q_m$, q_m is the digit at the m-th link, and $h_t(m)-h_t(m')\equiv h(m'm)$ is the height of the indicated segment. The graph in Fig. 1 refers specifically to a portion of the Escherichia coli DNA[4] with $f_1=-f_o$. In the simplest case of $b=0$ and of circular DNA the ground state coiled section, which borders on ground state helix segments, has [3,5] $h>V$ (so its helixing even amidst a helical environment increases the total energy by $h-V>0$). There is no segment with $h<-V$ inside it, otherwise the helixing of this segment

Fig. 1.

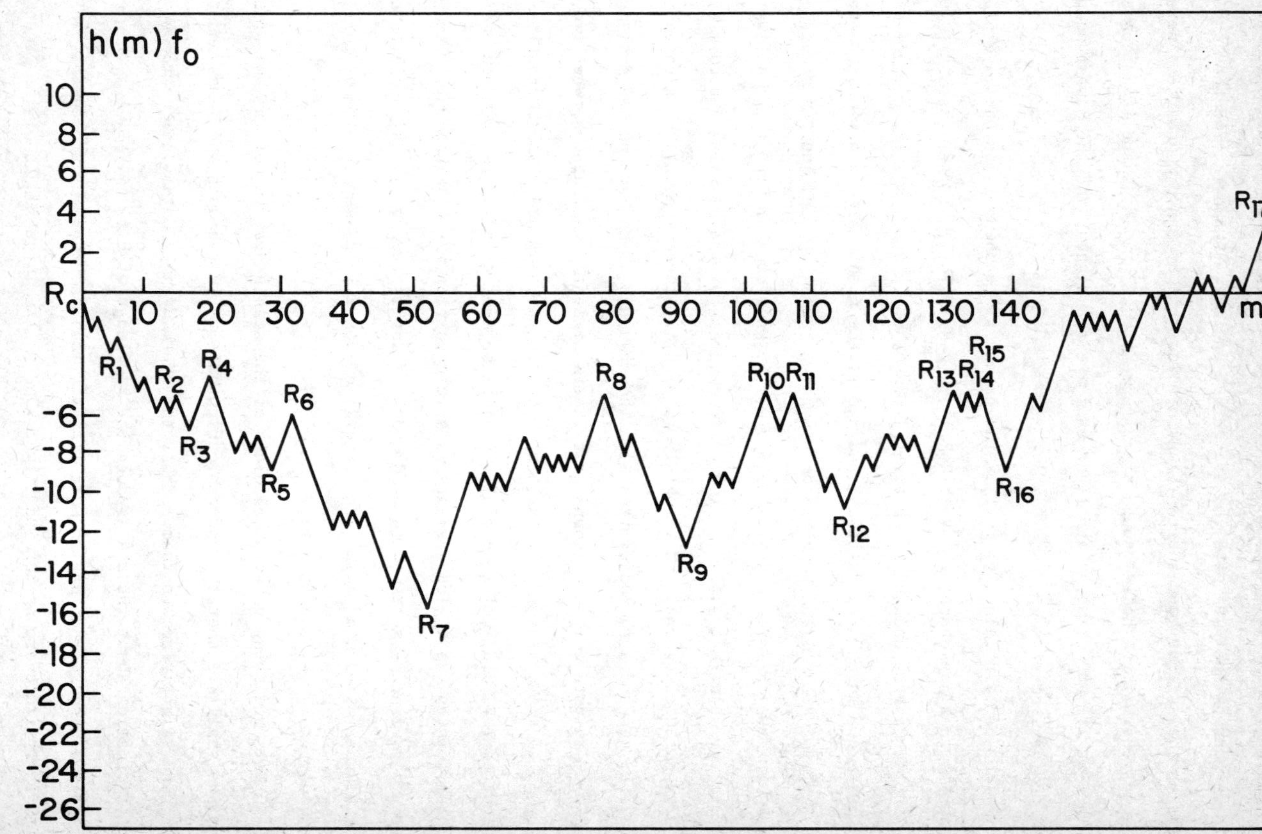
h(m) f$_o$
10
8
6
4
2
R$_c$
-6
-8
-10
-12
-14
-16
-18
-20
-22
-24
-26
10
20
30
40
50
60
70
80
90
100
110
120
130
140
m
R$_1$
R$_2$
R$_3$
R$_4$
R$_5$
R$_6$
R$_7$
R$_8$
R$_9$
R$_{10}$
R$_{11}$
R$_{12}$
R$_{13}$
R$_{14}$
R$_{15}$
R$_{16}$
R$_{17}$

would decrease the total energy. The left and the right ends of the coiled ground state segment on the plot of $h_t(m)$ are correspondingly the lowest and the highest points on the segment (and therefore any displacement of phase boundaries increases the total energy). A helix ground state segment has[3,5] $h<-W$, there is no segment with $h>W$ at it, and its left and right ends are the highest and the lowest points of $h_t(m)$ on the segment.

For certain values of f_o, f_1, V the ground state changes [in the case of Fig.1 it happens when consequently $V=|h(R_oR_7)|;|h(R_8R_9)|,|h(R_{10}R_{12})|=|h(R_{11}R_{12})|$, etc.], and then the energy changes by $\Delta E=f_o i+f_1 j+\frac{1}{2}Vn$ where i and j are the numbers of the newly helixed "zeros" and "unities" and $n\gtrless 0$ is the number of the new phase boundaries. Now, $\partial E/\partial f_o$, $\partial E/\partial f_1$, $\partial E/\partial V$ have corresponding jumps i, j, $\frac{1}{2}n$, and the second derivatives may have infinite peaks. Therefore, the knowledge of $E(f_o,f_1,V)$ allows us to draw (in variables $V/f_o, -f_1/f_o$), the phase diagram for the helixing and coiling of the ground state segments and to obtain i, j, n for these segments. In certain cases, the phase diagram determines also the sequence of the ground state segments of DNA. For instance, each ground state helix segment is situated between two ground state coiled segments. A helix segment appearing without new phase boundaries is adjacent to the "old" helix segment. A helix segment creating two new boundaries is situated between two coiled segments whose composition may be found out. Two simultaneously coiling segments (with $h_1>0$ and $h_2<0$ correspondingly), which create no new phase boundaries, are adjacent (for otherwise when V increases, h_1 cannot coil, and when V decreases, h_2 cannot coil).

This reasoning gives us the idea of the stable decoding of the DNA structure. Later, we shall see that thermodynamic fluctuations just smooth the jumps and the peaks, and slightly remove them. Thus, the thermodynamic data allow us to obtain the same information about the DNA as does the knowledge of $E(f_o,f_1,V)$.

Now suppose that $b\neq 0$ in Rule III, and thus the energies of segments which change their state are no more additive. Due to the large length of the ground state segments (which have $|h/>V>>kT,|f_o|,|f_1|)$, ℓnL depends slowly on L, and therefore in the approximation of neglecting terms of the order of 1/V this dependence can be ignored in comparison with the h's dependence on L. Thus, the previous determination of the ground state is applicable, if one uses the "renormalized"h's (in accordance with the coiling or helixing energy for b≠0). Obviously, a ground state segment coils in the helical environment, when

$$V = h - bkT\ell nL \equiv h^r; \tag{1}$$

h and L being the height and the length of the segment. A ground state segment helixes amidst the coiled environment when

$$V = - h - bkT\ell n \frac{L_1 L_2}{L_1 + L_2 + L} \equiv - h^r ; \tag{2}$$

where L and h are the length and the height of the helical segment, and L_1 and L_2 are the lengths of its adjacent coiled segments.

The reasoning for a non-circular DNA is quite similar, where one just has to take into account Rules (IV, V).

3. THE FREE ENERGY AND THE DECODING

Again we start with the case of b=0 and of circular DNA ($f_o>0>f_1$, as everywhere in this paper). Some new definitions are useful now. Let us denote segments which form the ground state for a (larger) boundary energy (V+ΔV) (where ΔV>>kT will be determined later) as "stable" segments, while their portions which change their ground state when the boundary energy decreases from (V+ΔV) to (V-ΔV) will be denoted as "unstable" segments. With the relative accuracy of the order of exp (-ΔV/kT) only the fluctuations which are related to the change of the state of unstable segments and/or to the displacement of the (V-ΔV) ground state phase boundaries, will contribute [6] to the free energy F. The unstable segments are either not adjacent to each other, or form groups of two, three, ... adjacent unstable segments. When we take into account these cases successively, we construct successive approximations (commonly of the order of increasing powers of 1/V). Each approximation determines $\Delta V \sim kT|\ell n\xi|$ self-consistently, where ξ is relative accuracy of the corresponding approximation. The approximation of separate unstable segments (i.e. of non-interacting fluctuations) is almost obvious:

$$\exp(-F/kT) \equiv Z = t^{n+} \prod_{+} x_{+} \prod_{-} (1+tx_{-}) \ ; \ t=\exp(-V/kT). \quad (3)$$

Here the subindexes "+" and "- " refer correspondingly to stable helix and unstable (both helix and coiled) segments; n_{+} is the number of stable helix segments; the products are performed over all indicated segments;

$$x_{\pm} = \Sigma \exp(|h_d|/kT) \equiv Q\exp(|h|/kT) \qquad Q \sim 1 \quad (4)$$

$$Q > Q_{min} = (u_1-u_o)/[(1-u_o)(u_1-1)] \quad (5)$$

where

$$u_o = \exp(-f_o/kT), \quad u_1 = \exp(-f_1/kT)$$

h is the height of the given segment, and h_d is its height when the segment phase boundaries are displaced. The summation is performed over all the possible displacements.

When applied to Gilbert's sequence[4] (Fig. 1), equations (3) and (4) provide an accuracy of 0.1-1% (the comparison is made with the exact numerical computation of F).

If $b \neq o$, then similar considerations prove that equations (3) and (4) remain valid with the same accuracy if, for a stable helix segment, h is replaced by $(h+bkT\ell nL)$, where L is the length of a left adjacent coiled segment. For an unstable segment h should be replaced by h^r according to equations (1) and (2). The resulting formula for F shows that $b>1$ implies a genuine phase transition. In a general case (in particular, for random and quasi-random sequences) F has an essential singularity (cf. Ref. 6): $\ell n|F| \alpha (T_c-T)^{-1}$, and the fluctuations are <u>less</u> pronounced the closer T is to the transition temperature T_c. In specific cases (e.g., for quasi-periodic sequence) the Poland-Scheraga-Applequist[7] singularity takes place: The transition is of the $(b-1)^{-1}$ order when $1<b<2$, but when $b>2$, it is of the first order and it is accompanied by the transition of the (b-1) order.

The integer b implies an additional logarythmic factor in $(T_c - T)$.

Equations (3) and (4) obviously relate the free energy to certain characteristics of the nucleotide sequence (cf. Ref. 3).

According to the equations (3) and (4) the DNA differential coiling curve $dN_c/dT \propto (\partial N_c/\partial f_o)_{V,f_o-f_1}$, where N_c is the total number of coiled links). This curve is expressed through the second derivatives of F, and it exhibits maxima. These maxima are related to the coiling of the ground state segments and therefore occur when $h \approx Vn/2$, where h is the height of the coiled segment and $n \lessgtr 0$ is the corresponding increase in the number of phase boundaries. For example in the casé of Gilbert's sequence[4], with b=0, the ground state changes at points $V/f_o=8.274$; $-f_1/f_o= 1.273$; $-f_1/f_o= 1$, while the numerically computed differential coiling curve shows maxima at $V/f_o=8.267$; $-f_1/f_o= 1.265$; $-f_1/f_o= 1.012$.

It is clear that using the DNA differential coiling curve we can determine h approximately for certain f_o, f_1 and thus, according to Section 2, it allows us to find out the composition of certain segments at DNA. Another approach for the (more detailed) DNA decoding is related to the representation of the experimentally obtained $Z=\exp(-F/kT)$ as a polynomial in $t=\exp(-V/kT)$ for fixed values of $f_o/kT, f_1/kT$. The lowest-power term and the roots (or the relation of the coefficients) of this polynomial allow us to obtain h's and therefore - the composition of the corresponding segments.

The accuracy and the stability of the decoding are determined by $Q/Q_{min} \sim 1$ as Q/Q_{min} slightly renormalizes $h \simeq V$. However, this accuracy[8] can be sufficient to obtain even the exact number of "zeros" and "unities". Again, in the case of the Gilbert's sequence[4] 40 arbitrarily chosen points (where F was numerically computed with the accuracy of three decimals) allowed us to obtain the composition of five consequent ground state segments. A larger number of experimental (or computed) points and their higher accuracy in this case provide 18 consequent segments: 0-3, 9-1, 13-19, 3-12, 2-3, 10-3, 9-5, 2-10, 10-2, 2-2, 1-7, 7-5, 4-0, 2-2, 0-4,

26-13, 5-5, 0-2, where commas separate segments, dashes separate i and j on the segment. According to the conditions which determine ground state segments, none of the segments with $j>i$ (or $i>j$) contains portions with $i>4+j$ (or $j>4+i$).

Of course, such a specific decoding is available only for sufficiently short sequences which contain few hundreds or thousands of nucleotides, and thus a relatively small number of ground state segments. If the sequence is very long, one can introduce (and obtain from the experimental data) the distribution function of the ground state segments whose height is h for given f_o, f_1. As it was for a short sequence, the stability and the accuracy of the decoding are related to small values of $kT/f_o, |f_o|/V$, and $|f_1|/V$.

To summarize: in the case of a long DNA, one must find out the DNA distribution function for the ground state segments, then to single out certain (sufficiently short) DNA portions and to find out the ground state segments which constitute them. Finally, it is necessary to separate these segments and obtain the exact sequence of nucleotides on each of them by the usual biochemical methods.

In order to use the experimental thermodynamic data such as the specific heat or the optical density (which are expressed[1] through F), one must replace the "theoretical" variables ($f_o/kT, f_1/kT, V/kT$) which enter the statistical sum, by "natural" experimental variables (T and the concentrations C_1, C_2 of the low molecular weight dissolved substances which renormalize[2,3,6] f_o, f_1 and V). This replacement is readily obtained from the thermodynamic formulas for homopoly-AT, homopoly-GC and homopoly-AU. For example, when b=0 we get

$$f_q/kT = \frac{c_q(T_q)}{T_q \theta'_q(T_q)}(T_q - T); \quad V/kT = 2\ln\left[\frac{c_q(T_q)}{4kT_q^2 \theta'^2(T_q)}\right] \qquad (6)$$

Subindex q indicates AT, AU or GC, T_q is the transition temperature (where a relative number of helix links θ_q equals 0.5), c_q is the heat capacity and θ_q is readily obtained[1] from the experimentally measured optical density. All these quantities are measured at $T=T_q(C_1,C_2)$ for given concentrations C_1 and C_2.

The F evaluation and the DNA decoding can easily be made more specific if one accounts for the interaction of nucleotide pairs.

Finally, let us discuss the value of b which enters Rule III. The coiled strands give rise to the entangling entropy. The condition that strands must meet each other at the ends of the coiled region decreases this entropy by ΔS. If we account for the elasticity of a strand by introducing a characteristic entangling angle $\varepsilon \ll 1$ between adjacent links and consider strands that are parallel at the starting and final points where strands are bonded, then simple consideration of the Brownian movement of strands containing L links gives $\Delta S = k \ln W$, where $W \sim 1$ if $L < \varepsilon^{-2/3}$; $W \sim \varepsilon^2 L^3$ if $\varepsilon^{-2/3} \lesssim L \lesssim \varepsilon^{-1}$; $W \sim \varepsilon^4 L^5$ if $\varepsilon^{-1} \lesssim L \lesssim \varepsilon^{-2}$, and $W \sim \varepsilon^{-3} L^{3/2}$ if $L > \varepsilon^{-2}$ (see Fig. 2). Therefore $T \Delta S$ slightly renormalizes the boundary energy and it implies that b changes slowly with L. In fact, considering the Brownian movement, when $L \gg \varepsilon^{-2}$, it is essential to take into account the impenetrability of strands and the appearance of knots. These knots can appear and disappear only through DNA ends, i.e. exponentially slowly. Various metastable states can therefore exist. But $L \gg \varepsilon^{-2}$ (commonly $L \gg 100$), can be essential only in the immediate vicinity of the genuine phase transition, and only for very long coiled regions whose relative contribution to F is small, and it can usually be ignored.

Thus, the F evaluation and the DNA decoding are irrelevant to any specific model assumptions and they are related only to a large value of V.

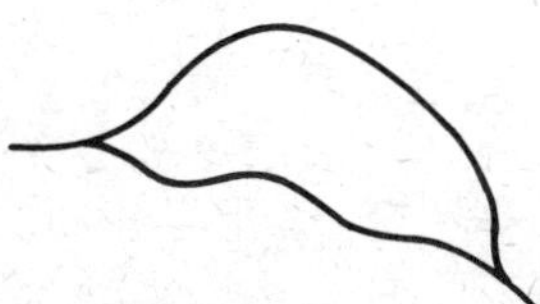

Fig. 2

References

1. Poland D., Scheraga H. A., Theory of H-C Transition in Biopolymers, Academic Press, New York, 1970.
   Wartell R. W., Montroll E. W., Adv. Chem. Phys., 22, 129 (1972).
   Lazurkin Yu. S., Frank-Kamenetsky M. D., Trifonov E. N., Biopolymers 9, 1253 (1970).
   Vedenov A. A., Dykhne A. M., Frank-Kamenetsky M. D., Uspehi Fiz. Nauk 105, 479 (1971); Sov. Phys. Uspehi 14, 715 (1972).

2. Azbel M. Ya., Pis'ma Zh.Eksp.Teor.Fiz. 16, 183 (1972); JETP Letters 16, 128 (1972).

3. Azbel M. Ya., Biopolymers, 12, 1591 (1973); Phys. Rev. Letts. 31, 589, 1592E (1973).

4. Gilbert W., private communication.

5. Lifshitz I. M., Zh.Eksp.Teor.Fiz. 65, 1100 (1973), Sov. Phys. JETP 38, 545 (1974).

6. Azbel M. Ya., J. Chem. Phys. 62, 3635 (1975).

7. Poland D., Scheraga H.A., J. Chem. Phys. 45, 1456, 1464 (1966); Applequist J., J. Chem. Phys. 45, 3459 (1966); 50, 600 (1969).

8. Due to the experimental inaccuracy some roots of Z(t) may be complex. In this case instead of complex roots x, x^* one should use $(x + x^*)$ and $xx^* / (x + x^*)$. One can see that this implies a relative error $\sim 1/V$, just as for the Q's.

CHAPTER 4

SOLUBLE MODELS ON THE TRIANGULAR AND OTHER LATTICES

R. J. Baxter

Research School of Physical Sciences,

The Australian National University,

Canberra, A.C.T., 2601 Australia

ABSTRACT

There are a number of two-dimensional lattice models for which the free energy can be calculated exactly, notably the Ising and dimer models on any planar lattice, the square lattice ice and eight-vertex models, and the pure three-spin triangular model. There are many common features in the solutions, in particular the occurrence of elliptic functions.

This suggests that there may be a more general model which includes all of these as special cases. The square lattice eight-vertex model itself contains the square Ising and ice-type models as special cases, and has recently been shown to also contain the three-spin model. However, it does <u>not</u> include the triangular and honeycomb Ising models.

I have constructed and solved a model which does include all of these as special cases, namely an eight-vertex model on the Kagomé lattice, satisfying certain restrictions on the vertex energies. (In general these restrictions are temperature-dependent.) In spin language, it is a triangular Ising model coupled to an honeycomb model via four-spin interactions. The two-spin interactions on the honeycomb lattice are related to those on the triangular by a generalized star-triangle transformation.

This model contains most of the previously solved planar models of systems with short-range forces. Apart from the spherical model, the only exceptions that come to mind are Ising models on irregular or non-translation - invariant lattices, and ice-type ferroelectric models in the presence of electric fields.

1. INTRODUCTION

Onsager[1] realized that the star-triangle relations for the Ising model implied certain commutation relations for transfer matrices. Here I wish to push this idea further and show that they can be used to solve a large class of planar eight-vertex models[2].

Construct a graph, or 'lattice' $\mathcal{L}$ by drawing straight lines in some convex planar region so that no three lines intersect at a common point. Two such lattices are shown in Fig. 1. A site is the intersection of two straight lines; each site is the endpoint of four edges.

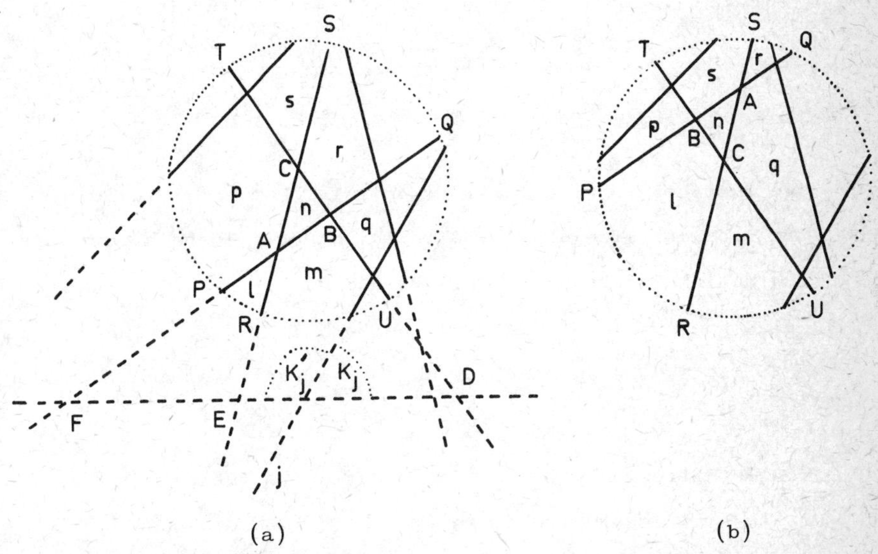

Fig. 1. Two irregular straight-line graphs or 'lattices' $\mathcal{L}$. They differ only in the position of the line PQ. The dotted circles are the boundaries. Also shown in (a) are: a base line and the extensions across it of the lattice lines (broken lines), and the coefficient K_j (and supplementary coefficient K'_j) of a typical line j.

With every angle PAR inside $\mathcal{L}$ associate a coefficient K_{PAR}. Give opposite angles at a site the same coefficient, e.g. $K_{PAR} = K_{QAS}$ in Fig. 1. Also fix a single 'four spin coefficient' K''. With each face ℓ associate an Ising spin σ_ℓ and define an 'eight-vertex' Hamiltonian $\mathcal{H}$ by

$$-\beta \mathcal{H} = \sum \left[K_{PAR}\sigma_\ell\sigma_n + K_{RAQ}\sigma_m\sigma_p + K''\sigma_\ell\sigma_m\sigma_n\sigma_p \right] . \qquad (1)$$

Here the summation is over all sites A of $\mathcal{L}$; ℓ, m, n, p are the four faces surrounding A, ordered anti-clockwise so that PAR is a corner of face ℓ, RAQ is a corner of face m, as indicated in Fig. 1a.

2. STAR-TRIANGLE RELATIONS

The two lattices in Fig. 1 differ only in that the line PQ has been shifted from one side of site C to the other. In both lattices A (B, C) is the intersection of lines PQ and RS (TU and PQ, RS and TU). Construct eight-vertex models as above on both lattices, using the same boundary conditions, the same set of angle coefficients, and the same K''.

The models differ only in the contribution to the partition function of the interactions between the spins in and around the triangle ABC, i.e. the spins on faces n, ℓ, m, q, r, s, p. Summing over σ_n gives a 'combined Boltzmann weight' W for the triangle ABC. In each case W is a function of $\sigma_\ell, \ldots, \sigma_p$.

For brevity, set

$$K_A = K_{SAQ}, \quad K_B = K_{PBT}, \quad K_C = K_{UCR},$$
$$K'_A = K_{RAQ}, \quad K'_B = K_{QBT}, \quad K'_C = K_{TCR}. \tag{2}$$

Then it is easy to show that the Boltzmann weights W are the same for both models, provided

$$\exp(2K'_B + 2K'_C) = \frac{\cosh(K_A + K_B + K_C - K'')}{\cosh(-K_A + K_B + K_C + K'')}, \tag{3}$$

plus two similar conditions obtainable by permuting A, B, C. When $K'' = 0$ these are the star-triangle relations[3].

Since all other interactions in the two models are identical, they have the same partition function if (3) is satisfied. Further all correlations not involving σ_n are the same for each model.

3. Z-INVARIANT EIGHT-VERTEX MODELS

Label the lattice lines j = 1, ..., N. Draw a 'base-line' below $\mathcal{L}$ and extend every lattice line to cross it. Assign coefficients K_j, K'_j to the intersection of line j with the base line, as indicated in Fig. 1a, for j = 1, ..., N.

Every site A of $\mathcal{L}$ is the vertex of a 'basic triangle' with one side the base line, e.g. AEF in Fig. 1a. I shall say that the eight-vertex model on $\mathcal{L}$ is 'Z-invariant' if the $\{K_j, K'_j\}$ can be chosen so that the star-triangle relations (3) are satisfied <u>for every basic triangle</u>.

Consider a typical triangle in $\mathcal{L}$, such as ABC in Fig. 1a. This is part of the quadrilateral ABCDEF, with fourth side the base line. The other three triangles AEF, BFD, CDE are basic, so the star-triangle relations (3) are satisfied for these. It follows that the star-triangle relations are also satisfied for ABC. (This can be proved either by brute force algebra, or quite neatly by elliptic functions.)

Thus a corollary of Z-invariance is that the star-triangle relations are satisfied for every triangle in $\mathcal{L}$.

Eliminating K_A, K_A', or K_B, K_B', between the three equations (3) gives

$$\Delta_A = \Delta_B = \Delta_C \quad , \tag{4}$$

where

$$\Delta_A = -\sinh 2K_A \sinh 2K_A' - \tanh 2K'' \cosh 2K_A \cosh 2K_A' \ . \tag{5}$$

Using (4) for all basic triangles, it follows that for all sites A of $\mathcal{L}$, and all sites on the base line, $\Delta_A = \Delta$, where Δ , like K'' is a single fixed numbers.

Such Z-invariant models certainly exist. One can construct them as follows: (a) Choose K'' , Δ , K_1, ..., K_N ; (b) Determine K_1', ..., K_N' from the equations $\Delta(K_j, K_j') = \Delta$; (c) Determine the K-coefficients for the angles at every site of $\mathcal{L}$ from the star-triangle relations for the corresponding basic triangle.

In particular, it is easy to show that the homogeneous square-lattice eight-vertex model is Z-invariant.

4. CORRELATIONS AND THERMODYNAMIC PROPERTIES

Hereinafter I consider only Z-invariant eight-vertex models. Their vital property is that correlations such as $\langle\sigma_\ell\sigma_n\rangle$ are unchanged by shifting lattice lines across sites, so long as no line is shifted across face ℓ or face n, and the order of lines at the boundary is preserved.

However, if ℓ and n are deep within $\mathcal{L}$, $\langle\sigma_\ell\sigma_n\rangle$ should not depend on what is happening at the boundary, so the second restriction can be ignored. (A fuller discussion of this point is given in ref. 2.)

By definition, $\langle\sigma_\ell\sigma_n\rangle$ is a function of K'', Δ and the 'line coefficients' $K_1, \ldots, K_N$. If line j does not pass between ℓ and n, then it can be shifted far away without changing $\langle\sigma_\ell\sigma_n\rangle$. But in this limit $\langle\sigma_\ell\sigma_n\rangle$ should be independent of K_j. Hence $\langle\sigma_\ell\sigma_n\rangle$ is a function <u>only of K'', Δ and the coefficients of the lines between ℓ and n</u>. The intervening lines can be re-arranged in any order, so it is a symmetric function of the coefficients of these lines, the same function for all lattices.

Further, for a given model on $\mathcal{L}$ the choice of $K_1, \ldots, K_N$ is not unique: any one can be chosen arbitrarily and the rest determined from the star-triangle relations. Thus $\langle\sigma_\ell\sigma_n\rangle$ is a function only of K'', Δ and the coefficients of the intersections of the lines between ℓ and n. Similarly for any correlation. In particular, if the boundary spins are always chosen to be up, then $\langle\sigma_\ell\rangle$ must be a function only of K'' and Δ, the same for all faces ℓ deep within $\mathcal{L}$, and the same for all lattices $\mathcal{L}$. Thus

$$\langle\sigma_\ell\rangle = M(K'', \Delta) , \tag{6}$$

where $M(K'', \Delta)$ is the magnetization of the homogeneous square lattice model, for which a very plausible conjecture is available[4].

Similarly, for any lattice $\mathcal{L}$ and any two adjacent faces ℓ and m,

$$\langle \sigma_\ell \sigma_m \rangle = P(K'', \Delta) \quad , \tag{7}$$

where $P(K'', \Delta)$ is the polarization of the square lattice model[5].

Further, if ℓ,m,n,p are the four faces round a site A, as in Fig. 1a, then any correlation of the corresponding four spins is a function only of K'', K_{PAR} and K_{RAQ}. (These determine Δ.) Any derivative of the total free energy F is a sum of such correlations. By integrating, maintaining Z-invariance and ignoring sites near the boundary, it follows that

$$F = \sum f(K_{PAR}, K_{RAQ}, K'') \quad , \tag{8}$$

where the summation is over all sites A and $f(K, K', K'')$ is the same function for all sites and all lattices. Thus f is the known free energy per site of the square lattice model[6].

This function f is analytic except when $\Delta = \pm 1$, so the same is true for F. Any Z-invariant eight-vertex model is therefore ordered if $|\Delta| > 1$, disordered if $|\Delta| < 1$, critical if $|\Delta| = 1$. Setting $t = 1 - \Delta^2$, it follows from the conjectured and known square-lattice results that when t is small and negative

$$M \propto (-t)^{\pi/16\bar{\mu}} \; , \; P \propto (-t)^{(\pi-\bar{\mu})/4\bar{\mu}} \; , \; F_{singular} \propto (-t)^{\pi/\bar{\mu}} \; , \tag{9}$$

where

$$\cos \bar{\mu} = \Delta^{-1} \tanh 2K'' \; , \; 0 < \bar{\mu} < \pi \; . \tag{10}$$

Thus the 'critical exponents' are also functions only of K'' and Δ, the same for all lattices.

5. HOMOGENEOUS KAGOMÉ LATTICE MODEL

An homogeneous Z-invariant model can be constructed on the Kagomé lattice of Fig. 2. Divide the sites into sub-lattices A, B, C as indicated. Assign coefficients K_A, K_B, K_C (K_A', K_B', K_C') to the 60° (120°) angles at all A, B, C sites, respectively. Fix K''. Choose these so that the three relations (3) are satisfied. Then there are four disposable parameters (e.g. K_A, K_B, K_C, K'') and the model is Z-invariant. It consists of a nearest-neighbour two-spin honeycomb Ising model with coefficients K_A, K_B, K_C, together with a model with coefficients K_A', K_B', K_C' on the dual (triangular) lattice. There is a four-spin interaction, with coefficient K'', on crossing edges. The coefficients are related by the generalized star-triangle relations (3).

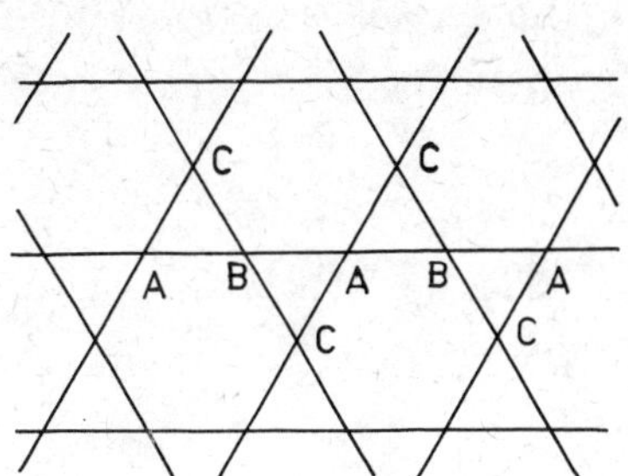

Fig. 2. The Kagomé lattice.

TABLE 1

Values of K_A, ..., K_C', K'' corresponding to previously solved special cases of the Z-invariant Kagomé lattice eight vertex model.

K_A	K_B	K_C	K_A'	K_B'	K_C'	$-K''$	Special Model
K	K'	∞	K'	K	0	0	Square Ising[1,7]
.	.	.	.	.	.	0	Honeycomb and triangular Ising[3]
K	K'	∞	K'	K	$-K''$	$-K''$	Square eight-vertex[6] and six-vertex[8]
K	K	K	0	0	0	-K	Triangular three-spin[9]
all infinite, differences finite							Restricted Kagomé six-vertex[10], and triangular twenty-vertex[11]. Critical triangular and square Potts models[10].

Even though the relations (3) are temperature-dependent, this model is still very interesting. It contains many previously solved models as special cases, as is indicated in Table 1.

Thus this new understanding of the significance of the star-triangle relations (3) unifies much previous work.

6. FORMULAE FOR f, M, P IN FERROMAGNETIC ORDERED PHASE

Consider again an arbitrary lattice $\mathcal{L}$, as in sections 1 to 4. Eq. (5) has the structure of a non-Euclidean trigonometry relation. As Onsager (ref. 7, p. 144) noted, such relations can be simplified by using elliptic integrals and functions. Suppose all the angle coefficients K_{PAR} are positive and that $\Delta < -1$. Then the system is ferromagnetically ordered. Define

$$k^{-2} = \Delta^2 \cosh^2 2K'' - \sinh^2 2K'' , \quad 0 < k < 1 , \tag{11}$$

$$g(\tau) = \int_{2\tau}^{\infty} (1 + k^2 \sinh^2\phi)^{-\frac{1}{2}} d\phi , \tag{12}$$

$$\lambda = g(-K'') , \quad \alpha_{PAR} = g(K_{PAR}) , \quad \mathcal{K}' = g(0) . \tag{13}$$

Then (5) and (3) imply that for all sites A and triangles ABC,

$$\alpha_{PAR} + \alpha_{RAQ} = \lambda , \quad \alpha_{CAB} + \alpha_{ABC} + \alpha_{BCA} = \lambda . \tag{14}$$

It is now easy to verify the quadrilateral property of the star-triangle relations that was used in section 3. If $\mathcal{K}$ is the complete elliptic integral of modulus k, then the known and conjectured results for f, M, P in (8), (6) and (7) are

$$q = \exp(-2\pi \mathcal{K}'/\mathcal{K}) , \quad x = \exp(-\pi\lambda/2\mathcal{K}) , \quad q < x^2 , \tag{15}$$

$$z_A = \exp\left[-\pi(\alpha_{PAR} - \alpha_{RAQ})/2\mathcal{K}\right] , \quad x < z_A < x^{-1} , \tag{16}$$

$$-\beta f(K_{PAR} , K_{RAQ} , K'') = K_{PAR} + K_{RAQ} + K''$$

$$+\sum_{n=1}^{\infty} \frac{x^{-n}(x^{2n}-q^{n})^{2}(x^{n}+x^{-n}-z_A^{n}-z_A^{-n})}{n(1-q^{2n})(1+x^{2n})}, \qquad (17)$$

$$M = \prod_{n=1}^{\infty}\left[(1-x^{4n-2})/(1+x^{4n-2})\right], \qquad (18)$$

$$P = \prod_{n=1}^{\infty}\left\{(1+q^{n})(1-x^{2n})/\left[(1-q^{n})(1+x^{2n})\right]\right\}^{2}. \qquad (19)$$

The forms (14) of the Z-invariance conditions clearly can be interpreted geometrically. An obvious (but not unique) solution is, for every angle PAR on a given lattice $\mathcal{L}$,

$$\alpha_{PAR} = (\lambda/\pi) \times \text{the angle PAR expressed in radians.} \qquad (20)$$

1. L. Onsager, in 'Critical Phenomena in Alloys, Magnets and Superconductors', eds. R.E. Mills, E. Ascher and R.I. Jaffee, McGraw Hill N.Y. (1971).
2. R.J. Baxter, 'Solvable eight vertex model on an arbitrary planar lattice', submitted to Proc. Roy. Soc.
3. R.M.F. Houtappel, Physica 16, 425 (1950), eq. 23.
4. M.N. Barber and R.J. Baxter, J. Phys. C 6, 2913 (1973).
5. R.J. Baxter and S.B. Kelland, J. Phys. C 7, L403 (1974).
6. R.J. Baxter, Ann. Phys. 70, 193 (1972).
7. L. Onsager, Phys. Rev. 65, 117 (1944).
8. E.H. Lieb, Phys. Rev. Lett. 18, 1046 and 19, 108 (1967).
9. R.J. Baxter and F.Y. Wu, Australian J. Phys. 27, 357 (1974).
10. R.J. Baxter, H.N.V. Temperley and S.E. Ashley, 'Triangular Potts model at its transition temperature, and related models', to be published in Proc. Roy. Soc.
11. S.B. Kelland, Australian J. Phys. 27, 813 (1974).

CHAPTER 5

RECENT ADVANCES CONCERNING THE d - 2 AND 4 - d EXPANSIONS

E. Brézin

Service de Physique Théorique, CEN-Saclay,

BP N° 2, 91190 Gif-sur-Yvette, France

ABSTRACT

Critical properties of classical n-vector models are studied near two dimensions. The basis of the expansion in powers of (d-2) is reviewed.

Large orders of the expansion of critical exponents in powers of 4-d are computable. For instance, if

$$\gamma = \sum_{0}^{\infty} \gamma_K \, \varepsilon^K$$

$$\gamma_K \underset{K\to\infty}{=} K!\left(\frac{-3}{n+8}\right)^K K^{4+\frac{1}{2}n} \, C\left(1 + O(\tfrac{1}{K})\right) \quad .$$

Various implications are discussed. Similarly three-dimensional calculations in which the same type of information is available are reported.

I. INTRODUCTION

In the critical region, the neglect of short range effects leads to the substitution of continuous spin distributions S(x), to the original lattice spin variables S_i, i.e. to a field theory. However the final continuous model depends on the analysis applied to the lattice Hamiltonian. If one performs an expansion around mean field theory, in the long distance limit this leads to the Landau-Ginzburg-Wilson model[1])

$$\beta H = \int d^d x \left\{ \frac{1}{2} \nabla S^2 + \frac{1}{2} r_o S^2 + u_o S^4 \right\} . \tag{1}$$

A low temperature expansion performed on a (classical) Heisenberg n-component Hamiltonian

$$H = \sum_{i,j} V_{ij} \vec{S}_i . \vec{S}_j , \tag{2}$$

leads, for n > 2 to the continuous model[2,3])

$$\beta H = \frac{1}{T} \int d^d x \left\{ \sum_{\alpha=1}^{n-1} (\nabla S_\alpha)^2 + \left[\nabla \sqrt{1-(S_1^2+\ldots+S_{n-1}^2)} \right]^2 \right\} , \tag{3}$$

which describes interacting spin waves. Perturbation theory is not directly applicable to the Hamiltonians (1) or (3), but the long distance properties are governed by the renormalization group analysis. The existence of a fixed point u_o^* of order ε in $(4-\varepsilon)$ dimensions for the Hamiltonian (1) leads to the ε-expansion of Wilson and Fisher[4]). Similarly a fixed point T^* of order ε' in $(2+\varepsilon')$ dimensions for the Hamiltonian (3) yields a systematic expansion in powers of ε' of all the critical properties for n larger than two[(3)].

Since the ideas underlying the ε'- expansion are not very familiar I shall outline the basic steps in a first part.

For numerical applications to three-dimensional systems the ε-expansion is quite accurate ... provided one does not include the order ε^3 [5]). It is now known that the ε-expansion has zero radius of convergence due to an essential singularity at $\varepsilon = 0$ [6]). However its divergence seems to be controllable in a way which would allow one to replace systematically the divergent ε-series by a convergent algorithm, (although a proof of the Borel summability of the ε-expansion seems to lie beyond the scope of the presently available techniques). These questions are discussed in the second part.

The last part is devoted to direct three-dimensional calculations by Callan-Symanzik equations. They rely on divergent expansions in powers of u_0. However this divergence is now completely characterized and this knowledge leads to powerful numerical techniques which give very accurate critical exponents[7,8].

II. LOW-TEMPERATURE EXPANSION NEAR TWO DIMENSIONS

Consider a classical n-component Heisenberg model

$$\begin{cases} H = \sum_{i.j} V_{ij}(\vec{S}_i\vec{S}_j - 1) \\ \vec{S}_i = (S_{i,1};\ldots;S_{i,n}) \\ \vec{S}_i^2 = 1 \end{cases} \tag{4}$$

with ferromagnetic, short-range, translationnaly invariant, interactions. Above two dimensions, at low enough temperature, the system is in an ordered phase, with a spontaneous magnetization pointing in a direction along which we put the last coordinate axis. Each spin $\vec{S}_i$ has thus a longitudinal component

$$\sigma_i = S_{i,n}$$

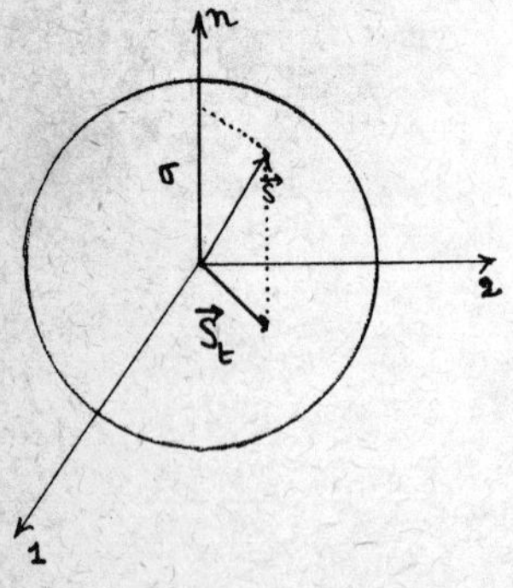

and (n-1) transverse components $\vec{S}_{i,t} = (S_{i,1};\ldots;S_{i,n-1})$. At low temperature most of the time the spins are pointing around the north pole of the sphere and the transverse components are typically very small.

In terms of the transverse variables the Hamiltonian is

$$H = \sum_{ij} V_{ij} \left(\vec{S}_{i,t} \vec{S}_{j,t} + \sqrt{1 - \vec{S}^2_{i,t}} \sqrt{1-\vec{S}^2_{j,t}} - 1 \right) \tag{5}$$

and the phase space measure for each spin is

$$\frac{d^{n-1} S_{i,t}}{\sqrt{1-\vec{S}^2_{i,t}}} \tag{6}$$

A systematic expansion in powers of T, may now be obtained in the following manner :

(i) the square roots in the Hamiltonian are expanded in powers of the $\vec{S}^2_t$

(ii) the quadratic part H_o of H

$$H_o = \sum_{ij} V_{ij} \left(\vec{S}_{i,t} \vec{S}_{j,t} - \frac{1}{2} \vec{S}^2_{i,t} - \frac{1}{2} \vec{S}^2_{j,t} \right) \tag{7}$$

is kept in exponential form as $\exp-\beta H_o$

(iii) interaction terms are generated by expanding $\exp-\beta(H-H_o)$ and the measure (6)

(iv) each component of $\vec{S}_{i,t}$ is integrated from $-\infty$ to $+\infty$. The error made in extending the integration range involves terms with a temperature dependence in $\exp-a/T$ which do not modify the expansion in powers of T.

Each term in this expansion may be represented by a diagram in which the propagator is the inverse of the quadratic form βH_o

$$G_{\alpha\beta}(\vec{q}) = \delta_{\alpha\beta} \frac{T}{\tilde{V}(o) - \tilde{V}(q)} \tag{8}$$ $\alpha, \beta = 1, \ldots, n-1$

$$\tilde{V}(q) = \sum_j V_{ij} \exp i \vec{q}\, \vec{r}_{ij} \tag{9}$$

The vertices are given by the rule (iii) ; they carry a factor 1/T (or 1). It is a straightforward exercise to show that the dependence of a given diagram in T is a power equal to the number of loops of the diagram. Let us for instance calculate the expansion in powers of T of the spontaneous magnetization

$$M = \langle \sigma_i \rangle = \langle \sqrt{1 - S^2_{i,t}} \rangle = 1 - \frac{1}{2} \langle S^2_{i,t} \rangle - \frac{1}{8} \langle (S^2_{i,t})^2 \rangle + \cdots \tag{10}$$

At lowest order the only contribution comes from $\langle (\vec{S})^2 \rangle$; there is one one-loop diagram

$$\text{(loop diagram)} = (n-1) \int \frac{d^d q}{(2\pi)^d} \frac{T}{\tilde{V}(0) - \tilde{V}(q)} \tag{11}$$

For a short-range interaction at small momentum

$$\tilde{V}(0) - \tilde{V}(q) \underset{q \to 0}{=} J q^2 \tag{12}$$

and the integral (11), over the first Brillouin zone is infra-red divergent when the dimension approaches two. A similar analysis performed on the higher orders terms leads to the following results

$$M = 1 - \frac{1}{2} \frac{(n-1)}{(d-2)} T/2\pi J - \frac{1}{8} \frac{(n-1)(n-3)}{(d-2)^2} (T/2\pi J)^2$$

$$- \frac{1}{48} \frac{(n-1)(n-3)(3n-7)}{(d-2)^3} (T/2\pi J)^3 + O(T^4) \tag{13}$$

in which, order by order in T, we have kept the most divergent part when d approaches two. The expression (13) is easily checked to be the expansion at order T^3 of the formula

$$M = \left[1 - \frac{n-2}{d-2} T/2\pi J \right]^{\frac{n-1}{2(n-2)}} \tag{14}$$

which predicts a critical temperature

$$T_c / 2\pi J = \frac{d-2}{n-2} + O(d-2)^2 \tag{15}$$

and a critical exponent

$$\beta = \frac{n-1}{2(n-2)} + O(d-2) \tag{16}$$

However this observation is certainly not sufficient to prove a formula like (14) which makes predictions on the leading divergence of any power of T. The proof is quite technical[9]. It makes use of the renormalization group analysis applied to the Hamiltonian (3). It is also possible, to go beyond the leading order of Eq.(13) and to obtain corrections to (15) and (16). The results for the critical exponents are[9]

$$\beta = \frac{n-1}{2(n-2)} - \frac{n-1}{2(n-2)^2}(d-2) + O(d-2)^2 \tag{17}$$

$$1/\nu = d-2 + \frac{(d-2)^2}{n-2} + O(d-2)^3 \tag{18}$$

The correction of order $(d-2)^2$ to the critical temperature (15) is not universal. The corrections to scaling may be analysed in this model and they are governed by the exponent

$$\omega = 2 - \frac{n}{n-2}(d-2) + O(d-2)^2 \tag{19}$$

At fixed temperature below T_c, the long distance behaviour of the correlation functions is governed by spin waves. Near T_c there is a cross-over to the critical behaviour which is described by the renormalization group.

The nature of the expansion in powers of (d−2) is unknown. It does not give very interesting numbers in three dimensions.

The theory that has been described is limited to n strictly larger than two. In two dimensions, Kostezlitz and Thouless have shown the existence of vortex excitations for n=2 from which results a low temperature phase corresponding to a line of fixed points[10]. D. Fisher and D. Nelson have recently argued than for d=2+ε', n=2 the critical exponent ν diverges as $1/\sqrt{\varepsilon'}$.

III. LARGE ORDERS OF THE ε-EXPANSION

The ε-expansion (ε=4-d) of Wilson and Fisher[4] has given a lot of informations on critical exponents, scaling functions, ratios of critical amplitudes etc ... . It normally starts into the right direction,gets rather close to the experimental value after one or two terms, and then begins to drift away from the correct value. As an example we take γ for n=1[5]

$$\gamma = 1 + 0.167\varepsilon + 0.077\varepsilon^2 - 0.049\varepsilon^3 + O(\varepsilon^4) \tag{20}$$

The omission of the ε^3 term gives a much better result. It is thus suspected that the true ε-expansion is asymptotic, but divergent in much the same way that it appears in Wilson's approximate recursion formula[1]. One could thus be tempted to conclude that, (i) equation (20) should indeed be stopped at order ε^2 , (ii) there is no possible use of the higher orders in ε and (iii) even as a matter of principle, the theory has reached a maximum accuracy with the order ε^2 and will never be able to go beyond that limit. However, these conclusions are certainly very pessimistic, though the ε-expansion is indeed divergent as we shall see below. Consider, as an illustrative example the much simpler following problem. Suppose that we want to calculate by perturbation theory the ground state energy of the one dimensional anharmonic oscillator

$$H = \frac{1}{2}(p^2 + x^2) + g x^4 \tag{21}$$

$$E(g) = \frac{1}{2} + g E_1 + \dots + g^K E_K + \dots \tag{22}$$

The table of the first E_K' s is

	0	1	2	3	4	5	6	7	8
E_K	0.5	0.75	-2.625	20.8125	-2.41289×10^2	3.58098×10^3	-6.39828×10^4	1.32973×10^6	-3.1448×10^7

and the E_K's keep growing very rapidly (E_{75} is of order 10^{144}). However the ground state energy is of order one for values of g of order one. Therefore the truncated perturbation theory is absurd. Can we still extract in a systematic way E(g=1) from the knowledge of the E_K's ? The answer involves several steps. The first one is the exact characterization of the E_K at large K[11,6) :

$$E_K \underset{K\to\infty}{\sim} K!\,(-3)^K K^{-1/2} \frac{6}{\pi^{3/2}} \left(1+O\left(\frac{1}{K}\right)\right) \tag{23}$$

which exhibits the divergence of the series (22) for any value of g. It is thus tempting to divide out the K! of Eq.(23) by the Borel summation procedure :

$$\sum_K g^K E_K \equiv \sum_K g^K \frac{E_K}{K!} \int_0^\infty e^{-t} t^K dt$$

and if summation and integration are interchanged

$$\sum_K E_K g^K = \int_0^\infty e^{-t} f(tg)\, dt \tag{24}$$

$$f(z) = \sum_0^\infty z^K \frac{E_K}{K!} \tag{25}$$

The proof of the legitimacy of this interchange lies outside perturbation theory[12).

From the $(-3)^K$ of Eq.(23) we see that f(z) has a singularity at z = -1/3, and from the $K^{-1/2}$ that this singularity is a square root branch point. Various numerical techniques may be used at this stage : for example (i) replace f(z) by a simple Padé approximant, (ii) make a conformal mapping of the cut z-plane into the interior of a circle. The Taylor expansion in the new variable converges inside the whole circle (if all the singularities of f(z) lie on the real negative axis $z \leq -1/3$). As a simple numerical experiment we can imagine that our knowledge of the E_K's is limited to $K \leq 6$. The Padé-Borel technique (i) gives a result which differs from the true ground state energy, which can be accurately calculated by variational methods, by less than 10^{-3} for g=1 . The error from the conformal mapping method (ii), with the same information $K \leq 6$ drops to 10^{-4}.

We thus conclude that, provided one knows the precise way in which the series diverges, one can transform the divergent perturbation series and deplace it by convergent algorithms dictated by the high order behaviour.

The equivalents of (23) for the ε-expansion are

$$\gamma = \sum_{K=0}^{\infty} \varepsilon^K \gamma_K\,, \quad 1/\nu = \sum_{0}^{\infty} \varepsilon^K \sigma_K \quad \omega = \sum_{0}^{\infty} \varepsilon^K \omega_K$$

$$\left\{\begin{matrix} \gamma_K \\ \sigma_K \\ \omega_K \end{matrix}\right. \underset{K\to\infty}{\sim} K!\left(-\frac{3}{n+8}\right)^K \left\{\begin{matrix} K^{3+n/2} \\ K^{4+n/2} \\ K^{5+n/2} \end{matrix}\right. \qquad (24)$$

again divergent series with signs alternatively positive and negative.

The derivation of (24) cannot be included in this report[6]. The method relies on saddle-point expansion of functional integrals. It was initiated by Lipatov[13] who studied the expansion in powers of g of the four dimensional $g\,\phi^4$ theory, but the method is quite general[14]. We are now in position to perform the same numerical analysis which was done for the anharmonic oscillator. Unfortunately our knowledge of the ε-expansion is limited at present to $K \leq 3$, and we cannot expect a very good accuracy at this stage. For instance for n=1 we obtain with these techniques a value for γ equal to 1.22, an error of a few percents which is to be expected with so few terms. It is very likely that with two or three more terms the ε-expansion would give results with an error smaller than 10^{-3}. Unfortunately this requires gigantic calculations.

IV. THREE-DIMENSIONAL CALCULATIONS

The N-point correlation functions in the Landau-Wilson model (1) satisfy the differential equation[15)]

$$\left[m\frac{\partial}{\partial m} + \beta(u)\frac{\partial}{\partial u} - \frac{N}{2}\eta(u)\right]\Gamma^{(N)} = \Delta\Gamma^{(N)} \tag{25}$$

in which m is the inverse of the correlation length ; this equation is valid order in order in a double expansion in powers of u and ε . The fixed point u^* is a zero of β

$$\beta(u^*) = 0 \quad \text{and} \quad \beta'(u^*) > 0$$

and the critical exponent are given by

$$\eta = \eta(u^*)$$

and there is a similar formula for ν . It has been suggested several years ago by Parisi[16)] that these equations could be used in a fixed dimension d=3 or d=2 through an expansion in powers of u alone. This has motivated very extensive calculations of the functions $\beta(u)$, $\eta(u)$, $\nu(u)$ in three dimensions up to six loops diagrams[17)]. The large order behaviour in this theory has been determined recently[6)] and gives for instance[18)] for $n = 1$

$$\beta_K \underset{K\to\infty}{\sim} \Gamma(K+9/2)(-0.14777422)^K\, 0.041175 \tag{26}$$

This information has been used for making elaborate numerical analysis of the u-series[7,8)]. Le Guillou and Zinn-Justin[7)] have made a very detailed use of the asymptotic information (26) in several ways which allowed them to reduce, the numerical uncertainties. They concluded that the three-dimensional critical exponents have the following best estimates

n	0	1	2	3
γ	1.1615 ± 0.0011	1.2402 ± 0.0009	1.3160 ± 0.0012	1.3866 ± 0.0012
η	0.0260 ± 0.0030	0.0315 ± 0.0025	0.0335 ± 0.0025	0.0340 ± 0.0025

REFERENCES

1. K.G. WILSON and J.B. KOGUT ; Phys. Rep. 12C, 75 (1974)
2. A.M. POLYAKOV ; Phys. Lett. B59, 79 (1975)
3. E. BREZIN and J. ZINN-JUSTIN ; Phys. Rev. Lett. 36, 691 (1976)
4. K.G. WILSON and M.E. FISHER ; Phys. Rev. Lett. 28, 240 (1972)
   K.G. WILSON ; Phys. Rev. Lett. 28, 548 (1972)
5. E. BREZIN, J.C. LE GUILLOU, J. ZINN-JUSTIN and B.G. NICKEL ; Phys. Lett. 44A, 227 (1973)
6. E. BREZIN, J.C. LE GUILLOU and J. ZINN-JUSTIN ; Phys. Rev. D15, 1544 and 1558 (1977)
7. J.C. LE GUILLOU and J. ZINN-JUSTIN ; Phys. Rev. Lett. 39, 95 (1977)
8. G.A. BAKER, B.G. NICKEL and D.I. MEIRON ; Saclay preprint DPh-T-77/39
9. E. BREZIN and J. ZINN-JUSTIN ; Phys. Rev. Lett. 36, 691 (1976), Phys. Rev. B14, 3110 (1976)
   E. BREZIN, J. ZINN-JUSTIN and J.C LE GUILLOU ; Phys. Rev. D14, 2615 (1976)
10. J.M. KOSTERLITZ and D.J. THOULESS ; J. Phys. C6, 1181 (1973)
    J.M. KOSTERLITZ ; J. Phys. C7, 1046 (1974)
11. C. BENDER and T.T. WU ; Phys. Rev . D7, 1620 (1972)
12. For this particular problem the Borel summability has been proven by S. GRAFFI, V. GRECCHI and B. SIMON ; Phys. Lett. 32B, 681 (1970)
13. L.N. LIPATOV ; JETP Lett. 24, 179 (796) ; 25, 116 (1977)
14. See for instance E. BREZIN. "The role of instantous in divergent perturbation series", Budapest Conference on Particle Physics, July 1977 (to be published)
15. For a review and references on the use of Callan-Symanzik in critical phenomena see : "E. BREZIN, J.C. LE GUILLOU and J. ZINN-JUSTIN in "Phase transitions and Critical Phenomena", vol.VI, edited by C. Domb and M.S. Green (Academic Press 1975)
16. G. PARISI, 1973 Cargese Summer School (unpublished)
17. G.A. BAKER, B.G. NICKEL, M.S. GREEN and D.I. MEIRON ; Phys. Rev. Lett. 36, 1351 (1976)
18. The K-independent coefficient 0.041175 has been calculated recently by G. PARISI and E. BREZIN (unpublished).

CHAPTER 6

SHAPE AND SIZE OF ISING CLUSTERS AND THE WHITNEY POLYNOMIAL

C. Domb

King's College, London WC2R 2LS, U.K*.
and Bar-Ilan University, Ramat-Gan, Israel

* Permanent address

ABSTRACT

In the earliest version of the droplet model of condensation no attempt was made to take account of the different possible shapes of droplet, all droplets being treated as spherical. Fisher's generalization of the model allowed for different surface conformations, but assumed that the droplets or clusters remained compact, i.e. that $s/n \to 0$ for large n, s being the number of surface molecules, and n the number of bulk molecules. Domb suggested that in the critical region ramified clusters of tree-like or sponge-like character could play an important role; for such clusters s/n tends to a finite limit for large n. Monte Carlo data for the two dimensional Ising model gave support to this view. Domb and Stoll used the cyclomatic number of a cluster as a measure of its degree of compactness or ramification, and used Monte Carlo data to estimate the average number of spins and average number of cycles per cluster as a function of temperature. At sufficiently high temperatures the coupling between spins is negligible, and results are relevant to random site percolation. At lower temperatures the results are relevant to site percolation with correlation, a problem which has attracted attention recently. The results are also related to the Whitney polynomial studied by Temperley and Lieb, and an exact calculation by these authors at the critical temperature enables the pattern of critical behaviour of the above cluster parameters to be conjectured.

1. INTRODUCTION

The droplet model of condensation was first introduced[1] in the 1930's, and served very effectively as a basis for explaining the phenomena of nucleation and metastability. The droplets were always taken to be spherical in shape, and although it was realized that other shapes ought to be taken into account[2], this was assumed to be a refinement of little significance to the kinetics of condensation.

In 1963 Essam and Fisher[3] first applied the ideas of the droplet model to critical behaviour, and the theory was later developed and amplified by Fisher[4]. The aim was to construct a mimic partition function, and changes of shape of the droplets were taken into account by including a surface entropy term. However the droplets were still considered to be 'compact' (Fig. 1a), i.e. the surface area s is proportional to n^σ for large n where σ is less than 1 .

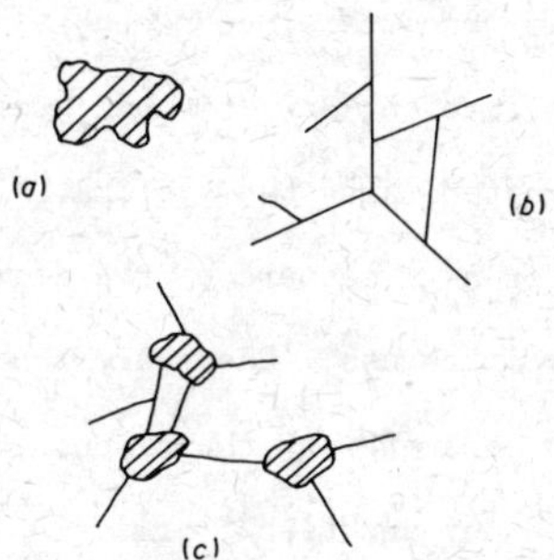

Fig. 1 Different types of droplet:
(a) compact, $\sigma < 1$;
(b) ramified, $\sigma = 1$;
(c) hybrid.

The present author suggested recently[5] that 'ramified' droplets of a tree-like or sponge-like character (Fig.1b) for which $\sigma = 1$ should be taken into account, and could play an important role in the critical region. In practice most droplets would possibly be hybrid (Fig.1c) with compact nuclei strung together in a ramified manner. A study of shapes of clusters in a Monte Carlo simulation of a lattice gas model[6] did indeed demonstrate such characteristics (Fig.2, it should be remembered that the maximum area in the isoperimetric problem on a sq. lattice corresponds to a square and not a circle.)

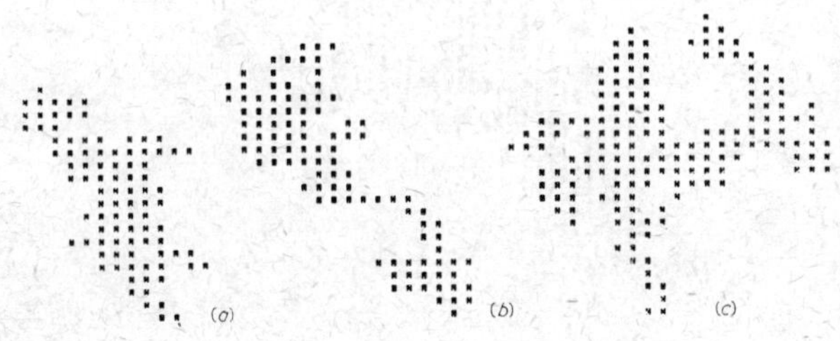

Fig. 2 Some typical clusters for $T < T_c$:
(a) $T/T_c = 0.958$;
(b) $T/T_c = 0.979$;
(c) $T/T_c = 0.99$.
The 'tree-like' character of the clusters should be noted, and the increase of ramification as T approaches T_c .

When $T > T_c$ the Fisher model breaks down, and although attempts have been made to modify the theory suitably[7], there is no longer a direct connection between the theory and cluster shapes. When T/T_c is sufficiently large the interaction between spins can be neglected and the system is one of random percolation. There is good theoretical support here for the suggestion that only ramified clusters play a significant role[8]. The problem of percolation in Ising systems, i.e. percolation with correlation, has recently attracted attention[9], and cluster shapes for $T > T_c$ show clear signs of ramification (Fig. 3a,b).

Fig. 3a $T/T_c = 1.1$

It is useful to quantify the above concepts of compactness or ramification, and Domb and Stoll[10] suggested using the cyclomatic number as a measure of the degree of compactness. They were led to this parameter by the discussions of Temperley and Lieb[11] on the

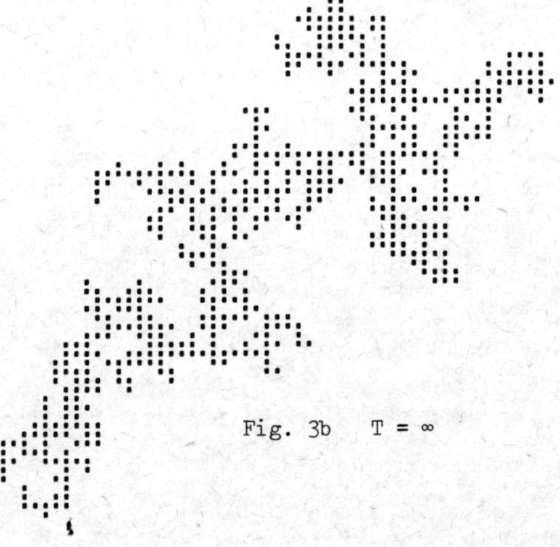

Fig. 3b $T = \infty$

Fig. 3 Cluster shapes for $T > T_c$ (from Monte Carlo simulation by Stoll and Schneider)

Whitney-Tutte polynomial* and its relationship to percolation and the Ising model. This is a polynomial in two variables q and x which directs attention to the cyclomatic numbers of different graphs. For $q = 1$ the polynomial is related to the bond percolation problem, and for $q = 2$ to the Ising model. The relation between the Potts model[12] and the percolation problem was dealt with in the pioneering work of Kasteleyn and Fortuin[13], who also established the relation between the Potts model and the Whitney-Tutte polynomial.

The only exact calculation available relating to shape and size of clusters applies to the critical point[14] of the sq. lattice. But Monte Carlo data obtained with a two-dimensional one-spin-flip Ising model (Glauber model) can be used to determine how the sizes and shapes of Ising clusters change with temperature. By varying the magnetic field at a sufficiently high temperature the same data can be obtained for a site percolation process as a function of concentration. The exact calculation provides a useful check on the Monte Carlo data, and helps in interpreting behaviour in the critical region.

2. STATISTICAL PARAMETERS†

Consider a cluster of n points and l bonds. The cyclomatic number $c(n,l)$ of the clusters is defined by

$$c(n,l) = l - n + 1 , \tag{1}$$

and represents the number of independent cycles in the cluster. For a tree $c = 0$,

*Professor P. W. Kasteleyn has suggested this terminology; although Whitney introduced the basic ideas, Tutte was the first to use the polynomial.

†The Monte Carlo data used in the following sections were obtained by E. Stoll and T. Schneider at IBM Zurich. The analysis of the data is due to C. Domb and E. Stoll.

for a simple polygon $c = 1$, and ramified clusters correspond to small c. The most compact cluster on a square (SQ) lattice is a square for which

$$c_{max} = (n^{1/2} - 1)^2 . \tag{2}$$

We therefore define the coefficient of compactness of a cluster, λ, as

$$\lambda = c/c_{max} = c/(n^{1/2} - 1)^2 . \tag{3}$$

This can go from 0 (completely ramified) to 1 (completely compact).

In analysing the statistics of clusters we must be careful to differentiate between 'per site' and 'per cluster' averages. The following definitions should clarify this point. Consider a lattice of N Ising spins a number of which are overturned; we are interested in the limit of large N. Let $z(n,N)$ be the number of clusters of n points, formed by the overturned spins. We then define $z(n)$, the probability per site of finding an n-cluster, as

$$z(n) = \lim_{N\to\infty} \frac{z(n,N)}{N} . \tag{4}$$

The probability that any given site belongs to an n-cluster of overturned spins is $nz(n)$. Hence if there is no infinite cluster, the average number of clusters per site is given by

$$\langle C \rangle_{site} = \sum_n z(n) , \tag{5}$$

and the average number of overturned spins per site by

$$\langle P \rangle_{site} = \sum_n nz(n) . \tag{6}$$

The latter is closely related to the magnetization.

In percolation theory we are often interested in statistics in the presence of an infinite cluster; for example, for the Ising model in three dimensions infinite clusters occur for most lattices even when $T < T_c$ [9]. Equation (5) does not have to be modified since the addition of one infinite cluster to $z(n,N)$ makes negligible difference asymptotically as $N \to \infty$. However, equation (6) must be modified to take account of the infinite cluster,

$$\langle P \rangle_{site} = \sum nz(n) + P_{\infty} \ , \tag{7}$$

where P_{∞} represents the probability per site of belonging to the infinite cluster; for a finite system we define a spanning cluster $S(N)$ as a cluster which extends to the boundaries of the system; if the number of spins belonging to the spanning cluster is $S(N)$ then

$$P_{\infty} = \lim_{N \to \infty} \frac{S(N)}{N} \ . \tag{8}$$

We also define the average number of cycles per site as

$$\langle c \rangle_{site} = \sum c(n) z(n) \ . \tag{9}$$

Passing now to cluster averages which characterise cluster sizes and shapes we define the average number of spins per cluster as

$$\begin{aligned} \langle n \rangle_{cluster} &= \frac{\text{Average number of overturned spins per site}}{\text{Average number of clusters per site}} \\ &= \frac{\Sigma nz(n)}{\Sigma z(n)} \ . \end{aligned} \tag{10}$$

Similarly the average number of cycles per cluster is given by

$$\langle c \rangle_{cluster} = \frac{\Sigma \bar{c}(n) z(n)}{\Sigma z(n)} \ , \tag{11}$$

where $\bar{c}(n)$ is the average of $c(n, l)$ over l .

3. ISING CLUSTERS

In Fig. 4 we show the behaviour of $\bar{c}(n)$ as a function of n for various values of T/T_c . Asymptotically the curves all become linear in n ; since by equation (2) c_{max} is asymptotically equal to n , we find that for large clusters, the compactness coefficient is given by

$$\lambda = \lim_{n \to \infty} (d\bar{c}/dn) \ . \tag{12}$$

By this means the curve in Fig. 5 has been derived showing the variation of λ with (T_c/T). At $T = \infty$ corresponding to random percolation at 5% of critical concentratio

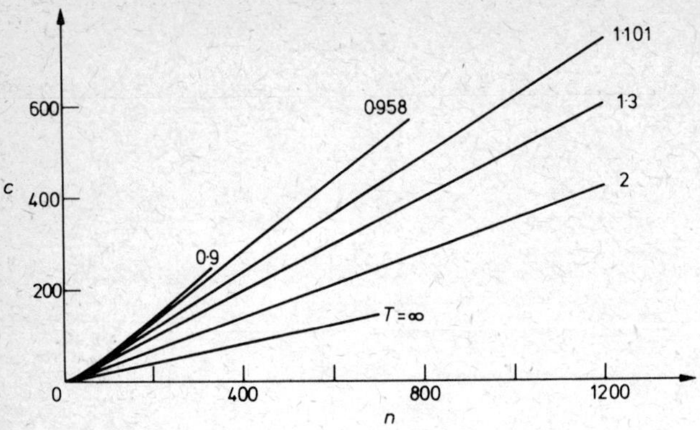

Fig.4 Cyclomatic number c as a function of n at different temperatures. For large n,c becomes linear, and the slope is a measure of the compactness of the cluster. Values of T/T_c are shown on the figure.

λ is about 0.2. As T decreases λ increases smoothly and steadily reaching a value of about 0.7 at T_c. Below T_c, λ increases quite rapidly to 1.

Fig. 6 shows the variation of the mean number of clusters per site $\langle C\rangle_{site}$ with temperature. The sudden increase as T goes below T_c seems surprising at first sight; but even though the total number of overturned spins drops very rapidly the fragmentation into small clusters is even more rapid, so that on balance $\langle c\rangle_{site}$ increases.

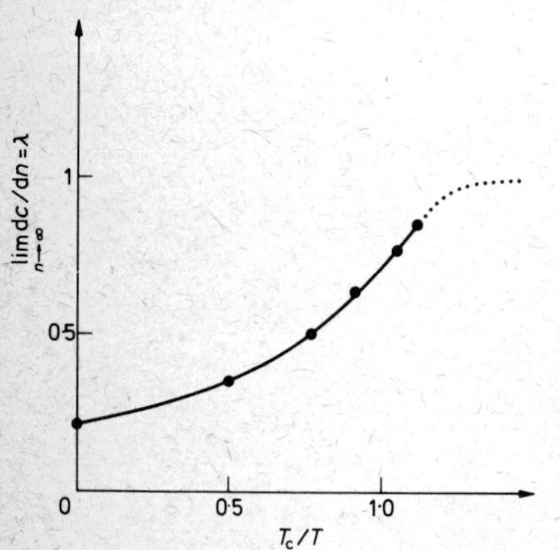

Fig.5 Coefficient of compactness λ of large clusters as a function of temperature.

Turning to statistics of individual clusters we first calculate the average number of spins per cluster $\langle n\rangle_{cluster}$ from equation (10). This is shown in Fig. 7 as a function of T_c/T; the number increases steadily as the temperature is lowered from about 7 at $T = \infty$ to 31 at $T = T_c$. Below T_c the number drops quickly and dramatically to 2 or 3.

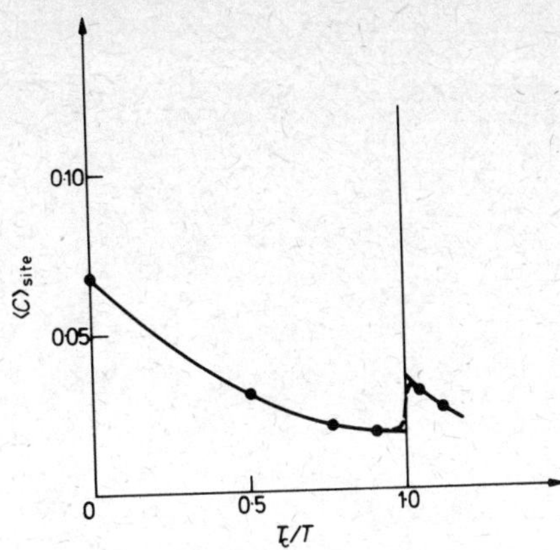

Fig.6 Mean number of clusters per site as a function of temperature. The broken curve represents conjectured behaviour in the critical region.

We now calculate $\langle c \rangle_{site}$ from (10) since it has a direct connection with the Whitney-Tutte polynomial, and it is shown graphically in Fig.8. The maximum value as $T \to T_{c+}$ is about 0.31, whereas for large clusters the value given by the asymptotic slope of $d\bar{c}/dn$ is about 0.36; we thus find that smaller clusters reduce the average of $\langle c \rangle_{site}$ since they have relatively fewer cycles. The value calculated by Temperley[14] and Lieb at $T = T_c$ is 0.128, and we have marked it on the figure. It is this value which indicates a rapid continuous variation near T_c, and suggests that the slope at T_c is probably infinite. The conjectured form of behaviour indicated in the critical region is based on this calculation.

The average number of cycles per cluster can be calculated from (11); the result is shown in Fig.9 and the behaviour parallels that of $\langle n \rangle_{cluster}$.

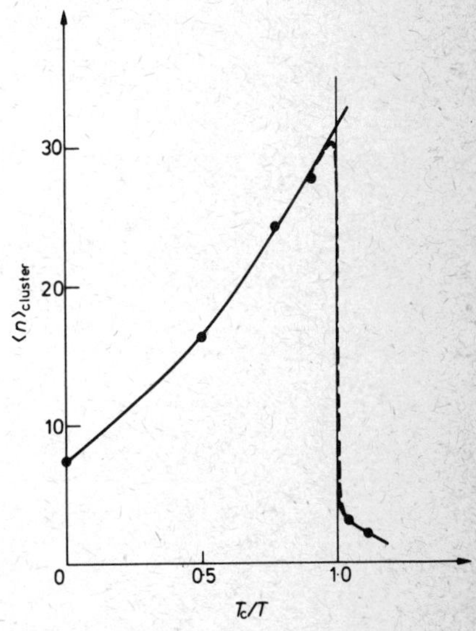

Fig.7 Mean number of spins per cluster as a function of temperature.

4. PERCOLATION CLUSTERS

The same calculations can be undertaken for random site percolation. Instead of changing the temperature we are now

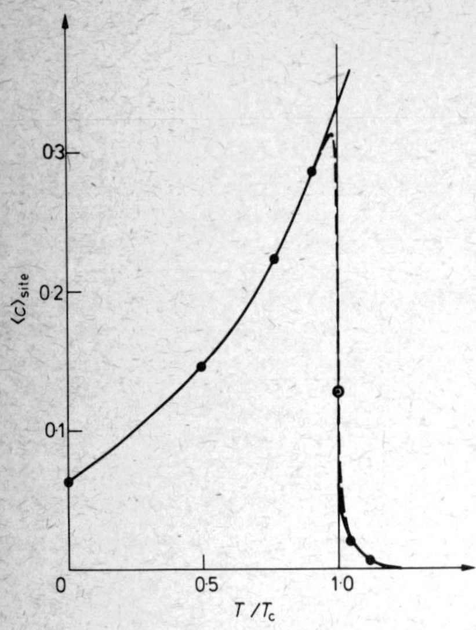

Fig.8 Mean number of cycles per site as a function of temperature. represents the exact calculation of Temperley and Lieb at $T = T_c$.

dealing with a fixed high temperature, and the variation of concentration is achieved by changing the magnetic field. We give a few typical illustrations. The variation of coefficient of compactness with concentration is shown in Fig.10 which should be compared with Fig.5 for Ising clusters. At critical concentration the percolation clusters are considerably more ramified than Ising clusters.

For p greater than the critical concentration p_c an infinite percolating cluster has formed and it is interesting to separate its properties from those of the finite clusters. This is indicated in Fig.11, where it will be seen that the percolating cluster becomes more compact as the concentration increases whereas the finite clusters experience little change.

In Fig. 12 we show the average number of spins per cluster and this should be compared with Fig. 7 for Ising clusters. The average number of spins per cluster at critical concentration is about 22.

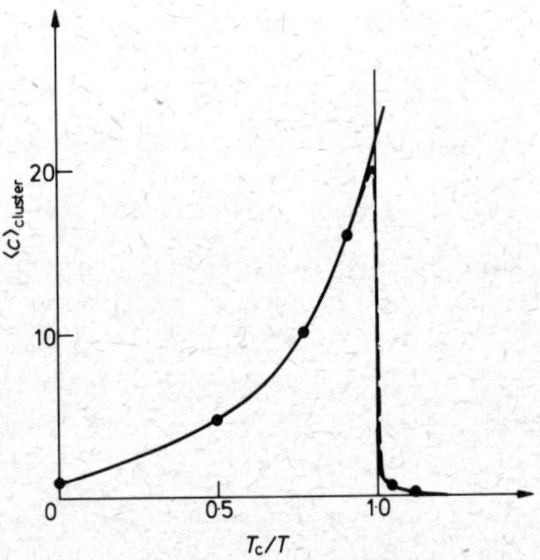

Fig.9 Mean number of cycles per cluster as a function of temperature.

5. WHITNEY-TUTTE POLYNOMIAL AND THE POTTS MODEL

During the lecture an account was given of the above relationship. Because of shortage of space this is not reproduced here; it

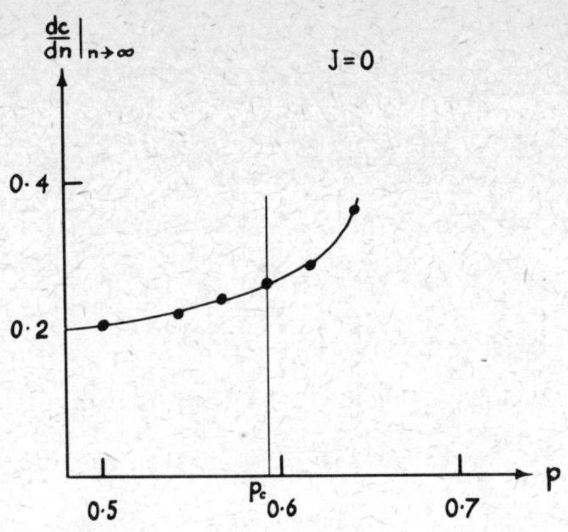

Fig.10 Percolation clusters ($T = \infty$) : coefficient of compactness vs concentration.

can be found in references 11-15.

6. NON-RANDOM PERCOLATION

Our study has been concerned with geometrical clusters i.e. with clusters formed by spins which are geometrically adjacent. We have derived some statistical properties of these clusters for the S.Q. lattice for the Ising model in zero field, and for the random site percolation problem. A natural generalization would be to the Ising model at fixed temperature in non-zero field which corresponds to percolation with correlation present.

The behaviour of the critical concentration is shown diagrammatically in Fig.13 for the S.Q. and S.C. lattices[15]. The most significant difference in three dimensions is the existence of a percolating cluster below T_c before phase separation occurs. This shows clearly that it is physical clusters of correlated spins rather than geometrical clusters which are relevant for condensation and critical behaviour[16].

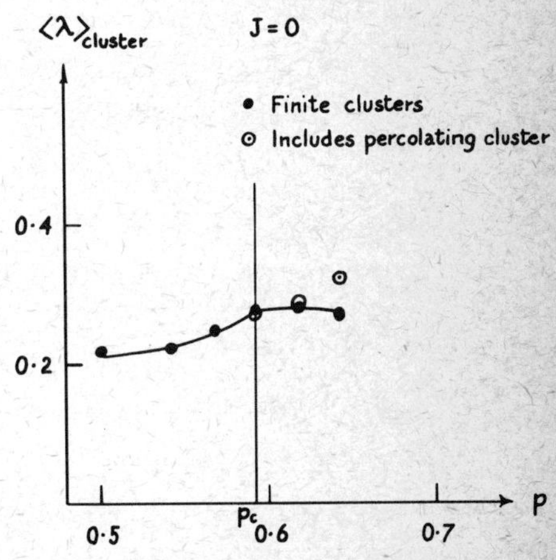

Fig.11 Percolation clusters: coefficient of compactness vs concentration Separation of infinite cluster

In two dimensions at temperatures below T_c percolation will normally correspond to phase separation

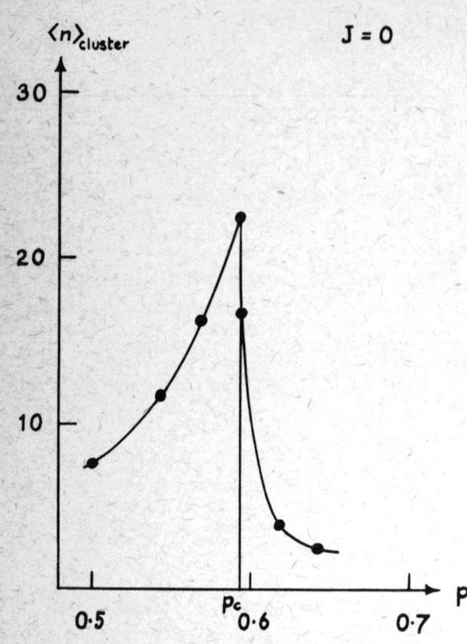

Fig.12 Percolation clusters: number of spins per cluster vs concentration.

This is the concentration expected for random percolation, and the clusters have a typical ramified shape; likewise B which represents a Pb-Ge alloy the critical concentration being 11% of Pb. But C represents an Aℓ-Ge alloy where there is strong interaction between the components, and the critical concentration is 68% of Aℓ; the shape of the clusters is now seen to be compact.

with $p_c = 0.5$. But under metastable conditions it is possible for p to exceed 0.5 without the formation of a percolating cluster as illustrated in Fig.14 (taken from the Monte Carlo kinetic Ising data of Schneider and Stoll[17]).

Finally I should like to show a beautiful experimental illustration of ramified and compact clusters in micrographs supplied to me by Dr. Michael Rapaport[18] of Tel Aviv University in Fig. 15. A represents an Indium-Germanium alloy and the critical concentration for percolation corresponds to 14% by volume of Indium.

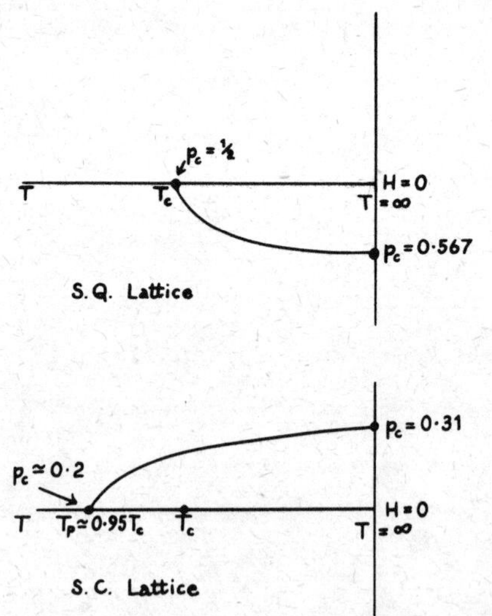

Fig.13 Percolation in Ising systems (i.e. with correlation).

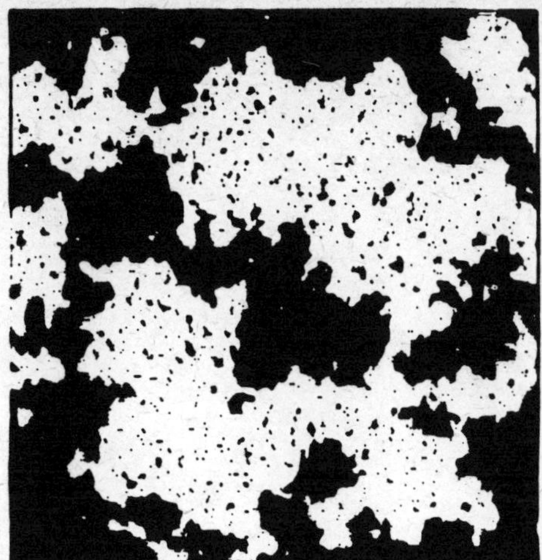

Fig.14 A metastable non-percolating configuration with p = 0.580 (T/T_c = 0.958).

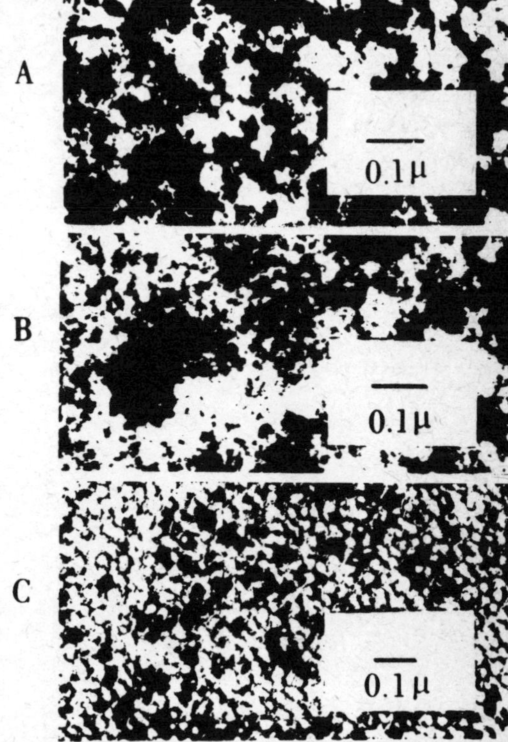

Fig. 15

[1]J. Frenkel, J.Chem.Phys. 7, 538 (1939), A. Bijl, Ph.D. Thesis, Leiden (1938), W. Bond, J.Chem.Phys. 7, 927 (1939).

[2]J. Frenkel, "Kinetic Theory of Liquids" Oxford 1946 p.382.

[3]J.W. Essam and M.E. Fisher, J.Chem.Phys. 38, 802 (1963).

[4]M.E. Fisher, Physics 3, 255 (1967).

[5]C. Domb, J.Phys.A: Math.Gen. 9, 283 (1976).

[6]C. Domb, T. Schneider and E. Stoll, J.Phys.A: Math.Gen. 8, L90 (1975).

[7]L. Reatto and E. Rastelli, J.Phys.C: Solid St. Phys. 5, 2785 (1975) and Ref.5.

[8]C. Domb, J.Phys.A 9, L141 (1976); P.L. Leath, Phys.Rev.B 14, 5046 (1976); D. Stauffer, Z.Phys. B 25, 391 (1976); and more recent preprints.

[9]H. Muller-Krumbhaar, Phys.Lett. 50 A, 27 (1974); A Coniglio, J.Phys.A 8, 1773 (1975).

[10]C. Domb and E. Stoll, J.Phys.A 10, 1141 (1977).

[11]H.N.V. Temperley and E.H. Lieb, Proc.Roy.Soc. A 322, 251 (1971).

[12]R.B. Potts, Proc.Camb.Phil.Soc. 48, 106 (1952).

[13]P.W. Kasteleyn and C.M. Fortuin, J.Phys.Soc. Japan Suppl. 26, 11 (1969), Physica 57, 536 (1972).

[14]H.N.V. Temperley, J.Phys.A 9, L113 (1976) and Ref.11.

[15]R.J. Baxter, S.B. Kelland and F.Y. Wu, J.Phys.A 9, 397 (1976).

[16]K. Binder, Ann.Phys. NY 98, 390 (1976).

[17]T. Schneider and E. Stoll in "Anharmonic Lattices, Structural Transitions and Melting" ed. T. Riste, Noordhoff-Leiden (1974).

[18]Work done in collaboration with G. Deutscher and Z. Ovadyahu.

CHAPTER 7

SOME FEATURES OF PHASE TRANSITIONS IN POLYMER SYSTEMS

P. G. de Gennes

Collège de France, 76231 Paris Cedex, France

ABSTRACT

We classify various transitions involving, long flexible chains, in solutions or in melts

1) phase separation between a) polymer and poor solvent
b) two molten polymers
c) two strongly incompatible polymers in good solvent

2) sol-gel transitions a) branching and cyclisation reactions based on small monomers
b) cross linking of long, preexisting chains.

We show that for the cases 1b and 2b the static critical exponents become classical for long chains. For all other cases listed here they are not.

I. INTRODUCTION

It is a remarkable achievement of chemistry that long polymer chains containing a number N of repetitive units (monomers) as large as 10^5 can be made without error (ie without any branching of the chain). The resulting objects, such as polyethylene

$$-\!\!\left[CH_2 \right]\!\!-_N$$

or polystyrene

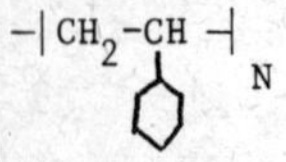

$$-\!\!\left[CH_2\text{-}CH \right]\!\!-_N$$

are often quite flexible when put in solution, and raise a number of interesting statistical problems. A beautiful presentation of the main ideas is found in classic book by Flory [1]. More recent advances, involving the more recent tools of statistical physics, are reviewed in ref. [2-3]. At present the statistics of long, flexible chains in a good solvent, or in a molten state, is reasonably well understood. In these cases we were dealing with linear chains with repulsive interactions.

We are coming now to a second generation of problems, where one or both of these features is modified :

1) what happens when certain interactions become attractive (and favor a demixtion)

2) what happens when a certain amount of branching is allowed during the synthesis.

In the present paper we shall present a few guidelines concerning these two types of problems. They are both very important in practice.

1) Polymer demixtion, or more generally polymer incompatibility, is the source of many local heterogeneities in industrial plastics, and may (or may not) be welcome depending of the case at hand. For instance many polymer glasses are naturally brittle, but their resistance to fracture is vastly improved by the incorporation of rubbers ; the two partners are generally incompatible and the rubber builds up small pockets which trap the craze or fracture tips.

2) The presence of branch points is quite common in industrial products such as commercial polyethylene. But in certain systems we purposely favor branching. An example would be the condensation of diacids with trialcohols : each alcohol molecule can initiate a chain in three directions and thus acts as a branching center. In such a reaction, at the early stages, all the molecules formed are "clusters" of finite size. But later, when the fraction of reacted bonds ρ crosses a critical value ρ_c, an infinite cluster appears. At this moment we change from a liquid phase ("sol") to a gel. These gelation transitions are not restricted to organic polymers : for instance the alumina gels are an important starting point for many catalysts and other chemical effectors.

Another important type of gel is found with the rubbers where long flexible chains are tied together by a relatively small number of cross links (sulphur atoms).

This cross linking (or vulcanization, as it is called for rubber) raises a number of interesting statistical problems, and will be discussed below in section III.

II. DEMIXTION TRANSITIONS INVOLVING POLYMERS

A) polymer + poor solvent

The simplest example of this class occurs with a single polymer species (chains of N monomers, number of monomers/cm^3 c) in a "poor" solvant. The word "poor" means that the monomers attract each other slightly. The mean field free energy per cm^3 for such a system has the form (1)

$$\frac{F}{T} = \frac{c}{N} \ln c + \frac{1}{2} v c^2 + \frac{1}{6} w^2 c^3 + \ldots \qquad (II.1)$$

where T is the temperature. The first term describes the translational entropy for chains of concentration $\frac{c}{N}$. The other terms represent interaction effects in the form of a low c expansion. The quantity v is the second virial of the monomer-monomer interaction, and depends on temperature like

$$v = \text{const. } a^3 \frac{T - \Theta}{\Theta} \qquad (II.2)$$

where a is a monomer size and Θ is the Flory compensation temperature. Above $T = \Theta$, v is positive (repulsive) and we have a good solvent. But we are concerned here with T slightly smaller than Θ (poor solvent). The constant w is also of order a^3.

From the mean field free energy it is easy to extract an osmotic pressure $\Pi = c^2 \frac{\partial}{\partial c} (\frac{F}{c})$ and to locate the critical point by the conditions $\frac{\partial \Pi}{\partial c} = \frac{\partial^2 \Pi}{\partial c^2} = 0$. The result is

$$v_{crit.} = - 2 w N^{-1/2} \qquad \text{(mean field)} \qquad (II.3)$$

$$c_{crit.} = w^{-1} N^{-1/2}$$

It is also interesting to construct a correlation length ξ : in mean field this has the form (4):

$$\xi = \xi_o \left(\frac{T_c}{\Delta T}\right)^{1/2} \qquad (II.4)$$

where $\xi_o \cong N^{1/4} a$. Not that ξ_o is <u>not</u> equal to the average size of one ideal chain ($R_o = N^{1/2} a$). But a good check on eq. (II.4) is obtained by taking $T = \Theta$ exactly. This corresponds to $\Delta T \sim \Theta N^{-1/2}$ and at $T = \Theta$, we get $\xi = R_o$, as expected for very weakly interacting objects.

In the limit $N \to \infty$ ξ_o is large and we might wonder whether this does not lead to a critical point with classical exponents, as is found for instance in superconductors or in superfluid He_3. The answer is no : the exponents are not classical. This can be shown through the use of a Ginsburg criterion (5), but it may also be understood in direct terms : the basic units are coils of large size R_o, but in critical conditions the average distance between them is also of order R_o : the interaction range is not larger than the interparticle distance, and no analogy can be drawn with superconductors.

B) two molten polymers

The situation becomes completly different if we deal with two polymers (A + B) with a weakly dominant AB attraction (+). Here, the Ginsburg criterion (5) shows that the critical fluctuations are not dangerous provided that we stay outside of a very small temperature region :

$$|\Delta T| > T_c N^{-2} \tag{II.5}$$

In this case the particle size is still R_o, but the interparticle distance is much smaller : the number of chains per cm^3 is $N^{-1} a^{-3}$ and their distance is $N^{1/3} a \ll R_o$.

C) two incompatible polymers in a good solvent

A very interesting case occurs when we have two polymers AB in a good solvent, but when the AB repulsion dominates. At low concentrations ($c_A = c_B = c$) we have separate coils of A and B. These coils are stronghly non ideal : they behave like self avoiding walks, and have a radius $R_F = a N^{3/6}$. At a concentration $c^* = N R_F^{-3}$ the coils begin to interpenetrate, and the AB repulsion is felt : very soon segregation takes place. Here, just as in case (a) the average distance between coils remains comparable to the interaction range, and the critical exponents should be non classical.

It may be worth while to point out that situation (c) occurs in the fabrication of high impact polystyrene - ie polystyrene strengthened by rubber.What is currently done is to start with a mixture of styrene monomer and rubber chains, and to polymerise the styrene. As soon as a weak polystyrene concentration is achieved ($c \sim c^* \sim 2\%$)

(+) *for simplicity we consider only the case* $N_A = N_B = N$, *and a critical point for equal concentrations. The situation* $N_A \gg N_B$ *will be discussed separatly.*

the medium becomes opalescent : the repulsions between polystyrene and rubber have driven a phase separation. Here the styrene monomer plays the role of a solvent.

III. GELATION

A) branching of small units

Let us consider for instance the branching reaction mentioned in the introduction with trialcohols and diacids as partners. An excellent introduction to the physics of the resulting gel point can be found in the papers of M. Gordon and coworkers [(6)]. However, I differ from the Gordon school for certain interpretations, to be described below.

Let us discuss the sequence of events occuring when the fraction ρ of reacted bonds increases.

1) In the sol phase ($\rho < \rho_c$), when $\rho \to \rho_c$ the clusters are finite but large. Their average molecular weight diverges like

$$\overline{M} \sim (\rho_c - \rho)^{-\gamma} \tag{III.1}$$

It would be also interesting to measure their gyration radius, but this requires blocking the reaction and subsequent dilution : in this process the sizes may be perturbed. However, if we ignore this difficulty, we should measure a correlation length

$$\xi \cong (\rho_c - \rho)^{-\nu} \tag{III.2}$$

2) In the gel phase ($\rho > \rho_c$) we can measure the fracture belonging to the infinite cluster (by washing out the finite clusters)

$$m \sim (\rho - \rho_c)^{\beta} \tag{III.3}$$

We can also measure the mesh size of the network after washing : if the washing procedure is not a dangerous perturbation , this could give us a correlation length ξ. Finally we can measure the elastic modulus E. This is delicate, because the gel is weak. The Gordon group used a small magnetic grain embedded in the gel (and driven by a weak field) to apply very small stresses [(6)]. We expect

$$E \sim (\rho - \rho_c)^{\mu} \tag{III.4}$$

Essentially all the litterature on the theory of gelation [1,6], is based on a Cayley tree approximation which ignores the possibility of cyclisation (or at most includes cycles below a certain fixed size). This led to rather good values for the threshold position ρ_c - just as in magnetic mean field calculations which give a good first approximation for the location of the Curie point. However, we expect the situation to be much worse as regards the critical exponents. The tree approximation predicts $\beta = \gamma = 1$, $\nu = 1/2$ and $\mu = 3$. It has been pointed out recently [7,8] that the problem is closely related to percolation. For one special model there is identity : namely when the monomers are dense, and located at the nodes of a periodic lattice, the nearest neighbor bonds then having a probability ρ of being linked, we have a bond percolation problem, and we expect (for $d = 3$) $\gamma = 1.8$, $\nu = 1.9$, etc.

It has also been shown [8] that the exponent μ of the elastic modulus coincides with the conducting exponent of a random network : thus we expect $\mu = 1.9$ in 3 dimensions.

There are two complications in practice

(i) the monomers are not on a lattice. We expect this type to be unimportant since the specific heat exponent α is negative for percolation transitions.

(ii) the monomers are not usually dense, but dilute. When we fix the fraction of reacted bonds, this creates spatial correlations between monomers. Thus we are led to a problem of percolation between correlated units (ie between + spins in an Ising ferromagnet at high but finite temperatures). I tend to think that whenever we maintain a homogeneous system (outside of the coexistence curve) we shall keep the same percolation exponents - but this will have to be checked.

To summarize : Stauffer and I expect values of the gelation exponents very different from the Cayley tree predictions ; this is in strong disagreement with most of the existing literature on gelation. The success of mean field theory for calculations of the threshold ρ_c has led to an overoptimistic view among the physical chemists, according to which the free approximation is sufficient. Many more experiments near the sol gel transition are required to clarify this point.

There is one case where the tree approximation is correct : namely for the cross linking of preexisting long chains in a dense state. This is interesting, because it corresponds exactly to the Goodyear process of vulcanization mentioned in the introduction. Here, again using a Ginsburg criterion [9], I find that classical behavior should hold, except in a relatively small region near critical, of magnitude

$$\frac{\Delta\rho}{\rho_c} \sim N^{-1/3} \qquad (d = 3) \qquad \text{(III.5)}$$

Thus in this case we need very large chains to restore the classical exponents.

References

(1) P. FLORY, Principles of polymer chemistry, Cornell Univ. Press, Ithaca (NY) (1971)

(2) M. DAOUD et al., Macromolecules 8, 804 (1975)

(3) P.G. de GENNES, Rivista del Nuovo Cimento (sept. 1977)

(4) P.G. de GENNES, Physics Lett., 26 A, 313 (1968)

(5) P.G. de GENNES, J. de Physique Lettres (to be published)

(6) M. GORDON, G. SCANTLEBURY, J. Chem. Soc., B, p. 1 (1967)
M. GORDON, R.G.C.P. "Plastiques" 4, 111 (1967)
See also "Polymer networks" (Chompff and Neuman, ets.) Plenum (1971)

(7) D. STAUFFER, J. Chem. Soc. (Faraday trans.) II, 72, 1354 (1976)

(8) P.G. de GENNES, J. de Phys. Lettres, 37, 1 (1976)

(9) P.G. de GENNES, J. de Phys. Lettres (to be published).

CHAPTER 8

ON THE DYNAMICAL THEORY OF IRREVERSIBLE PROCESSES AND THE MICROSCOPIC INTERPRETATION OF NON-EQUILIBRIUM ENTROPY †

A. P. Grecos & I. Prigogine *

Faculté des Sciences,

Université Libre de Bruxelles, Campus Plaine,

C. P. 231, 1050 Brussels, Belgium

†Partially supported by : R. Welch Foundation (Houston) and Belgian Government "Actions de Recherches Concertées" (Conv. 76/81 *II* 3).

*Also at Center for Statistical Mechanics and Thermodynamics, The University of Texas at Austin, Austin, Texas, U.S.A.

ABSTRACT

A survey of some recent work on dissipative processes in dynamical systems is presented. The role of non-unitary transformations, leading to a representation in which a non-equilibrium entropy can be defined, is emphasized. The formalism is applied to the derivation of linear irreversible thermodynamics for a neutral fluid.

1. INTRODUCTION

It is a fundamental problem of modern theoretical physics to investigate the relation between irreversible phenomenological laws and reversible (classical or quantum) dynamics. Irreversible processes occur in every branch of physics and consequently thermodynamic considerations play an ever increasing role in the description of the evolution of dynamical systems.

We are concerned here with the formulation of a dynamical theory of irreversible processes, having in mind Hamiltonian systems in which dissipative phenomena such as transport processes or decay of unstable states, take place. There are several aspects which need to be discussed in connection with such a program. However, here we will mainly center our attention to the possibility of a microscopic definition of a non-equilibrium entropy, i.e. a positive non-decreasing functional of the state of the system.

The purpose of this lecture is to present a survey of some recent work in this direction. Our point of view may be summarized as follows. The dynamical and thermodynamical descriptions are, in a certain sense, equivalent representations of the evolution of the system, connected via a non-unitary transformation. More precisely there exists a microscopic description of the evolution of the dynamical system which displays explicitly 'irreversibility' in the sense that we may associate to it a Liapounoff functional. Once this description is obtained from the usual formulation of dynamics through a non-unitary transformation, the macroscopic irreversible description follows. To construct such a transformation, we must examine the different dissipative processes which occur in the system. The analysis is based on the theory of subdynamics which we will summarize in the next section. In the third section we will indicate the main ideas of the transformation theory.

To illustrate the formalism we will consider the derivation of the (linearized) transport equations for a classical fluid near equilibrium. It is well known that the (complex) frequencies of the so-called hydrodynamical modes can be determined from a linear kinetic equation for the one-particle distribution function. However, the derivation of the macroscopic transport equations requires, in general, taking into account the effect of a class of correlations, which as we will show, are correctly predicted from the theory of subdynamics. Moreover, using the ideas of the transformation theory, we are able to deduce the Gibbs relation for the entropy and thus specify the domain of validity of linear irreversible thermodynamics.

2. KINETIC DESCRIPTION

For a Hamiltonian system, the time evolution of its state ρ is given by the Liouville (classical) or the von Neumann (quantum) equation which we write as

$$i\partial_t \rho = L\rho \; . \tag{2.1}$$

Here the operator L is defined by the Poisson bracket, $L\rho = i\{H,\rho\}$, or the commutator, $L\rho = [H,\rho]$, where H is the Hamiltonian of the system. The state ρ is a probability density or a density matrix. We may also consider the evolution of an observable A, which is given by

$$i\partial_t A = -LA \; . \tag{2.2}$$

Eq.(2.1) and eq.(2.2) are, respectively, the Schrödinger and the Heisenberg picture. They are connected by the equality of the expectation value of the observables in either picture

$$\langle A\rangle_{\rho}(t) = (A, \rho(t)) = (A(t), \rho) \; . \tag{2.3}$$

As usual (A,ρ) is a sesquilinear form defined by an integral over the phase space or a trace.

From eq.(2.1) we have that

$$\rho(t) = \exp\{-iLt\}\,\rho(0) \tag{2.4}$$

or in terms of the resolvent of L

$$\rho(t) = \frac{1}{2\pi i}\int_{\vec{C}} dz\, e^{-izt}\, \frac{1}{L-z}\,\rho(0) \; , \tag{2.5}$$

where the contour $\vec{C}$ is parallel and above the real axis. Clearly the above rélations, as well as the formulae which we will derive in the following, may be used to study the evolution of the observables provided that we replace L by $-L$, an operation which will be called 'L-inversion'.

To analyze the contributions to $\rho(t)$ in eq.(2.5) we follow the methods used in the derivation of generalized and asymptotic master equations[1]. Thus we decompose the state into several components and we examine the behaviour of each component separately. For instance, in the case of a quantum system we consider the diagonal part (with respect to some given representation) $\rho_0(t)$ of the state and the off-diagonal one $\rho_c(t)$. The formalism is quite simple when $\rho_0(t)$ may be written as $P\rho(t)$ where P is a projection[2]. Using the decomposition of the resolvent

$$\frac{1}{L-z} = [P + \mathcal{C}(z)] \frac{1}{PLP+\Psi(z)-z} [P + \mathcal{D}(z)] + \frac{1}{QLQ-z} Q \,, \tag{2.6}$$

where

$$\Psi(z) = - PLQ \frac{1}{QLQ-z} QLP \,, \tag{2.7}$$

$$\mathcal{D}(z) = - PLQ \frac{1}{QLQ-z} \tag{2.8}$$

and

$$\mathcal{C}(z) = - \frac{1}{QLQ-z} QLP \tag{2.9}$$

are, respectively the 'collision', 'destruction' and 'creation' operators, we obtain for the evolution of $P\rho(t)$ the so-called generalized master equation

$$i\partial_t \rho_0(t) + PLP\rho_0(t) = \int_0^t d\tau\, G(t-\tau)\rho_0(\tau) + \mathcal{F}(t, \rho_c(0)) \,. \tag{2.10}$$

The kernel $G(t)$ is the inverse Laplace transform of $\Psi(z)$, while the source term is the inverse transform of the $\mathcal{D}(z)$ acting on $Q\rho(0)$ $(Q=I-P)$.

The choice of the projections P and Q depends on the physical problem. In the fourth section we will consider a specific example. There are two points which we want to stress here : (1) Questions related to the problem of irreversibility cannot be discussed directly using the generalized master equation, because such an equation may be derived for any system, dissipative or not. (2) Any discussion limited to the P-component of the state is bound to be incomplete, or even incorrect, given that, in general, there are contributions from its Q-component which must be taken into account.

Irreversible kinetic equations are of the form

$$i\partial_t \tilde{\rho}_0(t) = \Theta \tilde{\rho}_0(t) \,, \tag{2.11}$$

where Θ is non-hermitian ($\Theta \neq \Theta^{+}$). For a homogeneous dilute classical gas or for a weakly coupled quantum system, the r.h.s. of eq. (2.11) is given by the asymptotic collision operator, $\Theta \approx \Psi(+i0)$, evaluated to first order in the density or to second order in the coupling constant (Boltzmann and Pauli equations, respectively). One may regard eq.(2.11) as an approximation to eq.(2.10). However, a different point of view may be pursued by considering this equation as describing exactly the evolution of a particular contribution, denoted by $\tilde{\rho}_0(t)$, to the P-component of the state. Using perturbation techniques, Prigogine and Résibois have shown[3] (see also ref. 1) that $\Theta = \Omega\Psi(+i0)$, where Ω is expressed in terms of the asymptotic collision operator and its derivatives. In fact, the collision term in kinetic equations proposed for moderately dense or strongly coupled systems is obtained as an appropriate approximation of $\Omega\Psi(+i0)$.

It is evident that the definition of $\Theta(=\Omega\Psi)$ through a series is, at least from a theoretical point of view, unsatisfactory. In several physically important problems the series does not converge and the meaning of eq.(2.11) becomes unclear. Therefore, it is interesting to consider the structure of Θ assuming certain properties for the collision operator. By definition $\Psi(z)$ is analytic for z non-real. Let us suppose now that the branch of $\Psi(z)$ defined for $\mathrm{Im}\, z > 0$ can be continued analytically in some domain below the real axis. Moreover, let us assume that there exists a set $\{z_n\}$ of complex numbers ($\mathrm{Im}\, z_n \leq 0$) and corresponding vectors $\{|u_n\rangle\}$ and $\{\langle v_n|\}$ such that

$$[PLP + \Psi(z_n)]\,|u_n\rangle = z_n |u_n\rangle \quad , \quad \langle v_n|\,[PLP + \Psi(z_n)] = z_n \langle v_n| \,. \tag{2.12}$$

The imaginary part of these numbers may be interpreted as the relaxation time of some collective motion described (partly) by the corresponding eigenvectors. Then, provided that the vectors $\{|u_n\rangle\}$ and $\{\langle v_n|\}$ form complete sets in the P-subspace, we get for Θ the expression

$$\Theta = \sum_n z_n\, |u_n\rangle\langle \bar{v}_n| \,, \tag{2.13}$$

where $\{\langle \bar{v}_n|\}$ is bi-orthogonal to $\{|u_n\rangle\}$. This result may be derived by considering the contribution to $\rho_0(t)$ due to $\{z_n\}$ which are poles of the analytic continuation of $P[L-z]^{-1}P$.

We observe now that the same poles contribute also to the Q-component of $\rho(t)$. The main idea of the theory of subdynamics[4,5] consists in treating together all these contributions which have a similar time dependance. It is an easy matter to see that from eqs.(2.5), (2.6) and (2.19) we obtain the contribution[6]

$$\tilde{\rho}(t) = [P + C]\; e^{-i\Theta t}\, A\,[P + D]\; \rho(0) \,. \tag{2.14}$$

Here Θ is given by eq.(2.13) and A, D and C are defined by

$$A = \sum_n [1-\mu_n']^{-1} |u_n\rangle\langle v_n| , \tag{2.15}$$

$$D = \sum_n |\bar{u}_n\rangle\langle v_n| \mathcal{D}(z_n) \tag{2.16}$$

and

$$C = \sum_n \mathcal{C}(z_n) |u_n\rangle\langle \bar{v}_n| , \tag{2.17}$$

where $\mu_n' = \langle v_n|\Psi'(z_n)|u_n\rangle$ and $\{|\bar{u}_n\rangle\}$ is bi-orthogonal to $\{\langle v_n|\}$.
Note that from eq.(2.14) we have

$$\tilde{\rho}(0) = \Pi\rho(0) , \qquad \Pi = [P+C] A [P+D] . \tag{2.18}$$

Formally we may show that Π is an idempotent but non-hermitian operator (if $\operatorname{Im} z_n \neq 0$ for some n) and that

$$\tilde{\rho}(t) = \Pi\rho(t) \quad , \quad t \geqslant 0 . \tag{2.19}$$

Comparing eq.(2.14) with eq.(2.11) we see that the later describes the evolution of the P-component of $\tilde{\rho}(t)$.

Several of the above mentioned assumptions can be relaxed with a corresponding extension of the domain of applicability of the formalism. These are essentially technical aspects which need not be discussed here. However, a few remarks should be made as to the implications of the assumption that eq.(2.12) admits complex roots. To have this possibility, we must require a Liouville-von Neumann operator with a continuous spectrum. In particular, a quantum system should be considered in some asymptotic limit, e.g. the thermodynamic limit.[7] It is only then that we may expect a sharp distinction between reversible and irreversible processes. Once we associate dissipative phenomena with the poles (or other singularities) of the analytic continuation (of matrix elements) of the resolvent of L, some interesting conceptual questions arise. They are related to the structure of the space in which act the operators that we introduced, e.g. the Π-projection.[8]

It is well known that difficulties arise when we attempt to derive thermodynamics from an equation which takes into account corrections to the 'Boltzmann approximation'. The reason is that the operator Θ in eq.(2.11) has no particular symmetry, cf. eq.(2.13). Consequently we are not able to define a functional which can be identified with a non-equilibrium entropy. This is a serious drawback of kinetic theory and the root of the difficulty may be traced to the fact that eq.(2.11) neglects entirely the evolution of the Q-component of the state. In the next section we will consider a transformation theory which permits a microscopic formulation of the second law.

3. TRANSFORMATION THEORY AND THE DEFINITION OF ENTROPY.

By construction the projection Π is similar to P, i.e. we can write

$$\Pi = \Lambda P \Lambda^{-1} \qquad (\Lambda^{-1} \neq \Lambda^{+}) . \tag{3.1}$$

From eq.(3.1) and eq.(2.17) we see that if $\chi_0 = P\Lambda P$ and $\chi_0^* = P\Lambda^{-1}P$, then

$$\chi_0 \chi_0^* = A . \tag{3.2}$$

It may be shown that χ_0 can be chosen so that Θ is transformed to a dissipative operator, namely

$$\Phi_0 = \chi_0^{-1} \Theta \chi_0 \quad , \qquad \frac{1}{2i}(\Phi_0 - \Phi_0^{+}) \leq 0 . \tag{3.3}$$

Thus if $\|\cdot\|$ is an appropriate norm, then $\|\chi_0^{-1}\rho_0\|^2$ is a Liapounoff functional which, cf. eqs.(2.11) and (3.3), decreases monotonically in time.

As one might expect, eq.(3.1) leaves $Q\Lambda Q$ essentially undetermined. This is due to the fact that we have considered only a particular class of contributions to eq.(2.5) arising from singularities of $P[L-z]^{-1}P$. To determine further the Λ transformation we need to analyze the singularities of other parts of the resolvent of L. We will not pursue this problem here (cf. ref.5). Instead we prefer to state briefly the main properties of such non-unitary transformations which lead to a representation in which the evolution is manifestly irreversible.

Consider the Liouville-von Neumann equation, eq.(2.1). Suppose that we change the representation using an invertible transformation Λ. We write

$$\overset{(P)}{\rho} = \Lambda^{-1} \rho \tag{3.4}$$

and from eq.(2.1) we get

$$i\partial_t \overset{(P)}{\rho} = \Phi \overset{(P)}{\rho} \quad , \quad \Phi = \Lambda^{-1} L \Lambda \ . \tag{3.5}$$

If Λ is unitary ($\Lambda^{-1} = \Lambda^{+}$), eq.(3.5) is simply the Liouville equation in a different (canonical) representation. On the other hand if Λ is non-unitary we may impose on Φ the generalized dissipativity condition

$$\frac{1}{2i} (\Phi - \Phi^{+}) \leqslant 0 \ . \tag{3.6}$$

A necessary condition for the inequality to be fulfilled is the continuity of the spectrum of L . In the case of (finite) classical systems more precise conditions, related to their ergodic properties, have been formulated .[9]

From eqs.(3.5) and (3.6) it follows that the Liapounoff functional

$$\Omega = (\overset{(P)}{\rho} , \overset{(P)}{\rho}) \tag{3.7}$$

is non-increasing

$$d\Omega/dt \leqslant 0 \ . \tag{3.8}$$

Consequently Ω may serve for the definition of a non-equilibrium entropy. The construction of the Liapounoff functional through a non-unitary transformation is of course a central point of this approach. However, it cannot be expected that there exists a unique transformation and, consequently, a unique functional. The requirement that Ω leads to the macroscopic entropy imposes supplementary conditions which will be considered in the fourth section.

There are several symmetry properties of the Λ transformation. Among them we mention that of 'star-hermiticity', i.e.

$$\Lambda^{-1}(L) = \Lambda^{+}(-L) \equiv \Lambda^{*}(L) \ . \tag{3.9}$$

The last equality defines the so-called 'star-conjugation', which is a combination of the hermitian conjugation and the L-inversion. It follows by imposing the equivalence of the Schrödinger and the Heisenberg pictures.

Let us finally remark that each class of singularities of the resolvent of L corresponds to a certain type of dissipative processes in the system. According to the theory of subdynamics we describe them by constructing an appropriate subspace, e.g. the Π-subspace. Each subspace, gives a specific contribution to the Ω-functional and thus to the non-equilibrium entropy. In principle, the Λ-transformation takes into account the effect of all possible processes. However, in certain problems we may limit our considerations to a particular class of processes and compute the corresponding entropy by calculating a part of Λ and not the whole transformation. This will be done in the next section where we discuss the linear transport theory of a classical fluid.

4. LINEAR IRREVERSIBLE THERMODYNAMICS

We consider a system of classical particles interacting with two-body, central, short-range forces. Its state is assumed to be slightly perturbed from the (absolute) canonical equilibrium ($\rho_{eq} = Z^{-1} \exp(-H/kT_0)$). Thus we may suppose that the evolution of the one-particle distribution function is essentially described by a linearized kinetic equation.

In the past few years there is a growing interest in the properties of the hydrodynamical modes, i.e. wave-like motions of long wave length[11,12,13]. An important aspect of the theory of these modes is the fact that their frequencies can be expressed in terms of the transport coefficients of the fluid. Moreover they can be used to investigate properties related to the so-called long-time tails of the Green-Kubo integrands (for a review see ref.14). Here we want to make a few remarks on the derivation of the linearized macroscopic transport equations and the Gibbs relation for the entropy. To formulate the problem in a way similar to that of the previous sections we introduce the projection $P_q (= P_q^2 = P_{-q}^+)$ defined as

$$\langle r, v | P_q | r', v' \rangle = \sum_{k=1}^{N} \rho_{eq} \, \varphi^{-1}(v_k) \, e^{iq\cdot(r_k - r_k')} \left[\ell_q(v_k, v_k') - \delta_q^{kr} \, \varphi(v_k) \right] \tag{4.1}$$

with

$$l_q(v,v') = \delta(v-v') - n_o \varphi(v) C_q . \tag{4.2}$$

Here $n_o = (\frac{N}{V})$ is the mean density, $\varphi(v)$ the Maxwellian velocity distribution and C_q the Fourier transform of the equilibrium direct correlation. Then from the Liouville equation we deduce the generalized master equation for $P_q \rho$. The linearized equation for $f_q(v,t)$, the (Fourier transform of the) one particle distribution function, is obtained from the equation for $P_q \rho$ by integrating over the positions and velocities of all particles except one, it reads

$$\partial_t f_q + iqvf_q - iqvC_q n_o \varphi(v) \int dv' f_q(v',t) = -i \int_0^t d\tau \int dv' G_q(v,v',t-\tau) f_q(v,\tau) + \mathcal{F}_q(t). \tag{4.3}$$

Obviously the kernel $G_q(t)$ and the 'source-term' $\mathcal{F}_q(t)$ are obtained from the corresponding terms of the master equation and it is not necessary to give their explicit form here.

The (complex) frequencies z_q of the hydrodynamical modes are solutions of the equation

$$[qv - z_q]\tilde{h}_q - qvC_q n_o \varphi(v) \int dv' \tilde{h}_q(v') + \int dv' \Psi_q(v,v',z_q)\tilde{h}_q(v') = 0 \tag{4.4}$$

which vanish for $q=0$. They give a class of exponentially varying contributions to the one-particle distribution function, which we denote by $\tilde{f}_q(v,t)$. The structure of the problem is identical to that of the second section, the only difference being that we have here the wave-vector q as a parameter. One computes the 'eigenfrequencies' and eventually the 'eigenvectors' of eq.(4.4) by perturbation. Assuming the existence of five collisional invariants we get five solutions corresponding to the two sound modes, the thermal mode and the (doubly degenarate) shear mode. Moreover one proves that the coefficients of the expansion

$$z_q^{(n)} = b^{(n)} q + d^{(n)} q^2 + O(q^{2+\epsilon}) \tag{4.5}$$

which are givenin terms of the matrix elements of the collision operator, may be identified to the transport coefficients and the thermodynamic derivatives of the fluid.

At this point we ask now the question whether using the hydrodynamical modes we can deduce the (linearized) hydrodynamical equations. This is the case in the Boltzmann approximation (cf.also ref.15). However, with a collision term depending on the frequency and the wave-vector such a derivation is impossible. It may be shown that the first five moments of $\tilde{f}_q(v,t)$ satisfy the set of equations

$$\partial_t n_q + iq u_q^{\parallel} = 0 \tag{4.6a}$$

$$\partial_t u_q^{\perp} + q^2 \eta/n \, u_q^{\perp} = 0 \qquad q \cdot u_q^{\perp} = 0 \tag{4.6b}$$

$$\partial_t u_q^{\parallel} + iq\left(\frac{\partial P}{\partial n}\right)_T \frac{n_q}{n} + iq\left(\frac{\partial P}{\partial T}\right)_n \frac{T_q^{K}}{n} + q^2\left(\frac{4}{3}\eta + \zeta + \mu\right)/n \, u_q^{\parallel} = 0 \tag{4.6c}$$

$$\partial_t T_q^{k} + iq \frac{T}{n c_v}\left(\frac{\partial P}{\partial T}\right)_n u_q^{\parallel} + q^2\left(\frac{K}{c_v} - \mu\right)/n \, T_q^{k} - q^2 \frac{\mu}{n}\left(\frac{\partial P}{\partial n}\right)_T\left(\frac{\partial T}{\partial P}\right)_n n_q = 0 . \tag{4.6d}$$

We use the standard notation and we have suppressed the index zero in the mean density and temperature. The coefficient μ depends on the second derivatives of $\Psi_q(z)$ and, in general, does not vanish (for more details see ref. 16).

Would the coefficient μ vanish, eqs. (4.6) reduce to the hydrodynamical ones. Clearly, in the general case, the difference stems from the fact that we have taken into account only the kinetic temperature. In the equations of hydrodynamics the total temperature appears, defined by

$$c_v T_q \equiv \frac{1}{N} \int (dp)^N (dr)^N H \rho = \frac{3k}{2} T_q^{K} + c_v T_q^{P} , \tag{4.7}$$

the second term depending on the correlations.

Using the theory of subdynamics (cf. the second section) we may predict that the correlations associated to the class of 'hydrodynamic' contributions to $P_q\rho$ are given by $C_q(P_q\rho)$. The operator C_q is expressed in terms of L, P_q and the frequencies $\{z_q\}$, by a formula analogous to eq. (2.16). It may be shown then that, to first order in q, there is a potential correction to the temperature, namely

$$T_q^{P} = -iqn\left(\frac{\partial P}{\partial T}\right)_n^{-1} \mu u_q^{\parallel} . \tag{4.8}$$

Taking into account this correction we may deduce the equations of hydrodynamics. They are identical to eqs. (4.6) with $\mu = 0$ and T_q^{K} replaced by T_q.

Another way of obtaining the correct hydrodynamical equations is by 'renormalizing' eqs.(4.6) using the ideas of the previous section. Let us call W_q the set $\{n_q, u_q, T_q^k\}$ and $\hat{W}_q$ the set $\{n_q, u_q, T_q\}$. We write eqs.(4.6) as

$$\partial_t W_q = \Theta_q^k W_q \; , \tag{4.9}$$

and the equations of hydrodynamics as

$$\partial_t \hat{W}_q = \Phi_q^k \hat{W}_q \; . \tag{4.10}$$

By construction the 5x5 matrices Θ_q^k and Φ_q^k have the same eigenvalues and thus there exists a transformation χ_q^k such that

$$\Phi_q^k = (\chi_q^k)^{-1} \Theta_q^k \chi_q^k \quad \Longleftrightarrow \quad \hat{W}_q = (\chi_q^k)^{-1} W_q \; . \tag{4.11}$$

A straightforward calculation gives

$$\chi_q^k = \begin{bmatrix} 1 & 0 & 0 & 0 & 0 \\ 0 & 1 & 0 & 0 & 0 \\ 0 & 0 & 1 & 0 & 0 \\ 0 & 0 & 0 & 1 & 0 \\ 0 & iq\alpha\mu & 0 & 0 & (3k/2c_v)^{1/2} \end{bmatrix} . \tag{4.12}$$

As we have already pointed out the effect of transformations of this type is to account for the correlations. Indeed one may verify that χ_q^k satisfies the relation

$$\chi_q^k (\chi_q^k)^* = A \; , \tag{4.13}$$

which is the analog of eq.(3.2).

It should be noticed that the matrix Φ_q^k is not dissipative and thus the Gibbs relation

$$\delta^2 S_q = \frac{1}{kT} \left[\left(\frac{\partial p}{\partial n}\right)_T \frac{|n_q|^2}{n^2} + |u_q|^2 + c_v \frac{|T_q|^2}{T^2} \right] \tag{4.14}$$

for the entropy is not simply $\|(\chi_q^k)^{-1} W_q\|^2$. As a matter of fact, we may construct another transformation,

$$\chi_q^e = \begin{bmatrix} \sqrt{kT}\left(\frac{\partial n}{\partial p}\right)_T^{1/2} & 0 & 0 & 0 & 0 \\ 0 & 1 & 0 & 0 & 0 \\ 0 & 0 & 1 & 0 & 0 \\ 0 & 0 & 0 & 1 & 0 \\ 0 & iq\alpha\mu & 0 & 0 & (3k/2c_v)^{1/2} \end{bmatrix} \tag{4.15}$$

which is also compatible with eq.(4.13). It includes a change of scale for the density and one may prove that

$$\delta^2 S_q = \| (\chi_q^e)^{-1} W_q \|^2 \quad . \tag{4.16}$$

These considerations will be developed in detail in a forthcoming paper.

5. CONCLUDING REMARKS

We have outlined here the main ideas of the theory of subdynamics and the transformation theory related to it. We think that the basic interest of this approach is that it leads to a formulation of dynamics in which the evolution is described by manifestly irreversible equations. This fact is expressed by the dissipativity of $\Phi = \Lambda^{-1} L \Lambda$. These considerations show that the 'dynamical' and 'thermodynamical' descriptions belong to different but equivalent representations.

Because of space limitations we have not discussed at all questions refering to the classes of states and observables to which our formalism applies. In several places we have assumed the possibility of analytic continuation of matrix elements of the resolvent of the Liouville-von Neumann operator, thus we have implicitly imposed restrictions on the admissible states and observables. These questions lead to rather interesting mathematical problems which appear also in other branches of mathematical physics[17].

Hydrodynamical modes have been used as an illustration of the formalism. In a relatively direct manner we have deduced the transport equations for a neutral fluid. The derivation of Gibbs formula for the entropy is a generalization of the work done by one of us (I.P.) in the case of a dilute gas [18,19]. What we have shown is that for a dense fluid the laws of linear irreversible thermodynamics cannot be obtained from the one-particle kinetic equation but effects of correlations must be included. Clearly these considerations can be extended to other types of physical systems, such as polyatomic fluids and ionized gases, where the structure of the macroscopic laws is not well established.

ACKNOWLEDGEMENTS

We acknowledge several fruitful discussions with our colleagues in Brussels and Austin. Especially, we wish to thank Dr. M. Theodosopulu who has obtained most of the results reported in the fourth section.

1. I. Prigogine, 'Non-Equilibrium Statistical Mechanics', Wiley-Interscience, New York, 1962.
2. In the context of non-equilibrium statistical mechanics, projection methods were introduced by Nakashima and Zwanzig; see : R. Zwanzig Physica 30, 1109 (1964)
3. I. Prigogine and P. Résibois, Physica 27, 629 (1961).
4. I. Prigogine, C. George and F. Henin, Physica 45, 418 (1969)
5. I. Prigogine, C. George, F. Henin and L. Rosenfeld, Chemica Scripta 4, 5 (1973)
6. A. Grecos, T. Guo and W. Guo, Physica 80A, 421 (1975)
   A. Grecos and M. Theodosopulu, Acta Phys. Polonica A50, 749 (1976)
7. In order to study rigorously a system in the thermodynamic limit, we should employ a more general mathematical framework. However, certain aspects of the problem of irreversibility can nevertheless be formulated at the level of the Liouville-von Neumann equation.
8. A simple example shows that, for a consistent interpretation of the formalism, a space with non-hermitian metric should be introduced; see A. Grecos in 'Proceedings of the International Conference on Frontiers of Theoretical Physics', Indian National Science Academy, New Delhi, 1977.
9. B. Misra, to appear.
10. I. Prigogine, C. George, F. Mayné and M. de Haan, to appear.
11. P. Résibois, J. Stat. Phys. 2,21 (1970)
12. D. Forster, Phys. Rev. 49, 943 (1974)
13. M.H. Ernst and J.R. Dorfman, J. Stat. Phys. 12, 311 (1975)
14. Y. Pomeau and P. Résibois, Physics Repts. 19c, 64 (1975)
15. M. Bixon, J.R. Dorfman and K.C. Mo, Phys. Fluids 14, 1049 (1971)
16. M. Theodosopulu, A. Grecos and I. Prigogine, to appear.
17. J.-M. Combes in 'Proceedings of the International Congress of Mathematicians', Vancouver, 1974.
18. I. Prigogine, Physica 15, 272 (1949)
19. S.R. de Groot and P. Mazur, 'Nonequilibrium Thermodynamics' , North-Holland, Amsterdam, 1962.

CHAPTER 9

NONEQUILIBRIUM QUANTUM STATISTICS OF LASERS AND NONLINEAR OPTICS

H. Haken

Institut für Theoretische Physik der Universität

Stuttgart, Germany

ABSTRACT

The fundamental difference between laser light and light from thermal sources rests in the statistical properties. Methods to cope with the laser problem and related ones of nonlinear optics are quantum mechanical Langevin equations, density matrix equations, and the method of quantum classical correspondence. Profound analogies between equilibrium and nonequilibrium systems exist.

1. INTRODUCTION

The laser is a light source whose remarkable physical properties, such as high intensity, spectral purity, high directionality etc. allow for numerous important and often novel applications. From a more fundamental point of view the question concerning laser light is this: Do its properties just present cases in which the corresponding properties of light from thermal sources are pushed to an extreme? Or are these properties fundamentally different? In the context of statistical physics it is most remarkable that the latter is true and that the radical difference between laser light and light from thermal sources rests in their statistical properties. As I predicted this fact sometime ago [1] my talk will necessarily have some historical traits. On the other hand concepts and methods of laser theory are still very lively and have impacts on many other fields.

2. LASER AND OPTICAL PARAMETRIC OSCILLATOR

In its typical form, a laser consists of a rod of an active medium which can be a gas, a solid, or a liquid. The laser-active atoms of the medium are "pumped" from the outside. Let us assume for simplicity two-level atoms. Their occupation number is inverted by the pump resulting in a Boltzmann distribution with negative temperature. The inverted atoms first emit light by spontaneous and later on by stimulated emission. This light (with frequency ω) is eventually absorbed by bodies with temperatures $T \ll \hbar\omega/k$ so that we may assume $T \cong 0$. In the sense of thermodynamics, laser light is coupled to two reservoirs: 1) the atoms with negative temperature, 2) the absorbers kept at temperature $T \approx 0$. Thus the whole system is far from thermal equilibrium. Furthermore, spontaneous and stimulated emission are quantum mechanical processes. What we are ultimately interested in are the physical properties of the emitted laser light [2] . Light from thermal sources is composed of wave tracks independently emitted by the atoms. The most complete description of this light can be obtained only by means of a statistical treatment. Thus in summary: an adequate theory of laser light requires <u>nonequilibrium quantum statistics</u>. Quite similar processes can be found in other fields of quantum optics, for instance in optical parametric oscillators [3]. A typical device consists of a so-called nonlinear crystal in which the incoming laser light (called pump) is split into the "idler" and the "signal". The idler may be a light wave or may consist of molecular or electric oscillations (Raman effect). The signal is a light wave whose properties we are interested in. Since the optical parametric oscillator can be treated in complete analogy to the laser I shall address myself to the latter. The physical properties of the emitted light, also called coherence properties [4], are usually described by correlation functions of the field strength

$$E(\underset{\sim}{x},t) = E^{(+)}(\underset{\sim}{x},t) + E^{(-)}(\underset{\sim}{x},t) \ . \tag{1}$$

In quantum electrodynamics the positive and negative frequency parts contain the photon annihilation and creation operators, respectively. The simplest correlation function

$$\langle E^{(-)}(\underset{\sim}{x}',t')\,E^{(+)}(\underset{\sim}{x},t)\rangle \qquad (2)$$

can be measured by Young's interference experiments or, in more modern experiments, by the measurement of beat frequencies. The next type of important correlation functions is given by

$$\langle E^{(-)}(\underset{\sim}{x}',t')\,E^{(-)}(\underset{\sim}{x},t)\,E^{(+)}(\underset{\sim}{x},t)\,E^{(+)}(\underset{\sim}{x}',t')\rangle \qquad (3)$$

which is accessible by the so-called Hanburgy Brown-Twiss experiment measuring photon-correlation. In (2) and (3) the E's are operators in the Heisenberg picture.

3. LASER MODEL. QUANTUM-MECHANICAL LANGEVIN AND DENSITY MATRIX EQUATIONS

Figure 1 explains our laser model.

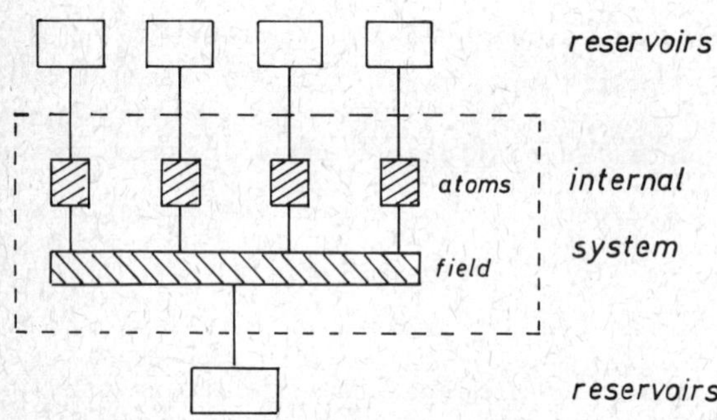

The Hamiltonian of the so-called internal systems within the dashed box consists of the Hamiltonians of the field, of the atoms, and of the interaction between field and atoms. The atoms are still subject to the external pump and interact with lattice vibrations in a crystal etc. These external effects are caused by reservoirs described by reservoir Hamiltonians H_R. The interaction between the atom μ and the reservoirs is represented by $H_{R,\mu}$. The interaction of the field with the mirrors or with the outer world is also described by reservoirs. This leads us to a total Hamiltonian

$$
\begin{aligned}
H_{tot} &= H_{field} + \sum_{\mu} H_{atom,\mu} + H_{int.} && \text{) internal) system} \\
&+ H_{f-R} + \sum_{\mu} H_{atom,\mu,R_\mu} && \text{interactions} \\
&+ H_R + \sum_{\mu} H_{R_\mu} && \text{reservoirs}
\end{aligned}
$$

As an example I treat the single mode laser. We decompose the field into standing waves assuming only one direction of polarization

$$E(\underset{\sim}{x},t) = \sum_k N_k \sin kx \left\{ b_k^+ + b_k \right\} \tag{5}$$

sinkx represents a standing wave, N_k normalization factors and b_k^+ and b_k are the creation and annihilation operators of a photon of mode k, respectively. In a single mode laser we keep only one k. This hypothesis is justified for well defined regions of laser action. The quantum statistical problem has been treated in either two more or less equivalent ways.

1) Quantum mechanical Langevin equations

We adopt the Heisenberg picture where the time dependence of an operator Ω is described by

$$\dot{\Omega} = \frac{i}{\hbar} \left[H_{tot}, \Omega \right] \tag{6}$$

The appropriate operators Ω for the laser are the creation and annihilation operators of the field mode, i.e. $\Omega = b, b^+$, and pairs of creation and annihilation operators of electrons in states k and j at atom μ : $\Omega = a_{k,\mu}^+ a_{j,\mu}$.

2)The density matrix equation

$$\dot{\rho} = - \frac{i}{\hbar} \left[H_{tot}, \rho \right] \tag{7}$$

In both cases we envisage a very large number of variables namely that of the field, the atoms and the reservoirs. On the other hand we are eventually interested only in correlation functions of the field, i.e. of the creation and annihilation operators b, b^+. This suggests to devise methods to first eliminate all unnecessary variables. I first describe the elimination of the reservoir variables coupled to the field. This problem of a quantum mechanical harmonical oscillator has been dealt with by Schwinger [5] and in a way especially suited for maser and laser physics by Senitzky [6] and others [2b]. The reservoirs cause damping. $-\kappa b$, and fluctuations, $F(t)$, so that the Heisenberg equation for b reads after the elimination procedure

$$\dot{b} = i\omega_o b - \kappa b + F(t) \tag{8}$$

Since the Heisenberg equations are time reversal invariant, whereas (8) is not, we see that the Markov approximation used in (8) brings in artificially irreversibility. However, one can show that this approximation is well justified in the laser process. The fluctuating force has the following correlation functions

$$\langle F(t) \rangle_r = 0, \ \langle F^+(t') \, F(t) \rangle_r = 2\kappa \, \bar{n}_{\omega_o} \, \delta(t-t').$$

Eq.(8) is, of course, strongly reminiscent of the Langevin equation of Brownian motion. We note, however, that both b and F(t) are operators. The noncommutivity of the F's secures that the commutation relation between the b's holds "on the average":

$$\langle b(t) \, b^+(t) - b^+(t) \, b(t) \rangle_r = 1 \tag{9}$$

Once we have found effective Langevin equations for the field mode alone it suggests itself to try the same trick with the atomic operators and this has indeed been done by Senitzky [7], Sauermann [2a], and others. There is, however, a more direct approach to find the correlation functions of the fluctuation forces in the following way [2a]. (Another approach gives Lax [2b]). First we substitute the pair of creation and annihilation operators by a projection operator $a_k^+ a_j \rightarrow P_{kj}$ allowing us to treat arbitrary quantum systems. Projection operators form an algebraic ring with the properties

$$P_{kj} \, P_{k'j'} = P_{kj'} \, \delta_{jk'} \tag{1o}$$

We now assume that the P's, when averaged over the reservoirs, obey the (most general) equations

$$\frac{d}{dt} \langle P_{kj} \rangle_r = \Sigma M_{kjmn} \langle P_{mn} \rangle_r \tag{11}$$

where the M's may be phenomenological damping and transition rates. The averaged projection operators do not obey the ring property (1o). However, we now require that (1o) is restored by the introduction of fluctuating forces $\Gamma_{mn}(t)$ into eq.(11) without averages. By this hypothesis the correlation function can be determined uniquely

$$\langle \Gamma_{m'n'}(t') \, \Gamma_{mn}(t) \rangle_r = G_{m'n'mn} \, \delta(t-t') \tag{12}$$

where the δ-correlation is a requirement which can be justified directly in many cases and the G's are functions of the damping and transition rates M. The laser Langevin equations in which the reservoir operators have been eliminated now read [2a]

$$\dot{b} = (-i\omega_0 - \kappa) b + i \sum_\mu g_\mu \, (a_1^+ a_2)_\mu + F(t) \tag{13}$$

$$(a_1^+ a_2)^{\bullet}_\mu = (-i\omega_{12} - \gamma) \, (a_1^+ a_2)_\mu - i g_\mu^* \, b(a_1^+ a_1 - a_2^+ a_2) + \Gamma_\mu(t) \tag{14}$$

and further equations for the atomic inversion operator. Now remember our goal. We want to derive correlation functions for the b's alone. This suggests to eliminate also the atomic operators. Imagine that we have resonance i.e. $\omega = \omega_{12}$ and that we eliminate the main time dependence by the transformation $b \rightarrow \exp(-i\omega t) b$ and a corresponding one for the dipole moment operators. This amounts to dropping the terms containing ω and ω_{12} in (13) and (14). If the damping of the field is much smaller than the damping of the atomic dipoles and the inversion the dipole moments and the inversion adjust adiabatically to the field. This allows us to neglect the time derivatives of the atomic operators in eq.(14) etc. and to express the atomic operators by the field. The result is the operator equation

$$\dot{b} = (G - \kappa)b - C\, bb^{+}b + F_{tot}(t) \tag{15}$$

F_{tot} is a superposition of the fluctuating forces of the field and of the atomic dipole moments. G, the so-called gain, depends on the strength of external pumping and can be changed appropriately. When the pump is small, i.e. $G < \kappa$, we can safely neglect the nonlinear term in (15). Since the noise term F_{tot} is Gaussian distributed also the field amplitude b is. This is the result to be expected for light from thermal sources. Let us now consider a region $G > \kappa$. Here the solution of (15) reads

$$b = (r_o + \rho(t)) \exp (i\varphi(t)) \tag{16}$$

i.e. the operator b now is decomposed into a c-number r_o and a residual operator ρ and the whole is multiplied by a phase operator. The correlation functions of (16) are entirely different from those of a Gaussian distribution. Figure 2 shows the prediction of that theory[1].

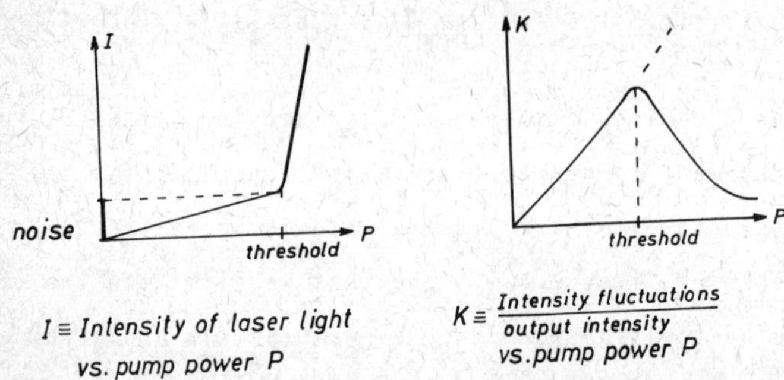

$I \equiv$ *Intensity of laser light vs. pump power P*

$K \equiv \frac{\text{Intensity fluctuations}}{\text{output intensity}}$ *vs. pump power P*

In a first exploration step we treat (16) by mean field theory which amounts to neglect F_{tot} and to replace b and b^+ by their expectation values $\langle b \rangle$. This procedure, also called "semiclassical approach", shows: For pump power small enough, i.e. $G < \kappa$, $\langle b \rangle = 0$. If $G > \kappa$, a steep increase of $\langle b \rangle$ as a function of pump occurs. The slope of the curve can by interpreted as efficiency. When we fully take into account fluctuations, we first ($G < \kappa$) find an increase of output intensity which is just the emission of usual lamps representing noise. The transition to the laser regime is rounded of. The change of the statistical properties of laser light versus light from thermal sources shows up when we plot the normalized fluctuations of the output versus the pump power (fig. 2). Below laser threshold, i.e. for $G < \kappa$, we obtain a linear increase well known from light from thermal sources while for $G > \kappa$ we obtain a sharp decrease. This is again a clear indication of the dramatic change of the properties of laser light. However, at that time I could not treat the immediate vicinity of the transition point which subsequently has been treated by Weidlich, Risken and Haken [2a] as well as by Scully and Lamb [2c] using a density matrix equation (the equivalence of both results has been recently demonstrated by Lugiato [8]) and in a classical domain (see below) by Risken [2a] and Hempstead and Lax [2b]. More recently Arecchi and Ricca [9] succeeded in solving the Langevin equation directly even close to and at the transition point.

The treatment of laser light statistics by means of density matrix equations also rests upon the idea of eliminating superfluous variables, especially to eliminate in a first step the reservoirs. This has been done in particular by Weidlich and Haake [2a]. A general frame for this is provided by Zwanzig's projection formalism [1o]. Lack of space does not permit to present details, but at least the general procedure is well known in the statistical physics community. Zwanzig's elimination procedure allows also to do away the atomic variables so that we are left alone with an effective equation for the field density matrix.

4. QUANTUM-CLASSICAL CORRESPONDENCE

I now want to discuss a rather new method called quantum classical correspondence, which I first explain by an example. In it we establish a one-to-one correspondence between the field operators b and b^+ with classical variables u and u^*. $b \longleftrightarrow u$, $b^+ \longleftrightarrow u^*$. This is achieved for instance by the Glauber-Sudarshan representation

$$f(u,u^*) = \pi^{-2} \int \exp(-i\beta u - i\beta^* u^*) \, \mathrm{Tr}\, (\exp(i\beta^* b^+) \exp(i\beta b)\rho) d^2\beta \quad (17)$$

Also other representations have been discussed in the literature.(17) establishes a correspondence between the density matrix ρ and the quasi-distribution function f(u,u). If ρ obeys a density matrix equation $\dot{\rho} = L\rho$ one can construct a (generalized) Fokker-Planck equation [2b] for the distribution function f. Furthermore all expectation and correlation functions between b's can be replaced by classical averages by a well defined description. Thus we can translate the original quantum mechanical problem into an entirely classical one. The principle of quantum-classical correspondence has been extended to arbitrary quantum systems described by projection operators [2a] or described by some other operators again having a ring property [11]. Due to this formalism we can construct, at least in a well defined approximation, a classical laser field equation

$$\dot{u} = (G - \kappa)u + Cu|u|^2 + F(t) \quad (18)$$

It is also possible to establish the Fokker-Planck equation connected with (18) and to write down its stationary solution [2a],[2b]

$$f(u) = N \exp(\alpha |u|^2 - \beta |u|^4) \quad (19)$$

5. ANALOGIES TO SYSTEMS IN THERMAL EQUILIBRIUM

This distribution function is strongly reminiscent of the Landau theory, if we draw an analogy between f and the distribution function of a system in thermal equilibrium and identify u with an order parameter [12]. This profound analogy is still more apparent for a continuum of laser modes. We obtain the Ginzburg-Landau equation in generalization of (18) and construct the Ginzburg-Landau functional as exact solution of the corresponding Fokker-Planck equation [13]. Its form

$$f(u(x)) = N \exp\left\{\int dx(\alpha|u(x)|^2 - \beta|u(x)|^4 + \gamma|u'(x)|^2)\right\} \qquad (2o)$$

needs no further discussion in this audience. It implies that laser light undergoes a pronounced disorder-order transition strongly reminiscent of a second order phase transition. Indeed there is a number of common features: form of stationary distribution function, critical slowing down, critical fluctuations, soft mode, long range order etc. On the other hand, the laser being a one-dimensional system, the characteristic singularities are now yet apparent. The time dependent correlation functions (2) and (3) of the single mode have been determined exactly by computer calculations solving the time-dependent Fokker-Planck equation [2a, b]. In the case of the continuous mode laser, the spatial correlation functions belonging to (2o) are known [14]. (2o) is still a special case: In a number of nonequilibrium situations a "potential" can be found analogous to thermodynamic potentials [15].

6. RECENT DEVELOPMENTS

In conclusion I mention some further developments of the quantum-statistics of the laser. Hepp and Lieb [16] discussed the problem if one finds also a phase transition if the coupling to external reservoirs is cut off so that one obtains the so-called Dicke model. Quite a number of subsequent papers has been devoted to this problem [17]. The answer is positive. The single mode laser model is presently used by a number of authors to test certain concepts, for instance Prigogine's entropy production principle or scaling theories [18]. Hasegawa [19] used it to define a new distribution function. In a very promising paper Sato [2o] showed that one can come close to the laser distribution function starting from the information entropy using a new type of constraints. Superradiance, superfluorescence [21] and nonequilibrium superconductivity are further examples for applications of the methods mentioned above. Another line of research leads into such different fields as chemistry, biology, ecology and sociology where one has to explain the spontaneous selforganized formation of macroscopic structure and order of complex systems composed of very many subsystems. These problems are now dealt with by a new interdisciplinary field called synergetics [22].

REFERENCES

1 H. Haken, Z.Physik 181, 96 (1964)
2 Laser theory (including statistical aspects) has been developed mainly by three groups and is correspondingly summarized in the following monographs which contain many further references.
a) H. Haken: Laser Theory, Vol.XXV/2c of Encyclopedia of Physics, Springer, New York 197o
b) W. Louisell, Quantum Statistical Properties of Radiation Wiley, New York, 1973, includes work of M. Lax
c) M.Sargent, M.O.Scully, W.E.Lamb: Laser Physics Addison-Wesley Publ.Comp. London 1974
3 see e.g. N. Bloembergen: Nonlinear Optics, W.A.Benjamin New York 1965
4 see e.g. L. Mandel and E. Wolf eds.: Coherence and Fluctuations of Light, Vol. I and II, Dover Publ. New York 197o
5 J. Schwinger, J. Math. Phys. 2, 4o7 (1961)
6 J.R. Senitzky, Phys.Rev. 119, 67o (196o); 124, 642 (1961)
7 J.R. Senitzky, Phys.Rev. 131, 2827 (1963)
8 F. Casagrande and L.A. Lugiato, Phys.Rev. A14, 778 (1976)
9 F.T.Areechi, M.Asdente and M.A.Ricca, Phys.Rev.A14, 383 (1976)
1o R. Zwanzig, Physica 3o, 11o9 (1964)
11 F. Haake, Springer Tracts in Physics 66 (1973) Springer, New York
12 H. Haken, in Advances in Solid State Physics, ed.O.Madelung, Vieweg, Braunschweig, 197o
V.DeGiorgio and M.O.Scully, Phys.Rev.A2, 117o (197o)
13 R. Graham and H. Haken, Z.Physik 237, 31 (197o)
14 see e.g. D.J. Scalapino, M.Sears, R.A.Ferrell, Phys.Rev. B6, 34o9 (1972)
15 R. Graham and H. Haken, Z.Physik 243, 289 (1971); 245, 141 (1971)
16 K. Hepp and E.H. Lieb, Ann.Phys. (N.Y. 76, 36o (1973); Phys.Rev. A8, 2517 (1973)
17 see for instance papers presented at the Conference on Coherence and Quantum Optics, Rochester N.Y. 1977
18 S. Suzuki, preprint 1976
19 H. Hasegawa, this conference
2o H. Sato, Progr. Theor.Phys. 57, 1115 (1977)
21 see for instance: Talks by R. Bonifacio, S. Haroche, L. Narducci at the conference 17 , and at the summerschool on Coherence in Physics, Versilia 1977
22 H. Haken: Synergetics - An Introduction, Nonequilibrium Phase Transitions and Self-Organization, Springer New York 1977

CHAPTER 10

ON SOME MATHEMATICAL MODELS EXHIBITING PHASE TRANSITIONS AND ON WHAT MIGHT BE LEARNED FROM THEM

Mark Kac

Rockefeller University, New York, N.Y. U.S.A.

ABSTRACT

Most (if not all!) of our understanding of phase transitions comes from the study of simplified models, and it is therefore useful to stop from time to time and to take stock of where we stand in this game. Such an exercise should be especially valuable in considering questions of universality of critical behaviour, the roles of dimensionality and symmetry, and of short versus long rangedness of interactions.

It will be argued that an intermediate spherical model ("soft-horn" model) introduced recently (Barrett and Kac, PNAS 72 (12), 1975, pp 4723-24; Barrett, J. Stat. Phys. 16(4), 1977, pp 371-94), is an appropriate vehicle for the discussion of these and related questions.

1. Most, if not all, of our understanding of the theory of phase transitions comes from studying models.

Models by their very nature have all sorts of artificial and unrealistic features built in and therefore caution is called for drawing physically valid conclusions from what after all are artfully contrived mathematical caricatures. And yet we are, for a long time to come, probably condemned to being dependent on models for insight and inspiration and we may as well try to develop certain systematic lines of inquiry.

2. In the sequel I shall consider only models of ferromagnets mainly for simplicity of presentation.

Here we have only three exactly soluble models:

(a) The mean field model

(b) The two-dimensional Ising model with nearest neighbor interactions and last (and probably least)

(c) The spherical model.

(I deliberately omit Gaussian models since they are closely related to (c) and since they lead to divergencies for low enough temperatures).

Of the three (b) is justly the most famous, most widely studied and probably the most realistic. Mean field theories are the simplest and remarkably good away from the critical point. Their fault is that they are based on the hypothesis of weak long range interactions which is untenable in the light of present day knowledge. The spherical models are the most "computable" but in a sense the least realistic. The lack of "realism" was however lessened considerably by Stanley's discovery (Stanley (1968)),(Kac and Thompson (1971)) that the spherical models are limiting cases of classical Heisenberg models based on n-dimensional (classical) "spins" in the limit $n \to \infty$.

A tempting strategy for further progress is to use the exactly soluble models as starting points for expansions in suitable chosen parameters.

An example of such a strategy is the familiar and much discussed $1/n$ expansion. Another example is the slightly modified "model A" of Helland and myself (Kac (1966)), i.e. a two-dimensional Ising model with interaction energy

$$(2.1) \qquad E = - \sum_{\substack{1<k<k'<N \\ 1\leq \ell < \ell' \leq M}} v(k,k,\ell,\ell')\mu_{k\ell}\,\mu_{k'\ell'}$$

where

$$(2.2) \qquad v(k,\ell;k',\ell') = J(e^{\gamma}-1)e^{-\gamma|k-k'|}\{\delta_{\ell,\ell'} + (2\cosh\gamma)^{-1}\,\delta_{\ell',\ell-1} + \delta_{\ell',\ell+1}\}$$

In the limit $\gamma \to \infty$ one obtains the classic Ising model with nearest neighbor interactions while in the limit $\gamma \to 0$ the model becomes a mean field model.

It can be argued that small γ is <u>qualitatively</u> the same as large γ and therefore use a small γ expansion to study qualitative features of the nearest neighbor model. (Unfortunately while this procedure confirms the universality of the <u>mathematical</u> mechanism underlying phase transitions -- i.e. the degeneracy of a principal eigenvalue it is not very well suited for studying critical behaviour.)

In my lecture today I should like to propose yet another model (or rather a class of models) whose discussion appears feasible and which might lead to better understanding of various features of the phenomenon of phase transitions.

3. The models I have in mind are based on the partition function

$$(3.1)\quad Q_N(\nu;\alpha) = \int\cdots\int_{\Sigma x_k^2=N} \exp\{\nu \sum_{i,j=1}^{N} \rho(i-j)x_i x_j\} \prod_{1}^{N}(1+\alpha x_k^2)\, d\sigma_{\sqrt{N}}, \alpha>0.$$

(I am using one-dimensional notation for simplicity in writing and am also considering first the case of zero external field).

If $\alpha=0$ we have the familiar spherical model. For $\alpha>0$ the weight $\prod_1^N(1+\alpha x_k^2)$ on the sphere $\Sigma x_k^2 = N$ has maxima at the 2^N points $(\pm1, \pm1,\ldots,\pm1)$ and thus the symmetry of the model is close to the symmetry of an Ising model. If N is kept fixed and $\alpha\to\infty$ one obtains (with a suitable normalization) the Ising model. However if the thermodynamic limit (i.e. $N\to\infty$) is taken first and then lets $\alpha\to\infty$ one is not (regretfully!) led to the Ising model. Still the new model (which I propose to call "soft horn" in contradistinction to the Ising model which has hard-delta function-"horns") must in some way be intermediate between the spherical and the Ising model and hence of interest.

4. Instead of (3.1) we could also consider

$$(4.1)\quad \overline{Q}_N(\nu;\alpha) = \int\cdots\int_{\Sigma x_k^2=N} \exp\{\nu/2 \sum_{i,j=1}^{N} \rho(i-j)x_i x_j\}$$

$$\prod_{k=1}^{N} \exp\{\alpha x_k^2-(\alpha^2/2)x_k^4\} \quad d\sigma_{\sqrt{N}}$$

which is obtained from (3.1) by truncating the formal expansion of $\log(1+\alpha x_k^2)$ at the second term.

The "soft horn" symmetry is preserved and the model looks like a discrete version of the popular ϕ^4 model except of course, for the spherical constraint $\sum_1^N x_k^2 = N$ whose role is neither wholly innocuous nor well understood.

Let me now sketch an argument that the model (4.1) exhibits a phase transition in two dimensions with nearest neighbor interactions.

First, we consider the auxiliary mean spherical version of (4.1) i.e. the model based on the partition function

$$(4.2)\quad \overline{Q}_N(\nu,s;\alpha) = \int_{-\infty}^{\infty}\cdots\int_{-\infty}^{\infty} \exp\{-s\sum_1^N x_k^2\}$$

$$\exp\{\nu/2\sum_{i,j=1}^{N}\rho(i-j)x_i x_j\}\prod_{k=1}^{N}\exp\{\alpha x_k^2-(\alpha^2/2)x_k^4\}\quad dx_1\cdots dx_N$$

where s is determined by the condition

$$(4.3)\quad -1/N\ \partial/\partial s\ \log\overline{Q}_N(\nu,s;\alpha) = 1$$

(because $\prod_{k=1}^{N} \exp\{-(\alpha^2/2)x_k^4\}$ is a very strong convergence factor, s can even be negative).

Assuming that we can prove that the limit

$$(4.4)\quad \lim_{N\to\infty} 1/N\ \partial/\partial s \log \overline{Q}_N (\nu, s; \alpha) = -\overline{f}(\nu, s; \alpha)$$

exists then since $\overline{f}(\nu, s; \alpha)$ is clearly a decreasing function of s, the equation

$$(4.5)\quad \overline{f}(\nu, s; \alpha) = 1$$

has an unique solution $\overline{s}(\nu;\alpha)$. It then follows by a standard argument that

$$\lim_{N\to\infty} 1/N \log \overline{Q}_N(\nu;\alpha) =$$

$$(4.6)\quad \overline{s}(\nu;\alpha) + \lim_{N\to\infty} 1/N \log \overline{Q}_N(\nu, \overline{s}(\nu;\alpha);\alpha)$$

and a phase transition would occur in the original model if for <u>every</u> ν $\overline{f}(\nu,s;\alpha)$ would have a break in analyticity in s at $s_o(\nu;\alpha)$ so that the graph of $\overline{f}(\nu,s;\alpha)$ (as a function of s) would be composed of two analytically distinct pieces.

By a result of G.S. Sylvester and H. Van Beijeren (to appear in the Journal of Functional Analysis) we can conclude that for <u>every</u> s there is a $\nu_c(s;\alpha)$ at which $\overline{f}(\nu,s;\alpha)$ has a break in analyticity and this ought to imply a break in analyticity in s.

The above argument needs a lot of tightening up but it does indicate how important the symmetry of a model is when it comes to exhibiting or not exhibiting a phase transition.

Unfortunately our argument is not immediately applicable to the simpler weight $\prod(1+\alpha x_k^2)$ since the Sylvester-Van Beijeren result is not strong enough (yet) to guarantee in this case the existence of $\nu_c(s;\alpha)$ (which would then have to be smaller than s to avoid divergence).

5. Returning to (3.1) it is easy to show (see Kac and Barrett (1975)) that

(5.1) $$\frac{Q_N(\nu;\alpha)}{Q_N(\nu;0)} = \exp\left\{\sum_{\ell=1}^{\infty} \frac{\alpha^\ell}{\ell!} \sum_{i_1,i_2,\ldots i_\ell=1}^{N} \chi_\ell^{(N)}(i_1,\ldots,i_\ell)\right\}$$

where the cluster functions $\chi^{(N)}$ are defined by the equations

$$1 = \langle x_i^2 \rangle_{(N)} = \chi_1^{(N)}(i)$$

$$\langle x_i^2\, x_j^2 \rangle_N (1-\delta_{ij}) = \chi_1^{(N)}(i)\chi_1^{(N)}(j) + \chi_2^{(N)}(i,j)$$

$$\langle x_i^2 x_j^2 x_k^2 \rangle_N (1-\delta_{ij})(1-\delta_{ik})(1-\delta_{jk}) =$$

$$\chi_1^{(N)}(i)\chi_1^{(N)}(j)\chi_1^{(N)}(k) + \chi_1^{(N)}(i)\chi_2^{(N)}(j,k) +$$

$$\chi_1^{(N)}(j)\,\chi_2^{(N)}(i,k) + \chi_1^{(N)}(k)\,\chi_2^{(N)}(i;j) +$$

$$\chi_3^{(N)}(i,j,k)$$

etc. and where $\langle\quad\rangle_{(N)}$ denotes the spherical average of the quantity inside < > e.g.

$$\langle x_1^2\, x_2^2 \rangle_{(N)} = \frac{\displaystyle\int\cdots\int_{\Sigma x_k^2=N} x_1^2 x_2^2 \exp\{\nu/2 \sum_{ij=1}^{N} \rho(i-j) x_i x_j\}\, d\sigma_{\sqrt{N}}}{\displaystyle\int\cdots\int_{\Sigma x_k^2=N} \exp\{\nu/2 \sum_{i,j=1}^{N} \rho(i-j) x_i x_j\}\, d\sigma_{\sqrt{N}}}$$

From now on we shall consider only the case in which the spherical model ($\alpha=0$) exhibits a phase transition at $\nu=\nu_c=\nu_c(o)$.

It is easy to conjecture that, at least for high enough temperatures, the cluster property holds, i.e.

(5.2) $$\lim_{N\to\infty} 1/N \sum_{i_1,\ldots i_\ell=1}^{N} \chi_\ell^{(N)}(i_1,\ldots i_\ell) = b_\ell(\nu)$$

exist.

It is also tempting to conjecture that the series

(5.3) $$\sum_{\ell=1}^{\infty} \frac{\alpha^\ell}{\ell!} b_\ell(\nu)$$

has a non-zero radius of convervence for $\nu<\nu_c$ where ν_c is the critical parameter (proportionate to the inverse temperature) for the spherical model. Dr. Groeneveld has communicated to me that he has indeed proved this statement.

What remains unresolved is whether

$$\lim_{\nu \to \nu_c} R(\nu) = 0 \qquad (5.4)$$

where $R(\nu)$ is the radius of convergence of (5.3).

In Barrett and Kac (1975) a conjecture was made that for sufficiently small $\alpha(>0)$ the critical temperature is the same as of the spherical model. This conjecture was made on the basis of the (erroneous!) belief that the cluster property holds even for low temperatures and that moreover $R(\nu) > 0$ even for $\nu > \nu_c$.

Recently Barrett (Barrett (1977)) calculated b_4 and the result for temperatures below the critical $(\nu > \nu_c)$ is

$$\begin{aligned} b_4 = - \Big[& \frac{105}{4} - 105\left(1-\frac{\nu_c}{\nu}\right)^2 + 112\left(1-\frac{\nu_c}{\nu}\right)^3 \\ & - 33\left(1-\frac{\nu_c}{\nu}\right)^4 - \frac{3}{\nu^4} \sum_{n=-\infty}^{\infty} \left(\frac{1}{2\pi} \int_{-\pi}^{\pi} \frac{\exp(in\theta)}{g(o)-g(\theta)} d\theta\right)^4 \\ & - \frac{12}{\nu^3}\left(1-\frac{\nu_c}{\nu}\right) \sum_{n+-\infty}^{\infty} \left(\frac{1}{2\pi} \int_{-\pi}^{\pi} \frac{\exp(in\theta)}{g(o)-g(\theta)} d\theta\right)^3 \\ & - \frac{4}{\nu^2}\left(1-\frac{\nu_c}{\nu}\right)^2 \sum_{n=-\infty}^{\infty} \left(\frac{1}{2\pi} \int_{-\pi}^{\pi} \frac{\exp(in\theta)}{g(o)-g(\theta)} d\theta\right)^2 \end{aligned} \qquad (5.5)$$

Here as usual

$$g(\theta) = \sum_{-\infty}^{\infty} \rho(n) \exp(in\theta) \qquad (5.6)$$

with the obvious modification in the multidenominal case

e.g. for the three-dimensional model with nearest neighbor interactions (on a cubic lattice)

$$g(\theta_1, \theta_2, \theta_3) = 1/3(\cos\theta_1 + \cos\theta_2 + \cos\theta_3) \tag{5.7}$$

We thus see that b_4 will diverge for temperatures below the critical temperature of the spherical model (three-dimensional model with nearest neighbor interactions on a cubic lattice) since

$$\frac{1}{(2\pi)^3} \int_{-\pi}^{\pi} \int_{-\pi}^{\pi} \int_{-\pi}^{\pi} \frac{d\theta_1 d\theta_2 d\theta_3}{[1 - 1/3(\cos\theta_1 + \cos\theta_2 + \cos\theta_3))]^2} = \infty. \tag{5.8}$$

Thus for short range interactions the cluster property fails to be valid for low enough temperatures and the series (5.3) has divergent terms.

For sufficiently long range interactions satisfying at least the condition

$$\int_{-\pi}^{\pi} \frac{d\theta}{(g(o)-g(\theta))^{\ell}} < \infty, \ \ell = 2,3,..., \tag{5.9}$$

the calculation of Barrett points toward the conclusion that the cluster property would in this case, hold even for $\nu > \nu_c$ and the conjecture that $\nu_c(\alpha) = \nu_c(o) = \nu_c$ (at least for sufficiently small α) becomes tenable. (However one may have to assume more than (5.9)).

For the mean field soft horn model i.e. for the model based on the partition function

$$(5.10) \qquad Q_N(\nu;\alpha) = \int_{\sum_1^N x_k^2 = N} \exp\left\{\frac{\nu}{2N}\left(\sum_1^N x_k\right)^2\right\} \prod_1^N (1+\alpha x_k^2)\, d\sigma_N$$

Barrett and Kac (1975) have shown that the transition temperature does not depend on α i.e. $\nu_c(\alpha) = \nu_c(o) = \nu_c$ for all $\alpha>0$.

It should be stated that we have not proved that models for which (5.9) holds for = 1 (i.e. the $\alpha = 0$ model has a phase transition) exhibit a phase transition but I don't think that anyone will doubt that this is indeed so.

6. I should finally like to mention a very interesting study of Kim and Thompson (1976) (to appear) which considered the soft horn version of the Dyson hierarchical model with interaction energy

$$(6.1) \qquad \sum_{i,j=1}^{N} \rho_{ij} x_i x_j = \sum_{p=1}^{2^L} 2^{-(1+\sigma)p} \sum_{r=1}^{2^{L-p}} S_{p,r}^2$$

where $N = 2^L$ and $S_{p,r} = \Sigma x_i$ for $(r-1)2^p + 1 \leq i < r2^p$.

The parameter σ measures the range of the interaction and σ^{-1} is in some sense an analogue of dimensionality.

For $0 < \sigma < 1$ the Dyson models with Ising spins have a phase transition and so do the corresponding spherical models. There is little doubt that the same holds for soft horn models ($\alpha>0$) but as far as I know this has not been proved.

The hierarchical models are of special interest since for them the iteration scheme which goes by the name of renormalization group is exact.

Kim and Thompson calculated (numerically) $\nu_c(\alpha)$ for $\sigma = 1/4$ for $\alpha = 0$, $\alpha = 1$, $\alpha = 10$, $\alpha = 10^2$, $\alpha = 10^3$ and $\alpha = 10^4$.

The results are

α	$\nu_c(\alpha)$
0	0.1166803
1	0.1166795
10	0.1166239
10^2	0.1164618
10^3	0.1163981
10^4	0.1163895

For $\sigma = 3/4$ which are much more short ranged than $\sigma = 1/4$ the results are

α	$\nu_c(\alpha)$
0	~1.07
10^{-1}	~1.052
1	0.99240
10	0.91566
10^2	0.88694
10^3	0.88283

One might expect that $\nu_c(\alpha)$ gets less and less dependent on α as $\sigma \to 0$. This would be in accordance with our discussion of §5.

7. It is not difficult to extend the calculations to the case of non-zero magnetic field.

The partition function is

$$(7.1)\qquad Q_N(\nu,\omega;\alpha) = \underset{\sum_1^N x_k^2 = N}{\int\cdots\int} \exp\left\{\frac{\nu}{2}\sum_{i,j=1}^{N} \rho(i-j)x_i x_j + \omega\sum_1^N x_k\right\} \prod_1^N (1+\alpha x_k^2)\, d\sigma_{\sqrt{N}}$$

where $\omega = H/kT$, H being the external magnetic moment.

Defining the free energy per site by the usual formula

$$(7.2)\qquad -\frac{\psi}{kT} = \lim_{N\to\infty} \frac{1}{N}\log\, Q_N(\nu,\omega;\alpha)$$

we obtain by a formal calculation

$$(7.3)\qquad -\frac{\psi}{kT} = \frac{1}{4}\,\frac{\omega^2}{s-\nu g(o)} - \frac{1}{2}\,\frac{1}{2\pi}\int_{-\pi}^{\pi} \log\,(s-\nu g(\theta))\,d\theta + \alpha - \frac{\alpha^2}{2}\mu_4 + \cdots$$

where s is determined from the equation

$$(7.4)\qquad \frac{1}{4}\,\frac{\omega^2}{(s-\nu g(o))^2} + \frac{1}{2}\,\frac{1}{2\pi}\int_{-\pi}^{\pi} \frac{d\theta}{s-\nu g(\theta)} = 1$$

and

$$(7.5)\qquad \mu_4 = 3 - 2\left(\frac{1}{2}\,\frac{\omega}{s-\nu g(o)}\right)^4$$

For multidimensional models the single integrals in the above formulas have to be replaced by appropriate multiple ones.

Thus for the three-dimensional model with nearest neighbor interaction on a cubic lattice the integral in (7.4) should be

$$\frac{1}{(2\pi)^3} \int_{-\pi}^{\pi} \int_{-\pi}^{\pi} \int_{-\pi}^{\pi} \frac{d\theta_1 \, d\theta_2 \, d\theta_3}{s - \nu/3 \, (\cos\theta_1 + \cos\theta_2 + \cos\theta_3} \tag{7.6}$$

The magnetization

$$m = \frac{\partial}{\partial\omega}\left(-\frac{\psi}{kT}\right) \tag{7.7}$$

turns to be given by the formula

$$m = \frac{1}{2}\,\frac{\omega}{S-\nu g(o)} + \alpha^2 \, \frac{\partial}{\partial\omega}\left(\frac{1}{2}\;\frac{\omega}{s-\nu g(o)}\right)^4 + \cdots \tag{7.8}$$

The first term in (7.10) is, of course, the magnetization of the spherical model which is an increasing function of ω. The α^2 correction is therefore positive in accord with the natural conjecture that the magnetization of soft horn models always is larger than the magnetization of the spherical model. (If this could be proved one would automatically have a proof that soft horn models exhibit a phase transition whenever the corresponding spherical model does).

The coefficient of α^2 in (7.10) is readily computable and the answer comes out to be

$$\frac{1}{4}\ \frac{\left(\frac{\omega}{s-\nu g(o)}\right)^4 \frac{1}{2\pi}\int_{-\pi}^{\pi}\frac{d\,\theta}{(s-\nu g(\theta))^2}}{\left(\frac{\omega}{s-\nu g(o)}\right)^3 + \frac{\omega}{2\pi}\int_{-\pi}^{\pi}\frac{d\,\theta}{(s-\nu g(\theta))^2}}$$

with the obvious modification in the multidimensional case.

As we let $\omega\to 0^+$ we discover a pronounced dependence on dimensionality.

In five and more dimensions the α^2 correction merely contributes a small correction to the coefficient of "classical"

$$\sqrt{1-\frac{\nu_c}{\nu}}$$

for the zero field magnetization.

In three dimensions because of the divergence of the integral

$$\int_{-\pi}^{\pi}\int_{-\pi}^{\pi}\int_{-\pi}^{\pi}\frac{d\theta_1 d\theta_2 d\theta_3}{[1-\frac{1}{3}(\cos\theta_i+\cos\theta_2+\cos\theta_3)]^2} \tag{7.9}$$

the α^2 term does not behave "classically". I do not know what the right behavior is but it is surely not "classical". Similarly in four dimensions I would not expect "classical" behavior but it is a borderline case since the integral corresponding to (7.9) diverges only logarithmically.

Unrigorous as this discussion is (though not more so than ε expansions in $4-\varepsilon$ dimensions so readily accepted by the community) it contains at least a germ of understanding of some features of universality.

8. In conclusion let me state a problem which I think is both important and perhaps broachable.

It is to prove (or disprove) that the specific heat for a three-dimensional soft horn model (based on nearest neighbor interactions on a cubic lattice) becomes infinite at the critical point from both the low and the high temperature side.

Am I optimistic? Moderately so for the high temperature side. When it comes to the low temperature side the complexity of Barrett's (1977) calculations are enough to discourage even the most incurable optimist. Still perhaps at Statphys 13 + n (n>0) answers to this and related questions will be forthcoming.

REFERENCES

1. Wayne W. Barrett, J. Stat. Phys. 16 (4) (1977).
2. Wayne W. Barrett and M. Kac, Proc. Nat. Ac. of Science., USA, 72 (12) (1975).
3. M. Kac, Brandei's Lectures (1966) Vol.1.
4. M. Kac and Colin J. Thompson, Phys. Norv.5 (1971).
5. H.E. Stanley, Phys. Rev. 176 (1968).

CHAPTER 11

A MODEL FOR INTERACTING QUARKS AND STRINGS *

Leo P. Kadanoff

Department of Physics and Materials Research Laboratory,
Brown University, Providence, Rhode Island 02912 U.S.A.

*Supported in part by the National Science Foundation under Grants DMS 7680560 and DMR 73–04886.

ABSTRACT

A model for an interacting system of quarks and strings is described. This model is unrealistic in many respects, but it illustrates how instantons can produce quark trapping in a relatively simple and direct way.

1. INTRODUCTION

According to one line of recent work in particle physics[1], strongly interacting particles like π mesons and protons may be considered to be made up of quarks, which interact via a gauge theoretical mechanism. This interaction is mediated by "strings" which are identified as vector mesons obeying an exact color gauge symmetry. This form of interaction is then used to derive (or at least argue for) the two most evident qualitative features of these particles, namely:

1. Asymptotic Freedom: That at short distances (i.e., high momentum transfers) the quarks behave like non-interacting particles, and

2. Quark Trapping: That at large distances, the interactions between quarks are so strong that we can only observe their color-singlet bound states. These bound states include a quark-antiquark multiplet, which includes among other things the π-meson, and also a three quark multiplet, exemplified by the proton.

In this note we describe a two dimensional model which seems to include in a simplified form the possibility of this qualitative behavior. The model described here does, however, differ most markedly from the usual colored quark-string theory, in that in this case there are "instantons"[2]-topological patterns of field excitation-which can trap the quarks. In the usual theory (reviewed in reference 1) the trapping mechanism arises from the non-linearities inherent in the non-abelian nature of the color symmetry.

To get a simple statistical mechanical picture of the quarks, represent them on a lattice in two dimensional space by creation and annihilation "operators" $\bar{\psi}(x)$ and $\psi(x)$ and use a Hamiltonian

$$H = \sum_{x} \bar{\psi}(x)\, \psi(x) - \sum_{\langle xx' \rangle} \bar{\psi}(x)\, \Sigma_o\, (x,x')\, \psi(x') \tag{1}$$

Here the first term simply describes the normalization of the quark variables; the second describes how the quarks may hop from site to site in the lattice. We know how to handle the fermion nature of the quarks by letting $\bar{\psi}$ and ψ describe a Grassmann algebra. However, for simplicity of presentation, we shall consider here boson quarks in which $\psi(x)$ is a complex variable which is integrated over the entire complex plane and $\bar{\psi}(x)$ is its complex conjugate. In that case, the Hamiltonian (1) describes non-interacting bosons, which have a propagator

$$\begin{aligned} g(x,x') &= \langle\, \psi(x)\, \bar{\psi}(x')\, \rangle \\ &= [\delta(x,x') - \Sigma_o(x,x')]^{-1} \end{aligned} \tag{2}$$

Here the inverse is to be interpreted in a matrix sense, so that the Fourier transform of the propagator is

$$g(\vec{p}) = [1 - \Sigma_o(\vec{p})]^{-1} \tag{3}$$

By choosing the hopping term correctly we can fit desired forms of particle propagation. For bosons, the right choice is one for which at small p

$$1 - \Sigma_o(p) \rightarrow (\text{const})(p^2 + m^2) \tag{4}$$

so that m is the boson mass.

So far we have a theory of non-interacting quarks. Next we construct very simple theory which have a gauge symmetry, namely an independent symmetry operation at every point in space. This theory serves to define the "strings." Then we couple the quarks and strings and see what arises.

Our string theory is closely related to the two-dimensional planar model. This is a model of two component vectors of fixed length lying at each point on the lattice. These vectors can be described by an orientation angle $\phi(\vec{x})$, which defines the inclination of these vectors from a standard direction. Of course any Hamiltonian for such a model must include a periodicity conditio[illegible] it is invar[illegible]nt under

$$\phi(x) \rightarrow \phi(x) + 2\pi\, n(x) \quad , \tag{5}$$

where n(x) are any set of integers varying from point to point on the lattice.

We employ a particular form of the planar model due to Villain[3] and José *et al*[4], and discussed in more detail by myself elsewhere in these proceedings. This model, which is believed to have behavior qualitatively similar to that of the planar model, employs in addition to the orientation variables $\phi(x)$ a set of link (or string) variables $\ell(x,x')$. These are a set of integers which are defined on every pair of nearest neighbor sites of the lattice and obey

$$\ell(x,x') = -\,\ell(x',x) \tag{6}$$

The planar model is then described by a Hamiltonian including nearest neighbor interactions among the ϕ's and an additional interaction among the ℓ's:

$$H_{string} = \frac{K}{2} \sum_{<xx'>} (\phi(x) - \phi(x') - 2\pi\ell(x,x'))^2 + H_o[\ell] \tag{7}$$

The planar model is invariant under the transformation (5). But for the Hamiltonian (7) this becomes invariant under the combined operations

$$\phi(x) \to \phi(x) + 2\pi n(x)$$

$$\ell(x,x') \to \ell'(x,x') = \ell(x,x') + n(x) - n(x') \qquad (8)$$

This invariance will be realized in the string Hamiltonian if $H_o[\ell'] = H_o[\ell]$. In turn, H_o will obey this symmetry if it depends upon the ℓ's only in the combinations $m(\vec{R})$, which are the sums of ℓ about closed loops. In particular (see Fig. 1), we define

$$m(\vec{R}) = \ell(\vec{x}_1,\vec{x}_2) + \ell(\vec{x}_2,\vec{x}_3) + \ell(\vec{x}_3,\vec{x}_4) + \ell(\vec{x}_4,\vec{x}_1) \qquad (9)$$

For simplicity, we choose H_o to be a quadratic form

$$H_o[\ell] = J \sum_R [m(\vec{R})]^2 \qquad (10)$$

Physically, $m(\vec{R})$ represents the vorticity at the site $\vec{R}$. To show this, we display in Fig. (2) the pattern of $\phi(x)$ which produces the minimum energy when there is a single non-zero $m(\vec{R})$ - equal to one - in the system.

To follow up on the idea of Polyakov, we wish to couple the vortices to the quarks. This can be done by replacing the quark Hamiltonian (1) by a "minimal coupling" between quarks and the vortices, namely

$$H_{quark} = \sum_x \bar{\psi}(x)\,\psi(x) - \sum_{\langle xx'\rangle} \bar{\psi}(x)U(x,x')\Sigma_o(x,x')\psi(x') \qquad (11)$$

Here $U(x,x')$, which represents a bit of string, is taken to be

$$U(x,x') = \exp[2\pi i\ell(x,x')/3] \qquad (12)$$

The reason for the "3" will appear in a moment.

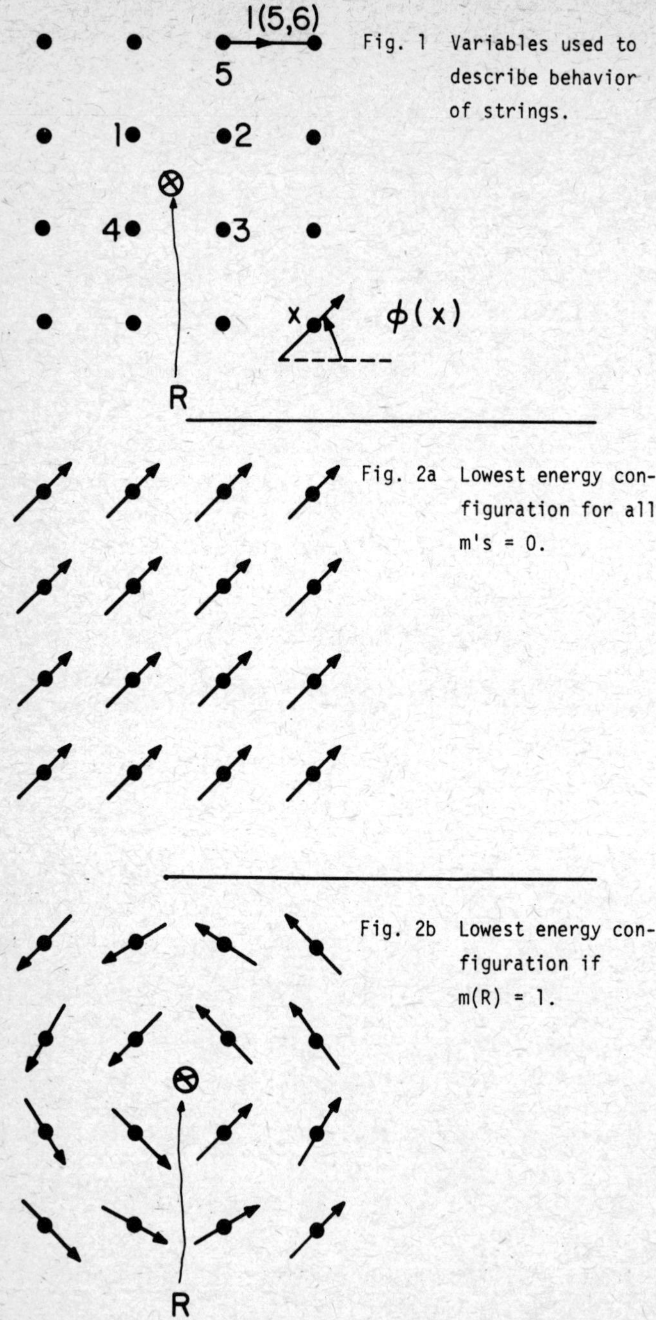

Fig. 1 Variables used to describe behavior of strings.

Fig. 2a Lowest energy configuration for all m's = 0.

Fig. 2b Lowest energy configuration if m(R) = 1.

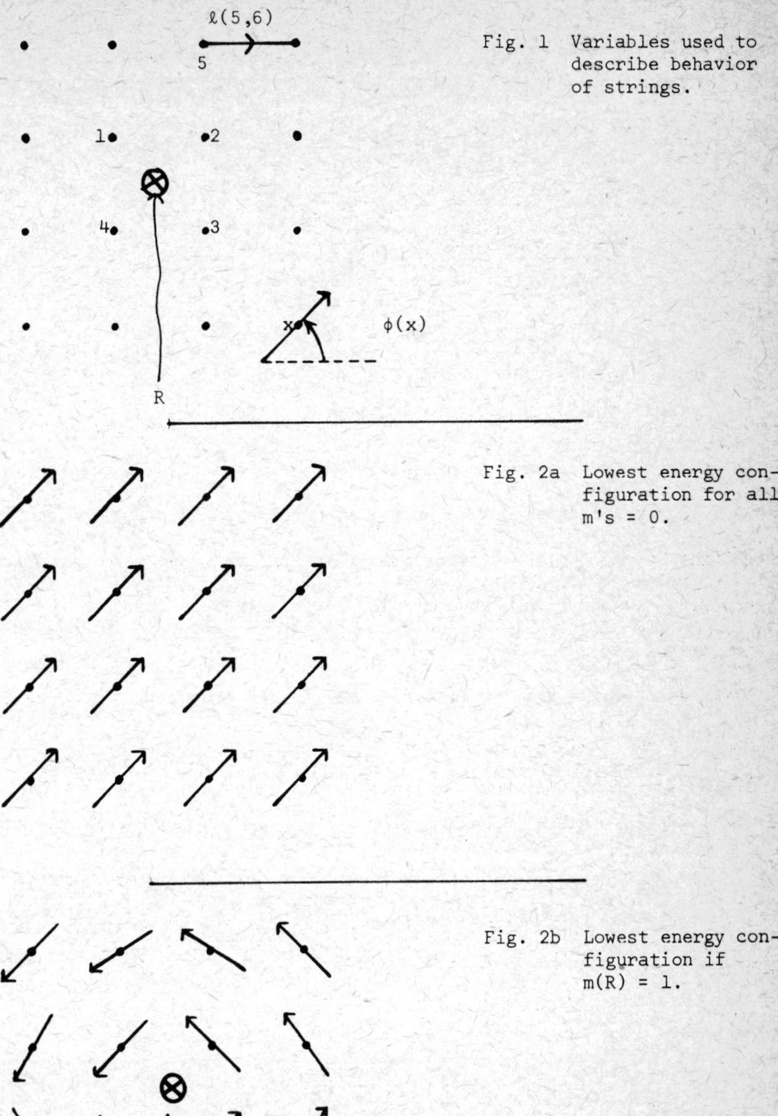

Fig. 1 Variables used to describe behavior of strings.

Fig. 2a Lowest energy configuration for all m's = 0.

Fig. 2b Lowest energy configuration if m(R) = 1.

The entire system is represented by the sum of the Hamiltonians (7) and (11). To make the transformations (8) a full invariance of the system we must also vary the quark operators according to

$$\psi(x) \to u(x)\, \psi(x)$$

$$\bar{\psi}(x) \to \bar{\psi}(x)\, u^*(x) \qquad (13a)$$

with

$$u(x) = \exp\,[\tfrac{2\pi i}{3}\, n(x)] \qquad (13b)$$

If it is true that only gauge invariant objects are observable, then the quarks themselves - which are preseneted by ψ or $\bar{\psi}$ - are unobservable. But the combinations $\bar{\psi}(x)\psi(x)$ (the "π-meson") and $\bar{\psi}(x)\bar{\psi}(x)\bar{\psi}(x)$ (the "proton") are gauge invariant and can be observed. Thus our simple abelian picture is - in some respects - analogous to the usual theory of colored quarks.

Next, we would like to examine the qualitative properties of the quarks in two limiting cases and show that

1. As $J \to \infty$ the quarks behave in exactly the same way as free particles, and
2. As $K \to 0$, the quarks are trapped.

To see this, examine a typical process in which a quark-antiquark pair is formed at one point, separates, and recombines as shown in Fig. (3). The amplitude for the process is roughly proportional to

$$A = \langle \prod_{\text{loop}} \Sigma_o(x,x')\, U(x,x') \rangle$$

$$= \langle\, (\prod_{\text{loop}} \Sigma_o(x,x')\, \exp[\tfrac{2\pi i}{3} \sum_{\text{loop}} \ell(x,x')] \rangle$$

$$= \langle \exp[\frac{2\pi i}{3} \sum_{\substack{\text{area} \\ \text{inside loop}}} m(\vec{R})] \rangle$$

$$\langle \prod_{\text{loop}} \Sigma_o(x,x') \rangle \tag{14}$$

The examination of the two limits is now trivial. As $J \to \infty$, expression (10) shows that all the vortex quantum numbers are forced to zero. Then all reference to the strings drops out of Eq. (14) so the fermions behave in exactly the same manner as if $U(x,x')$ were unity. Hence $J \to \infty$ implies free particle behavior for quarks.

On the other hand, if $K \to 0$, then all the different vortices are uncoupled. Thence, the right hand side of expression (14) is proportional to

$$A \sim \prod_{\substack{\vec{R} \\ \text{inside loop}}} \langle e^{\frac{2\pi i m(R)}{3}} \rangle$$

$$\sim [\langle e^{\frac{2\pi i m(R)}{3}} \rangle]^{XT} \tag{15}$$

$$\sim \exp[-\text{(const)}\ X\ T] \tag{15}$$

Here X and T are the two dimensions of the loop. But, if some amplitude is of the form

$$A \sim e^{-VT} \tag{16}$$

We interpret V as the portential or the energy involved in the process. (This is the map between the Euclidian world of statistical mechanics and the Minkowski world of field theory). Thence the potential for separating a quark-antiquark pair is

$$V = \text{const } X$$

Since the potential grows linearly, the quarks are infinitely strongly bound.

To get the standard picture of asymptotic freedom plus trapping, we must then get the renormalization group to give us renormalized interactions which show $K \to 0$ as the length scale goes to infinity and $J \to \infty$ as the scale goes to zero. Can this be?

Figure (4) shows the results of Kosterlitz and Kosterlitz and Thouless[6] for the renormalization group flow diagram of the planar model. Reference 4 shows how this particularly applies to the Hamiltonian (7). Assume that the analysis of this case is not changed in any substantial way by the presence of the quarks.

Now look at the flow in region I of Fig. 4. These flows point toward $K = 0$; hence they show trapping. They start from $J = \infty$; hence it includes asymptotic freedom. All flows in region III, on the other hand, go to $J = \infty$. Hence in this region, the theory flows toward a theory of non-interacting quarks. Finally, in region II, we have neither asymptotic freedom nor trapping.

The different regions of the flow diagram thus show physically very different behavior. The trapping and freedom of region I is the kind of behavior which we hope will emerge in the true four dimensional theory of colored quarks and strings. The free particle behavior of region III is a trivial result which can emerge from many field theoretical models. Finally, quantum electrodynamics is a gauge theory, which (like region II?) shows neither trapping nor asymptotic freedom.

This type of model may well be worthy of further investigation. In particular, the analysis given here is only heuristic in character because none of the effects of the quarks upon the strings have been taken into account.

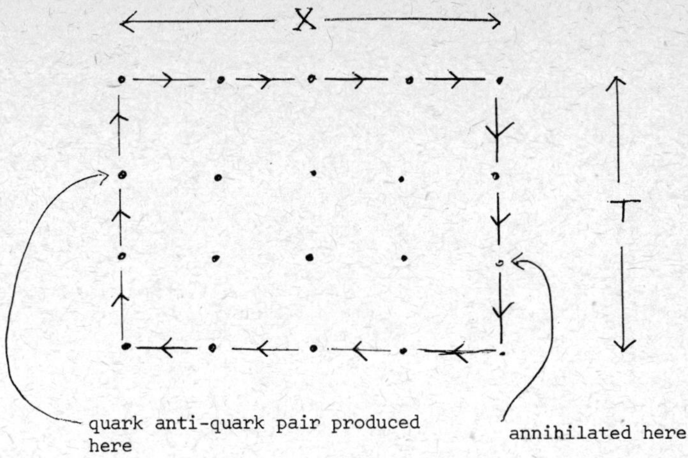

Fig. 3 A typical quark process.

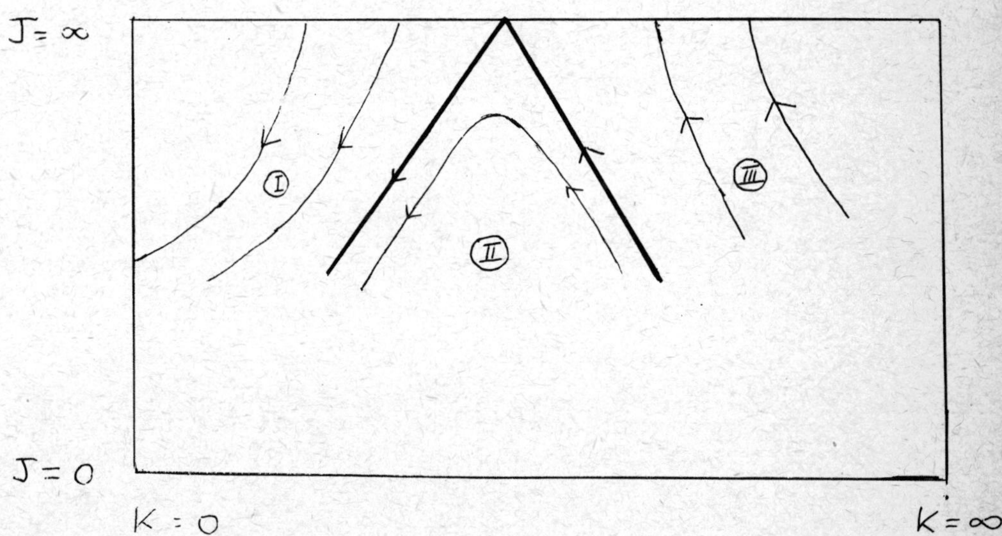

Fig. 4 Flow diagram. The heavy lines are separatrices which divide the picture into regions I, II and III.

FOOTNOTES

1. This work is reviewed in L. Kadanoff, Rev. Mod. Phys. 49, 267 (1977), where many additional references to the original way may be found.

2. A. M. Polyakov, Phys. Lett. B59, 79 (1975).

3. J. Villain, J. Phys. 36, 581 (1975).

4. J. José, L. Kadanoff, S. Kirkpatrick and D. Nelson, Phys. Rev. B (to be published).

5. See the review article of Kosterlitz and Thouless, J. Low Temperature Physics (to be published) for a full discussion of these trajectories.

CHAPTER 12

EXACTLY-SOLUBLE MODELS IN NONEQUILIBRIUM STATISTICAL MECHANICS

H. J. Kreuzer

Theoretical Physics Institute and
Department of Physics, University of Alberta,
Edmonton, Alberta, Canada

ABSTRACT

Recent progress in the construction of exactly soluble models in NESM is reviewed. We construct the exact time and space dependent density evolution in a gas (single- and multi-component) in which at time t=0 external potentials are switched on. The redistribution of matter in weak as well as strong potentials causing boundstates and resonances is illustrated as 3-dim. perspective views of these quantities over the (infinite) space-time plane.

1. INTRODUCTION

Exactly soluble models in non-equilibrium statistical mechanics (NESM) are constructed for a variety of reasons: (1) to exhibit the great complexity in the initial statistical regime of the time evolution of large systems (extending in a gaseous system over times of the order of the duration of a binary collision). For this, exact solutions of the Liouville or von Neumann equation are needed. The bulk of this paper is devoted to this problem; (2) to provide a theoretical "system" in which general approximation techniques and ideas in NESM, e.g. in the derivation of kinetic equations and hydrodynamic balance equations, can be tested. In our exact models we comment on some methods to extract the large time behaviour from pertubation theory; (3) to shed some light on the eternal problem in NESM, namely the origin of irreversibility. In this connection, we want to recall that one of the first exact models in NESM was Schrödinger's chain of mass points, with nearest neighbour harmonic coupling[1]. Having initially one mass displaced from its equilibrium position, the system will respond with damped oscillations (Besselfunctions) spreading the finite elastic energy stored in the initial displacement over infinitely many degrees-of-freedom. (Irreversibility due to infinite volume limit! The finite chain can also be evaluated, establishing very nicely the Poincaré recurrence cycle.) The simplest exact model in NESM is an ensemble of systems, each consisting of a free particle in a box with elastic walls. Here irreversibility comes about through statistical initial conditions.[2] Both mechanisms will be recovered in the more complicated models discussed later where we will also be able to identify momentum mixing in collisional processes as a third and very important source of irreversibility. In passing we should mention that recently master equations have been constructed which exhibit exponential decay to equilibrium.[3]

Exact (microscopic) models in the initial statistical regime of the time evolution of a large system have as ingredients: (a) a Hamiltonian $H = H_o + \theta(t)V$, where V is most likely an external potential; (b) the von Neumann equation $i\hbar\dot{\hat{\rho}} = [\hat{\rho}, H]$ for the statistical operator $\hat{\rho}$; and (c) initial conditions $\hat{\rho}_o = \hat{\rho}\,(t=0) = \exp[-\beta H_o]$ which must be known exactly. We outline our computional method with the calculation of the local time dependent mass or particle density $\rho(\vec{r},t) = \mathrm{Tr}[\psi^+\psi\hat{\rho}_t]/\mathrm{Tr}\hat{\rho}_t$ where ψ^+ and ψ are field operators (in second quantization) and $\hat{\rho}_t = \exp[-iHt/\hbar]\,\hat{\rho}_o \exp[iHt/\hbar]$ is the formal solution of the von Neumann equation. Shifting the time evolution operators onto ψ^+ and ψ , we get

$$\rho(\vec{r},t) = \mathrm{Tr}\,[\psi^+(\vec{r},t)\,\psi(\vec{r},t)\hat{\rho}_o\ /\mathrm{Tr}\,\hat{\rho}_o. \qquad (1)$$

The construction of exact models in NESM is thus split into two steps: (a) the calculation of the exact dynamics in $\psi(\vec{r},t)$, a problem in quantum mechanics; and (b) the trivial averaging of the operator products $\psi^+(\vec{r},0)\psi(\vec{r},0)$ at time t=0 over an exactly known initial distribution $\hat{\rho}_o$.

2. RESPONSE OF AN IDEAL GAS IN A SEPARABLE POTENTIAL [4]

We consider an ideal gas initially in equilibrium (any statistics) and uniformly distributed throughout a volume Ω which is subjected at time t=0 to an external potential. We want to calculate the resulting time and space dependent redistribution of matter. The Hamiltonian is

$$H = \sum_{\vec{k}} \varepsilon_{\vec{k}}\, a_{\vec{k}}^+ a_{\vec{k}} + \Theta(t)\Omega^{-1} \sum_{\vec{k},\vec{k}'} V_{\vec{k}\vec{k}'}\, a_{\vec{k}}^+ a_{\vec{k}'} \qquad (2)$$

($\psi^+(\vec{r},t)=\Omega^{-1/2}\Sigma \exp[i\vec{k}\cdot\vec{r}]\, a_{\vec{k}}(t)$). The dynamical single-particle problem can be solved explicitly for only a limited number of potentials $V_{\vec{k}\vec{k}'}$, e.g. Coulomb, square well, Bargmann, and separable potentials. We choose here a separable ansatz $V_{\vec{k}\vec{k}'} = g\, v_{\vec{k}} v_{\vec{k}'}$ acting in s-waves only. (Generalization and justification can be found in ref. 4). Heisenberg's equation-of-motion can be solved to yield

$$a_{\vec{k}}(t) = e^{-i\varepsilon_{\vec{k}} t} a_{\vec{k}}(0) + \Omega^{-1} \sum_p F_{\vec{k}\vec{p}}(t)\, a_{\vec{p}}(0) \qquad (3)$$

where the expicit form of $F_{\vec{k}\vec{p}}(t)$ depends on the form factor $v_{\vec{k}}$ which in our numerical examples is chosen to be $v_{\vec{k}}=(\vec{k}^2+\gamma^2)^{-1}$ well-known in nuclear physics.[5]

Inserting (3) into (1), taking the large volume limit and

introducing dimensionless variables $\lambda=\gamma^{-1}$, $V_o=g\lambda$, $\sigma=g/|g|$, $\lambda_o=\hbar/(4mV_o)^{1/2}$, $\upsilon=\lambda_o/\lambda$, $\delta=V_o/k_BT$ and $\tau=2V_ot/\hbar$, we find

$$\rho(r,t)/\rho(0) = 1 - \frac{\sigma}{\pi\sqrt{\pi}}\, {}^{\lambda}0/\lambda\ ({}^{\lambda}0/r)^2\ \delta^{3/2} \int_0^\infty \frac{y}{y^2+\nu^2}\, e^{-\delta y^2} * \tag{4}$$

$$* \left(2\sin(yr/\lambda_o)\ \mathrm{Re}[G] - \frac{\sigma}{16\pi}\frac{\nu y}{y^2+\nu^2}|G|^2\right) dy$$

where

$$G = e^{iy^2\tau}\sum_{k=1}^{5} A_k\, e^{ic_k^2\tau}\left[e^{-r/\lambda}\mathrm{Erfc}(e^{i\pi/4}c_k\sqrt{\tau})\right. \tag{5}$$

$$\left. - e^{c_kr/\lambda_o}\ \mathrm{Erfc}\ (e^{i\pi/4}c_k\sqrt{\tau} + e^{-i\pi/4}r/(2\lambda_o\sqrt{\tau})\right]$$

with c_k and A_k known functions of the parameters above. This is the exact time- and space-dependent density evolution in an ideal gas in which at time t=0 an external separable potential is switched on. Now we give a number of numerical examples of the density evolution plotted

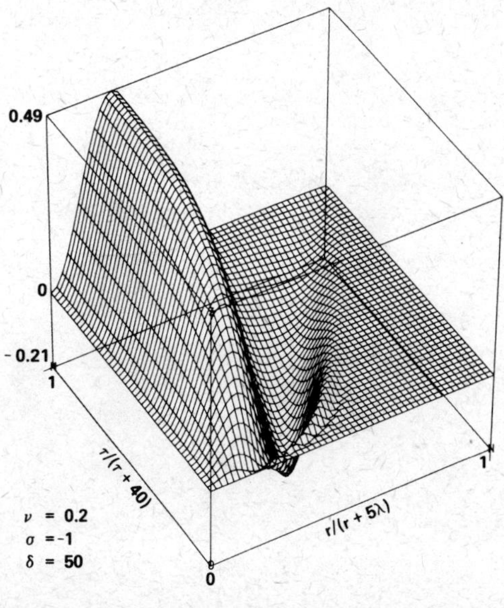

← Fig. 1: Density evolution of a Maxwell-Boltzmann gas in a weak attractive potential. Mass is drawn into the potential centre and a weak disturbance travels diagonally across the (r,τ)-plane as a density wave.

as the surface $(r/\lambda)^2(\rho(r,\tau)/\rho(0)-1)$ over the entire space-time plane, rescaled with the transformations $\tau \to \tau/(\tau+\tau_0)$ and $r \to r/(r+r_0)$ with appropriate scale constants τ_0 and r_0. The discussion is given in the figure captions.

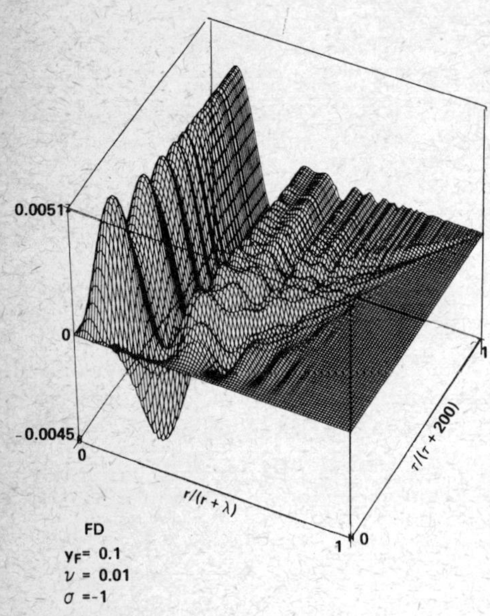

⇐ Fig. 2: Density evolution of a Fermi-Dirac gas at T=0K (normalized Fermi-momentum $y_F = (6\pi^2\lambda_0^3\rho(0))^{1/3}$ in a strong attractive potential, so strong that gas particles can be trapped into a bound-state.
Notice: (a) the time-dependence along the boundstate-ridge. The frequency of the oscillation is proportional to the boundstate energy. (b) For $t\to\infty$: the stationary mass distribution in this potential is given by the boundstate wavefunction in the range of the potential. The stationary density ripples further out can be traced to the continuum.

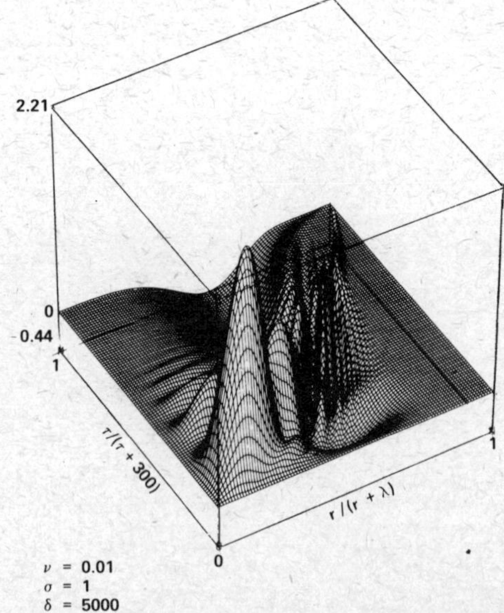

Fig. 3: ⇒
Density evolution of a Maxwell-Boltzmann gas in a strong potential causing gas particles to be trapped in resonances.
Notice: (a) resonating mass enhancement peak. (b) shockwave-like disturbance running diagonally across the (r,τ)-plane, best seen at large r and τ: hydrodynamic regime. (c) strong dispersion of waves for intermediate times (centre of the plot).

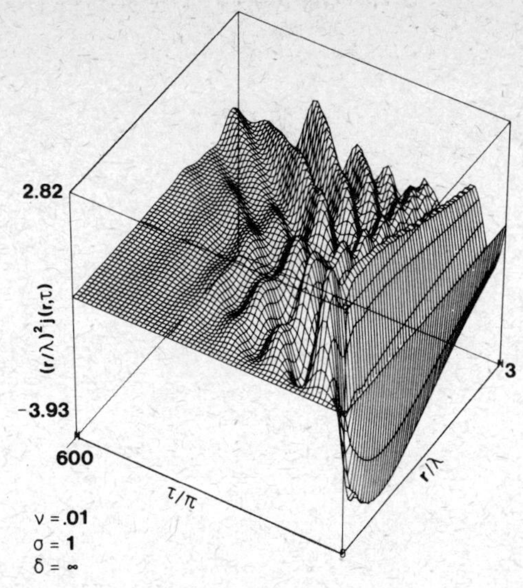

⇐ Fig. 4: The mass current accompanying the mass redistribution in Fig. 3. Care must be excercised in calculating the current in a nonlocal potential. The "nonlocal" contribution is negligible compared to the "local" or Schrödinger current for weak potentials. In strong potentials the "nonlocal" current is dominant for short times.[6]

Comments: (1) Small time behaviour in weak potentials:

$$\rho(r=0,\tau)/\rho(0) \approx 1 - \frac{\sigma}{A}/\pi - \delta[1-v\sqrt{\pi\delta}\ \ e^{-\delta v^2}\mathrm{Erfc}(v\sqrt{\delta})]\ \underline{\underline{\sqrt{\tau/2\pi}}}$$

$$\rho(r\neq 0,\tau)/\rho(0) \approx 1 + A(r,v,\delta)\underline{\underline{\tau^2}} + \ldots$$

(2) Large time limit: For weak potentials in the exact result:

$$\lim_{\tau\to\infty} \rho(r=\text{const.},\tau)/\rho(0) > 1 ,$$

indicating an apparent violation of mass conservation. This is due to the fact that as $\tau\to\infty$ the mass conserving depletion layer (for an attractive potential) will have moved past any finite r. For any finite τ we always get $\int d^3r\ \rho(r,\tau)=\rho(\tau=0)$ i.e. mass conservation.

(3) 2nd order pertubation theory and large time limit ($\lim_{\tau\to\infty} \sin^2\omega t)/\omega^2 =$ $= t\pi\delta(\omega)$)yield for the time dependent occupation function in momentum state $\vec{k}$:$n_k(t)\approx n_k(0) + f_k^{(2)}t/\Omega$. The large time limit in the <u>exact</u> solution yields a similar result (with a different f_k). Interpreta-

tion: an external potential of finite range cannot and does not lead to a finite change in $n_k(t)$ which, after all, is an average quantity over the whole infinite system.[7] To obtain meaningful results one has to study local quantities as done above.

Possibilities: (1) Extension to superposition of several separable potentials and to several scattering centers has been done. (2) A similar 1-dim. model has been used to study the adsorption kinetics of a gas at a solid surface.[3] (3) Extension to multi-component systems in which a molecule of type "i" is catalytically converted into a molecule "j" in the potential $V_{kk'}$ (e.g. spin-flip scattering at a magnetic impurity!). For a 2-component system the Hamiltonian reads

$$H = \sum_k \left(\varepsilon_k^{(1)} a_k^+ a_k + \varepsilon_k^{(2)} b_k^+ b_k + \Theta(t)\Omega^{-1} \sum_{kk'} V_{kk'} (a_k^+ b_{k'} + b_k^+ a_{k'}) \right) \quad (6)$$

Numerical work with Yamaguchi form factors in a separable potential $V_{kk'} = g v_k v_{k'}$ has been done.

3. TIME EVOLUTION OF A TWO-COMPONENT SYSTEM WITH CATALYTIC REACTIONS IN A RANDOM DISTRIBUTION OF STATIC SCATTERING CENTERS[9]

The interaction in the Hamiltonian (6) has to replaced by

$$V_{kk'} \rightarrow V_{kk'} \sum_i \exp\left[-i (\vec{k}-\vec{k}')\cdot\vec{R}_i\right]$$

where $\vec{R}_i$ are the random positions of the scattering centers. The additional difficulty is obviously the averaging process over the random positions $\vec{R}_i$. A diagrammatic method is used. Conclusions: (1) There is a subset of topologically simple graphs which can be summed explicitly yielding the "diagonal" model, i.e. a system in which the conversion of "a" into "b" particles takes place in an averaged external potential which is constant in coordinate space and thus diagonal in momentum space, i.e. its Hamiltonian is [10]

$$H_d = \sum_k [\varepsilon_k^{(1)} a_k^+ a_k + \varepsilon_k^{(2)} b_k^+ b_k + V_o (a_k^+ b_k + b_k^+ a_k)]$$

which can be diagonalized by an orthogonal transformation to quasi particle operators $\alpha_k = \cos\theta_k a_k + \sin\theta_k b_k$ and $\beta_k = -\sin\theta_k a_k + \cos\theta_k b_k$ with $\tan 2\theta_k = 2V_o/(\varepsilon_k^{(1)} - \varepsilon_k^{(2)})$. The "a" particle density is given by

$$\rho_1(t) = \rho_1(0) + \frac{1}{(2\pi)^2} \int_0^\infty k^2 \frac{V_o^2}{V_o^2 + \frac{1}{4}(\varepsilon_k^{(1)} - \varepsilon_k^{(2)})^2} \times$$

$$\times (1 - \cos(\lambda_k^{(1)} - \lambda_k^{(2)})\frac{t}{\hbar}) (n_k^{(2)}(0) - n_k^{(1)}(0))dk .$$

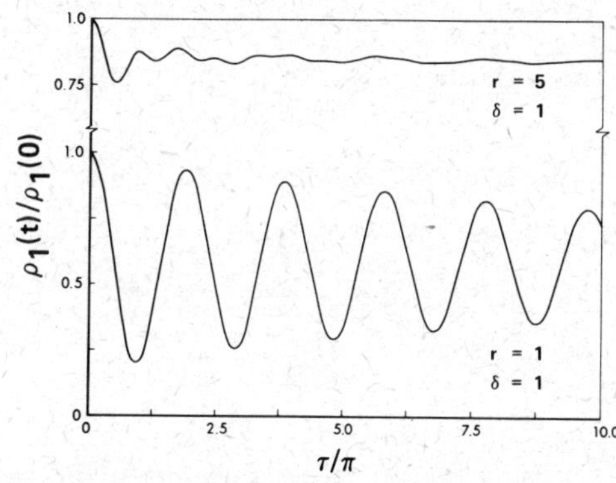

← Fig. 5: Diagonal model: 2-component Maxwell-Boltzmann gas. $\varepsilon_k^{(1)} = \hbar^2 k^2/2m_a$, $\varepsilon_k^{(2)} = \hbar^2 k^2/2m_\sigma + \varepsilon_o$; $\tau = \pi\hbar/V_o$; $\delta = V_o/k_\beta T$; $r = \varepsilon_o/V_o$.

Observe the fast decay of the high frequency oscillation in upper graph and slow decay of the asymptotic frequency (ω=1) oscillation in lower graph. The large time behaviour is given by $\rho_1(t)/\rho_1(0) \approx 1 - A\cos(\tau + \pi/4)/\sqrt{\tau}$ similar to the Schrödinger chain!

Comments on the "diagonal" model[10]: (1) Irreversibility is due to (a) large volume limit and (b) statistical initial conditions which eventually cause the oscillations to get out of phase. (2) The final state of the systems depends on the initial conditions, i.e. no approach to equilibrium takes place due to the lack of momentum mixing (3). If at time t=0 both gases "a" and "b" are present but at different temperatures, a beat pattern will emerge in their density evolution. (4) If initially the "a" particle gas is weakly degenerate, the amplitude of the oscillations compared to a Maxwell-Boltzmann gas is smaller

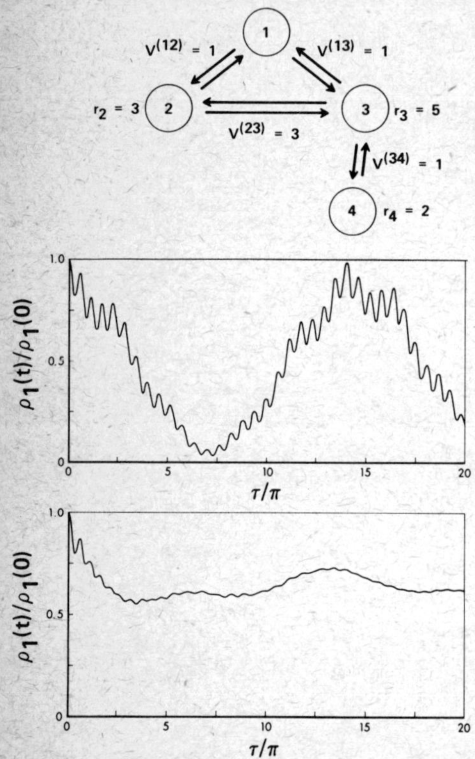

← Fig. 6: Diagonal model: 4-component Maxwell-Boltzmann gas. $V^{(ij)}$ are the interchannel coupling constants for the transformation of "i" into "j" particles. r_i are the normalized threshold energies. The parameters for the lower graph are given in the coupling scheme. In addition: $\delta=1$ and $m_1/m_2=m_1/m_3=m_1/m_4=0$. For the upper graph: $r_2=2$, $r_3=4$, $r_4=1$ and $m_1=m_2=m_3=m_4$. (For details see ref. 10).

for Fermi-Dirac particles (because the Pauli exclusion principles acts to prevent further creation of "b" particles) and is larger for a Bose-Einstein gas (because B-E statistics favors a larger occupation in a given energy state). (5) Separable Two-body forces can also be included[11].

The general problem of a two-component gas interacting catalytically in a random distribution of static scattering centers has been formulated and solved formally.[9] Numerical work is in progress.

REFERENCES

[1]E. Schrödinger, Ann. Physik 44, 916 (1914).
[2]A. Hobson, Phys. Lett. A 26, 649 (1968).
This system has nothing to do with an ideal gas in a box, see H.J. Kreuzer and R. Teshima, Can. J. Phys. 55, 189 (1977).
[3]D.P. Vasholz, Physica 74, 577 (1974); M. Razavy, Intern. J. Theor. Phys. 13, 237 (1975).
[4]H.J. Kreuzer and R. Teshima, Physica A (1977)(in press).
[5]Y. Yamaguchi, Phys. Rev. 95, 628 (1954).
[6]H.J. Kreuzer, Microscopic hydrodynamics in a nonlocal potential, preprint (University of Regensburg).
[7]I. Prigogine, in Topics in Nonlinear Physics, N.J. Zabusky, ed. (Springer, Berlin, 1968) pp. 239-246, discusses the external field problem in pertubation theory as an illustration of the methods to be used in his general theory of Non-equilibrium Statistical Mechanics. In view of our results it seems that its usefulness is somewhat reduced; in particular, the assertion that the linear time evolution for large times is due to "the scattering of the incoming beam" is less convincing.
[8]H.J. Kreuzer and G. Obermair, to be published.
[9]H.J. Kreuzer and Y. Kurihara, Physica 87A, 94 (1977).
[10]H.J. Kreuzer and K. Nakamura, Physica 78, 131 (1974) and H.J. Kreuzer and E. Hiob, Phys. Rev. A 14, 2321 (1976).
[11]H.J. Kreuzer and H.Z. Capri, Physica 81A, 411 (1975).

CHAPTER 13

STATISTICAL MECHANICS OF GRAVITATING SYSTEMS

Myron Lecar

Center for Astrophysics, Harvard College Observatory,

Smithsonian Astrophysical Observatory, Cambridge, Mass., U.S.A.* and

Racah Institute of Physics, Hebrew University of Jerusalem, Israel

* Permanent Address

ABSTRACT

Within our own galaxy, there are systems of a million stars called Globular Clusters. These clusters are sufficiently dense that star-star scatterings relax the system in the same way that collisions relax gaseous systems. However, because Globular Clusters are held together by their own gravity, the consequences of relaxation are unusual. At some stage, they become thermodynamically unstable and evolve so as to increase their central density without limit.

In Galaxies, which contain about a million-million stars, encounters between stars are unimportant. 'Mean Field' relaxation is responsible for the fact that they have smooth and similar shapes. Unlike collisional relaxation, which leads to a unique equilibrium state (if one exists), a variety of equilibrium states are possible when only the 'Mean Field' is operative.

In the Universe, there are about a million-million galaxies. They tend to occur in clusters, and possibly clusters-of-clusters. If, in the beginning, galaxies were distributed at random, clustering would have been induced by their mutual gravitational attraction. The rate at which clustering developed was influenced by the amount of matter in the Universe. The present density of the Universe is uncertain to within a factor of one hundred, and thus it is not known whether the Universe will continue to expand indefinitely, or will someday collapse. A considerable effort is underway to measure the amplitude of the correlations in the distribution of galaxies, and to understand the growth of correlations in expanding Universes of different densities.

Globular Clusters are relatively dense, spherical star systems of $10^5 - 10^6$ stars. There are on the order of two hundred of them in our galaxy, and they are also seen in nearby galaxies. They have been relaxed by star-star encounters in much the same way as gaseous systems relax by molecular collisions. Thus, equilibrium statistical mechanics might be applicable and such an analysis of their stability is reviewed here.

Galaxies, which contain $10^{11} - 10^{12}$ stars have symmetrical shapes and look relaxed. However, in this case, star-star encounters are completely negligible. Collective ("Mean Field") interactions relax these systems, but such processes conserve phase density, and thus Entropy, along trajectories. The statistics have to be worked out for a "coarse-grained" equilibrium state. The result resembles Fermi-Dirac statistics.

The Universe contains about $10^{10} - 10^{12}$ galaxies. In the large, they are distributed uniformly, but their positions are observed to be correlated up to scales of 10^8 light years (e.g., of the order of 1% of the radius of the universe). Numerical simulations indicate that initially uncorrelated galaxies will develop correlations similar to those observed, for a wide variety of initial conditions and in both open and closed universes. Our understanding of this process is still rudimentary.

I will discuss systems of stars (and galaxies) that are held together by their mutual gravitational attraction. While gravity is the weakest of the forces,

$$\frac{Gm_p^2}{\hbar c} \cong \frac{1}{10^{38}} \qquad \text{while} \qquad \frac{e^2}{\hbar c} \cong \frac{1}{137}$$

it is long range, unshieldable, and thus dominates for large enough bodies. Most bodies in the solar system have densities on the order of 1 gm/cm^3. Thus their radii are approximately given by $(3M/4\pi\rho)^{1/3}$ and their gravitational binding energy (per nucleon) is about

$$\epsilon_G = \frac{1}{2}\frac{G m_p M}{R} \cong 6\times 10^{-20}\,\rho^{1/3} M^{2/3}\ \mathrm{eV/nucleon}$$

where m_p is the proton mass, M the total mass, and ρ the density. The gravitational binding energy is plotted against the mass in figure 1. Gravitation contributes only 10^{-16} eV per nucleon to the forces that hold a man together, which is clear from his unsymmetrical figure. Bodies held together by gravity tend to have simple, symmetrical shapes, without protrusions. Even for the asteroids (the largest of which is the size of the British Isles), gravity contributes little to their binding energy. I have taken the break-even point, where gravity takes over from chemical forces as 10^{-3} eV per nucleon (or about 30 calories per gram). For bodies as large as the moon, gravity is the dominant force.

The Energy of a gravitating system is

$$E = 1/2\ M V^2 - G M^2/R$$

where M = Nm, N being the number of stars and m the mass of a typical star. It is critical to notice that the energy is not proportional to the number of particles. This will have the consequence that the various statistical ensembles (the Grand Canonical, the Canonical and the Micro-Canonical) describe different physical systems and give different results.

Gravitating systems in dynamical equilibrium (i.e., with no mass flows) satisfy the Virial Theorem
$M V^2 = G M^2/R$ so that $E = -1/2\ M V^2 = -\ 1/2\ G M^2/R$. Identifying the temperature by $kT/m = 1/3\ V^2$, we notice that $E \sim -T$ so that the heat capacity is negative. This implies that should you try to cool the system down by removing heat, they respond by heating up. For example, the kinetic temperature of an orbiting earth satellite is momentarilly lowered when the satellite collides with air molecules. But as it loses velocity, it falls, and as it falls, it speeds up. Since $V^2 \sim 1/R$, the more heat is removed, the faster (hotter) it gets.

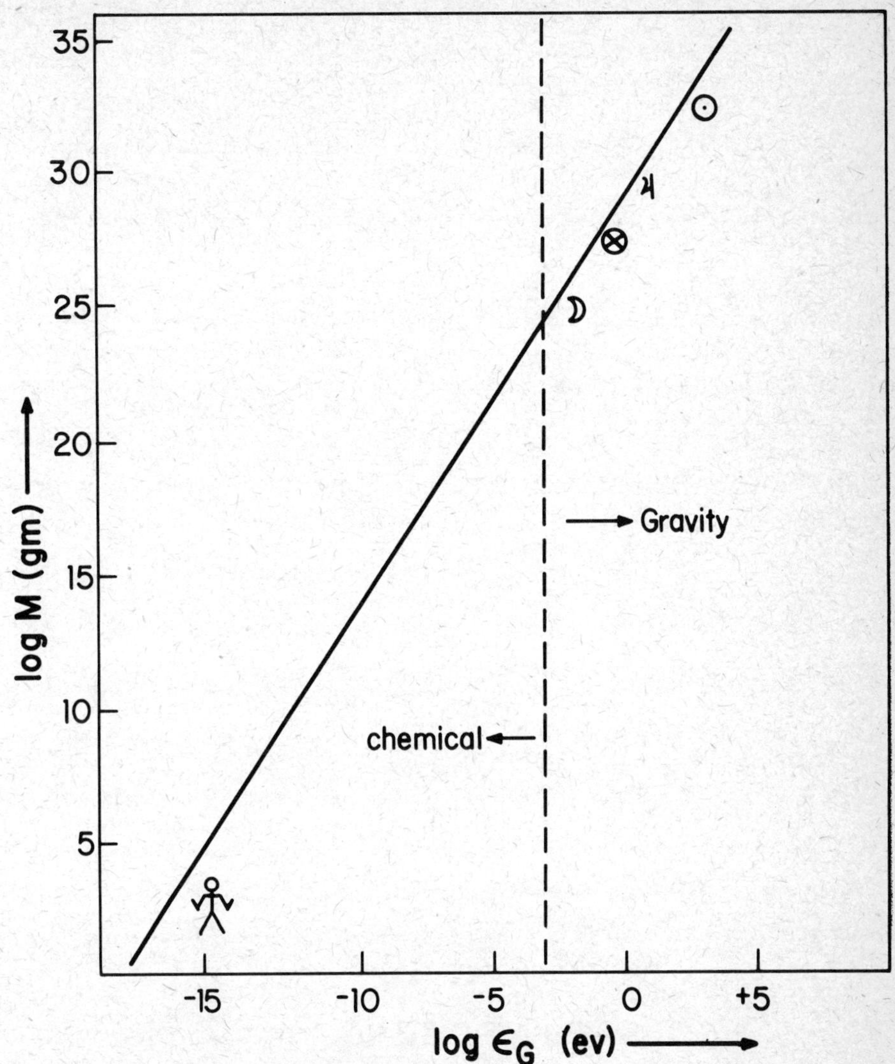

Fig. 1. Gravitational binding energy per nucleon

In large gravitating systems composed of many stars, an individual (test) star feels the average force of a large number of stars, and its' trajectory is relatively smooth. However, on occasion it does pass close to another star and a transfer of energy results. The time for such encounters to perturb the energy by an amount comparable to its initial energy is called the relaxation time, T_r. We can estimate T_r as follows. In order to have a large energy transfer, the stars will have to pass by each other sufficiently close so that their potential energy is comparable to their kinetic energy. If b denotes their separation, then $Gm/b \sim V^2$. Therefore the cross-section for strong encounters, $\sigma = \pi b^2$. If the cluster contains N stars in a volume R^3, then in a crossing of the cluster, the test star sweeps out a volume b^2R which contains, on the average, $b^2R \times N/R^3$ stars; or the chance of having a close encounter in one crossing is $N \times (b/R)^2$. Now, $b \sim Gm/V^2$, but the Virial Theorem, gives $V^2 = GNm/R$, so $b = R/N$. Thus the number of strong encounters per crossing is 1/N, or it takes N crossings to exchange a sizeable fraction of one's initial energy. As the crossing time is R/V, $T_r = N\ R/V$.

This is approximately correct, but, in fact, weak encounters dominate over strong encounters by a factor Log N, so that

$$T_r \sim N/\text{Log } N \times R/V.$$

Typical dimensions and relaxation times for Globular Clusters and Galaxies are :

	M ($M_\odot$)	R(pc)	V(Km/sec)	T_r(yrs)
Globular Cluster	10^6	20	15	$< 10^9$
Galaxy	10^{11}	10^4	250	$> 10^{14}$

where $M_\odot = 2\times10^{33}$ gms and pc = 3×10^{18} cm = 3 light years.

As the Universe is thought to be about 10^{10} years old, Globular Clusters are well relaxed by encounters, while Glaxies are not.

GLOBULAR CLUSTERS :

We know quite a bit about the dynamical evolution of globular clusters, mostly from numerical simulations. Figure 2, taken from the work of Henon, shows the radius containing a fixed fraction of the mass as a function of time. The outer layers expand, while the inner layers contract and then expand. The region that continually contracts is called the core, and the core radius is indicated by the dashed line. Note that the core does not contain a fixed mass. For the densest globular clusters (core density about $10^5\ M_\odot/pc^3$), the core mass is presently about 10% of the cluster mass. In time, the core density will increase but the core mass will shrink. From the work of Larson, we learn that the core density tends to infinity in a finite time, t_c. It is convenient to count time from t_c. Let $\tau = t_c - t$, so that as $t \to t_c$, $\tau \to 0$. Then, in a typical case, Larsons results can be expressed as follows :

$$\rho \sim \tau^{-1.5}, \quad v^2 \sim \tau^{-0.3}, \quad M \sim \tau^{+0.3}, \quad r \sim \tau^{+0.6}$$

$$E \sim Mv^2 \sim \frac{GM^2}{r} \sim \tau^0$$

Lynden-Bell and Wood (1968) studied the stability of gravitating systems with fixed energy and mass; i.e., the Micro-Canonical Ensemble. Recently, Horowitz and Katz rediscussed this problem using more powerful and rigorous methods, and in addition, looked at the Canonical and Grand Canonical Ensembles. Katz and I, motivated by the behavior of globular cluster cores, studied the stability of a system with fixed energy but variable mass (the "Grand Micro-Canonical Ensemble").

In order to obtain an equilibrium solution at all, it is necessary to enclose the system in a box. Otherwise, the entropy diverges, reflecting the tendency of the system to evaporate stars to fill the infinite configuration space. Secondly, one has to introduce a short range cutoff in the force law to prevent the formation of binaries. Observations do indicate a lack of binaries in globular clusters, but we don't understand the reason. It is easy to estimate that the time to make a binary with a close three-body encounter is approximately N relaxation times, but what is surprising is that there are no initial binaries (formed out of the same

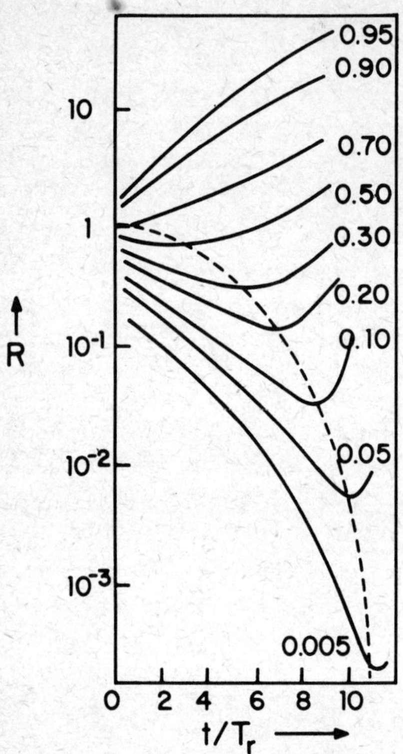

Fig. 2. Radius containing a given fraction of the mass

lump of gas). In the field outside of globular clusters, most stars are members of an initial binary.

Thirdly, one investigates the stability of equilibrium isothermal spheres. That is, the equation of state is

$$P = \rho \frac{kT}{m} = \frac{\rho}{\beta}$$ where P is the pressure.

Then hydrostatic equilibrium requires that

$$\frac{dP}{dr} = -\frac{GM}{r^2}\rho$$

where M is the mass interior to r, and we also have the relation

$$\frac{dM}{dr} = 4\pi\rho r^2$$

We require that the density be finite at the origin. For reference, the singular solution is $\rho \sim r^{-2}$, $M \sim r$.

The non-singular solution approaches the singular solution at very large r (well outside of the range of stability for any of the ensembles).

Solution of the isothermal sphere are tabulated by Chandrasekhar and Wares, in the following variables.

$$\rho = \rho_c e^{-\psi}, \quad \xi = kr, \quad k^2 = 4\pi G\beta\rho_c$$

$$\frac{1}{\xi^2}\frac{d}{d\xi}\left(\xi^2\frac{d\psi}{d\xi}\right) = e^{-\psi}$$

$$\text{or,} \quad u = \frac{d\ln M}{d\ln r} = \frac{\xi e^{-\psi}}{\psi'}, \quad v = -\frac{d\ln\rho}{d\ln r} = \beta\frac{GM}{r} = \xi\psi'$$

$$\frac{u'}{v'} = \frac{u}{v}\frac{(3-u-v)}{(u-1)}$$

In this way, one generates a family of equilibrium solutions. It turns out that all of these solutions are stable to collective modes, which go on the crossing time scale. What is investigated here thermodynamic instabilities which operate on the relaxation time scale.

The steepest descent analysis of Horowitz and Katz looks at the sign of the coefficient of the second order term in a Taylor expansion of the entropy around its' extremal value. Their analysis justify the following prescription.

We use the notation $f = Ae^{-\beta\epsilon}$, $\rho = \left(\frac{2\pi}{\beta}\right)^{3/2} Ae^{-\beta\phi}$

so that A is the natural logarithm of the chemical potential. The prescription is to construct a dimensionless function, F, from the gravitational constant G, E or β (depending on whether energy or temperature is fixed) and M or A (depending on whether mass of chemical potential is fixed). The (unique) functions, and their expressions in terms of the isothermal variables u and v, are :

$$F_{GCE} = 8\pi G\,(2\pi^3/\beta)^{1/2} R^2 A = uve^{-v}$$

$$F_{CE} = \beta GM/R = v$$

$$F_{GMCE} = (1/2)(8\pi^2 G)^5 EA^4 R^7 = (uve^{-v})^4 v(u-3/2)$$

$$F_{MCE} = ER/GM^2 = (u-3/2)/v$$

The turning points of F indicate the passage from a region of stability to a region of instability (or vice versa). It is convenient to plot F vs the density contrast (ρcenter / ρboundary). It is clear physically that for uniform density, the systems behaves like a perfect gas and is stable. Therefore the first turning point indicates the onset of instability. These occur at the following density contrasts:

Ensemble	Density contrast
Grand Canonical (GCE)	1.58
Grand Micro-Canonical (GMCE)	1.66
Canonical (CE)	32.1
Micro-Canonical (MCE)	709.

Thereafter, for higher density contrasts, the ensembles remain unstable, except for the Grand Micro-Canonical which has one additional region of stability for density contrasts from 11.6 to 92.6. The justification for disregarding higher turning points comes from the analysis of Horowitz and Katz.

The grand ensembles, which allow diffisuve contact with a reservoir, are clearly the most unstable. Unfortunately, I find them difficult to apply to astronomical systems. The most direct interpretation of the thermodynamics is to imagine the system under consideration to be in diffisive contact with a reservoir at fixed chemical potential. However, since the chemical potential, $\ln A = \ln(\rho\beta^{3/2}) + \beta\phi$ contains the gravitational potential ϕ, it is difficult to keep the chemical potential of the resevoir fixed (i.e., to shield it from the test systems). For example, we can understand the Grand Canonical Ensemble by imaging many clusters, well seperated in space so that they do not perturb each other gravitationaly; all at the same temperature and chemical potential, and connected by pipes. For (very) low density contrasts, an increase in mass is accompanied by an increase in chemical potential. Since matter wants to flow higher chemical potential to lower chemical potential, the systems gaining mass will return it. When the systems go unstable, and increase in mass is accompanied by a decrease in chemical potential, so that once a system gets a taste of a little mass, it gobbles up all the other clusters (who, when they are unstable, raise their chemical potential when they give away mass). However, when Katz and I tried to wrap the resevoir around the test system, the chemical potential of the resevoir was perturbed just enough to stabilize the system.

The Micro-canonical and Canonical ensembles are easier to mimic, say, in a computer experiment. The Micro-canonical ensemble is simulated by a system in a box with reflecting walls, and the stability has been checked by a computer experiment by Larson.The instability can be described as follows. If the center is allowed to contract slightly, it heats up. The heat flows down the temperature gradient to the boundary, where it is trapped. For low density contrasts, the boundary temperature increases fast enough to reverse the temperature gradient and re-establish equilibrium. For density contrasts greater than 709, the boundary never catches up, and the central temperature (and density) increase without bound (just as a system without a wall, would). From this description, it is easy to see why the Canonical ensemble is less stable, as there the wall is kept at a fixed temperature, and if the stars that evaporate to the wall, are hotter than the wall, they cool off. The functions, F, for the Canonical and Micro-canonical ensembles are plotted in figure 3.

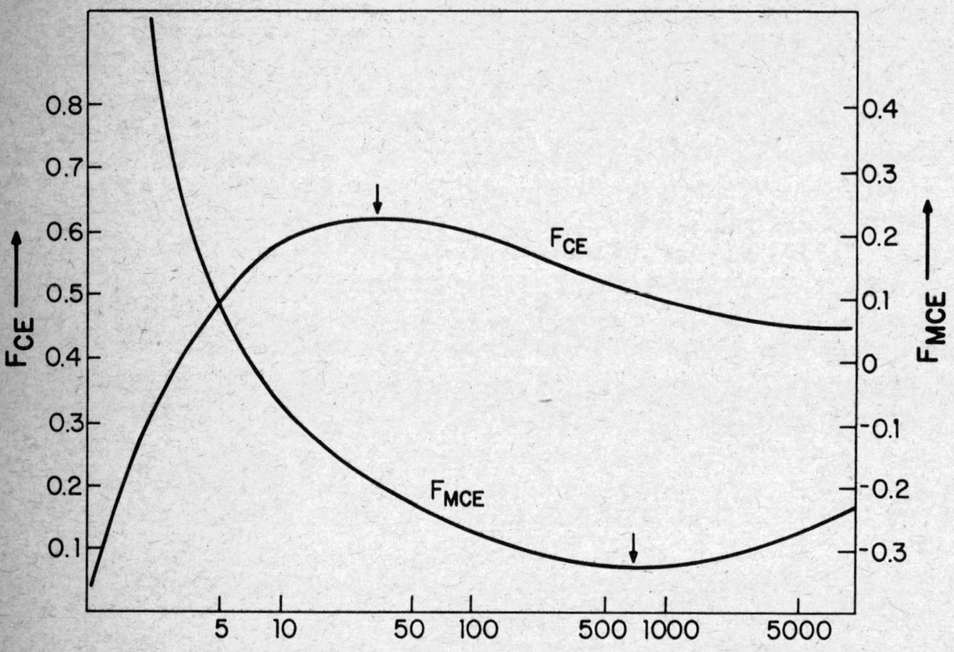

Fig. 3. Canonical and microcanonical ensembles

GALAXIES:

Galaxies are collionless systems. As we saw earlier, their relaxation time is of the order of 10^{14} years. Their dynamical evolution is governed by the Collisionless Boltzmann Equation (also called the one-particle Lioville Equation or the Vlasov Equation).

If f is the phase density, so that $\int f d\vec{v} = \rho, \int f d\vec{v} d\vec{x} = M$

Then $$\frac{\partial f}{\partial t} + \vec{v}\cdot\frac{\partial f}{\partial \vec{r}} + \vec{a}\cdot\frac{\partial f}{\partial \vec{v}} = \frac{D}{Dt} f = 0$$
$$\frac{\partial \vec{a}}{\partial \vec{r}} = -4\pi G\rho$$

This equation says that f is constant along particle trajectories, or that f flows like an incompressible fluid. The relaxation is accomplished by "Phase Mixing" which is a result of particles having different frequencies, and by "Dynamical Mixing" which reshuffles the frequencies. The $\vec{a}\cdot\frac{\partial f}{\partial \vec{v}} \sim f^2$ term is non-linear in f and exchanges energy. "Landau Damping" is an example of a resonant collective interaction. In general, if a particle oscillates out of phase with the bulk of the system, it will find more or less mass interior to it when it falls in, than when it pops out the other side, and thus it will gain or lose energy in the oscillation. Numerical simulations of galaxies forming in free fall collapse confirm this notion of collective relaxation. Usually, this phase of violent relaxation lasts a few free-fall (crossing) times.

Because f is conserved microscopically, equilibrium is possible only in a coarse grained sense. Lynden-Bell has worked out the statistics for this case, and finds that the coarse grained f evolves to a Fermi-Dirac distribution. Imagine that we started the system with blobs of $f = \eta$, and f = 0. In time, the average phase density, $\bar{f}$ will lie between 0 and η. A given arbitrary mass m will occupy an area (in a one-dimensional system) $f\,\Delta x.\Delta v$. As $f \longrightarrow \bar{f}$ the area occupied by m will be

$$\Delta x\cdot\Delta v = m/\bar{f} > m/\eta = \text{constant } (\hbar ?)$$

Of course, m/η is an arbitrary constant, but still the incompressibility of f acts like an exclusion principle. Thus, it is not surprising that $\bar{f}$ is given by

$$\bar{f} = \frac{\eta}{e^{\beta(\epsilon-\mu)}+1}$$

I, and my colleagues, have verified by numerical simulation that gravitating systems do behave this way. However, usually the relaxation is incomplete, because, in the first bounce, the system often throws off a halo, which thereafter only weakly interacts with the remainder of the system. Note that for large energies, $\bar{f}$ behaves like the usual Maxwell-Boltzmann isothermal distribution. This was once thought to be a major defect of this theory as it is well known (and we have seen earlier) that for an isothermal distribution, M r, so the mass is unbounded. However, recently, radio observation of disk-like galaxies have indicated that M does increase proportional to r out to quite large distances, and many of us now think that galaxies do look like isothermal spheres, until such distances that they interact tidally with a neighboring galaxy.

THE UNIVERSE :

The key observational facts of the universe are:

1. It is isotropic

   The 2.7°K Black Body radiation (the adiabatically cooled radiation from the Big Bang) is constant to 10^{-3} on scales greater than 10°.

2. The density of matter is uniform

   The number of galaxies out to a distance R is proportional to R^3 out to a distance of 10^9 light years (about 10% of the radius of the universe).

3. It is expanding, with an expansion velocity that is proportional to the distance.

   A galaxy at a distance R recedes from us with a velocity V where V = HR and H depends on time but not on R. Currently H = .05 c/10^9 light years or $H^{-1} = 1.8 \times 10^{10}$ years.

If we reversed the velocities, the galaxies would collapse to a point in less than H^{-1} years; less than, because gravity would speed up the collapse. Thus the age of the universe can be taken as somewhat less than H^{-1} years.

Because of the homogeneity, the dynamics of the expanding universe can be described rather simply. Imagine a sphere of radius R bounded by marked galaxies. We now follow that R of the marked galaxies. To a good approximation, we can neglect diffusion across R, so that the mass M, interior to R, is a constant.
Then, we have :

$$\ddot{R} = -\frac{GM}{R^2} \quad , \quad \frac{1}{2}\dot{R}^2 - \frac{GM}{R} = E$$

E less than zero implies a bound (closed) universe, while E greater than zero implies an unbound (open) universe. For the border-line case, E = 0, we have

$$\frac{1}{2}\dot{R}^2 = \frac{GM}{R}$$

Substituting $\dot{R} = HR$ and $M = \frac{4\pi}{3}\rho_c R^3$

we find $\rho_c = 3H^2/8\pi G = 5.7 \times 10^{-30}$ gm/cm^3 $= 3 \times 10^{12}$ $M_\odot/(10^7$ lys$)^3$.

If $\rho = \rho_c$, E = 0 and Ht = 2/3

If $\rho > \rho_c$, E < 0 and Ht < 2/3 $\qquad 2/3\ H^{-1} = 12 \times 10^9$ years

If $\rho < \rho_c$, E > 0 and 2/3 < Ht < 1

There are a number of programs directed to the determination of ρ . The one that will concern us relates to the clumpiness of the universe. On scales of a fraction of the radius of the universe, galaxies are seen to be clumped. It seems reasonable to assume that this tendency to group is due to gravity, and that the more matter there is, the more this tendency is encouraged.

We want to distinguish between the two configurations pictured in figure 4. More quantitatively, let dP_{12} be the joint probability of finding a galaxy in volume dV_1, and a second galaxy in dV_2. If there are no correlation $dP_{12} = n^2\ dV_1 dV_2$, where n is the mean number density. To take into account correlations, we introduce and write $dP_{12} = n^2[1 + \xi(r_{12})]dV_1 dV_2$. We observe that goes as

$$\xi = \frac{40}{r(\text{Mpc})}^{1.8}$$

for 100 Kpc < r < 40 Mpc.

The upper limit may be due to the paucity of cataloged galaxies and should, not at this time, be taken as a break in the power law behavior of the correlation function (although, should the universe be open, a break will have to occur before r = c/H).

A calculation, which makes plausible, the notion that density features can develop even in an open universe, is presented below.

Look at the linearized equations of continuity and motion (conservation of mass and momentum). The zero order equations are :

$$\frac{\partial \rho}{\partial t} + \nabla . (\rho \vec{u}) = 0$$

$$\frac{\partial \vec{u}}{\partial t} + (\vec{u} . \nabla)\vec{u} = -\nabla \phi$$

$$\nabla^2 \phi = 4\pi G \rho$$

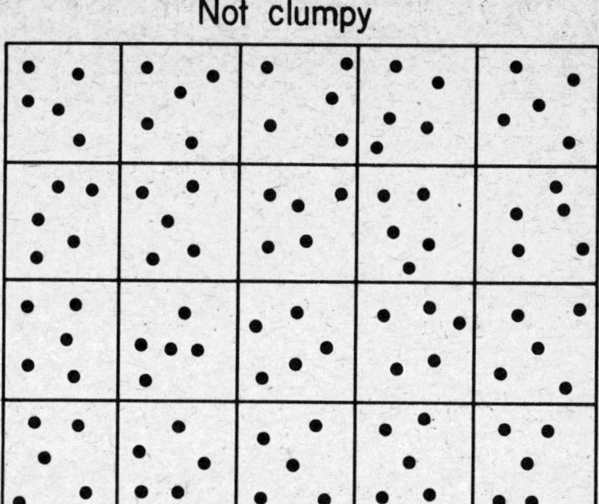

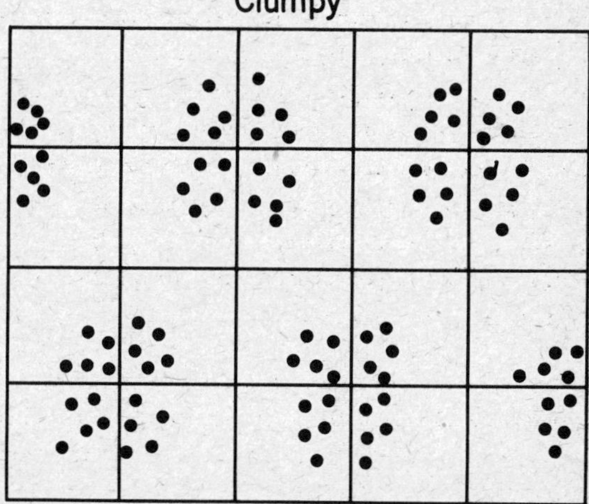

Fig. 4. Correlations
n = 5 per box

Let

$$\rho = \rho_0(t)\,[1+\delta(r,t)]\,,\quad \vec{u} = H\vec{r} + \vec{v}(r,t)$$

where δ and v are first order perturbations to the density and velocity fields. Note that δ measures the contrast relative to background density.

The equation for δ is

$$\ddot{\delta} + 2H(t)\dot{\delta} = 4\pi G\rho_0(t)\,\delta$$

If E = 0, H = 2/3 1/t and $4\pi G\rho_0 = 3/2\ H^2$. Making these substitutions, we find that there is a growing solution $\delta \sim t^{2/3}$. δ grows relative to ρ, but as $\delta \sim 1/R^3 \sim 1/t^2$, in absolute terms, δ also decays.

The forgoing picture of correlations is almost entirely the work of Peebles and his group at Princeton. Recently, Aarseth has done a variety of numerical simulations of 4,000 galaxies in an expanding universe. He starts with uncorrelated galaxies, and observes the correlations develop. In both open and closed universes, and with a variety of initial perturbations (e.g., white noise) ξ develops a power law behavior similar to that observed. White the rate of growth does depend on the density of the universe (closed universes develop correlations much faster than open universes), the r-dependence finally looks the same. As we don't have an independent measure of the age of the universe, this, as yet, does not distinguish between universes. However, it is a powerfull comfirmation that the general idea of gravity-induced correlation is correct.

We hope that as a theory devèlops for the growth of correlations, it will discriminate between open and closed universes. Such a theory, however, is still wanting.

REFERENCES:

Globular Clusters:

D. Lynden-Bell and R. Wood, 1968, Monthly Notices of the Royal Astronomical Society, 138, 498

G. Horowitz and J. Katz, 1977, Astrophysical Journal, in press

Galaxies:

D. Lynden-Bell, 1967, Monthly Notices of the Royal Astronomical Society, 136, 101.

Numerical Experiments relating to the above two topics are reviewed in :

M. Lecar, 1972, Gravitational N-Body Problem, D. Reidel Publishing Co., Dordrecht, Holland

S.J. Aarseth and M. Lecar, Computer Simulations of Stellar Systems, Annual Reviews of Astronomy and Astrophysics, vol. 13, 1975

Cosmoloty :

P.J.E. Peebles, Physical Cosmology, Princeton University Press, Princeton, N.J., 1971

P.J.E. Peebles, 1974, Astrophysical Journal Supplement, 253, 37

The numerical simulations of Aarseth have not as yet been published.

CHAPTER 14

POSITION-SPACE RENORMALIZATION APPROACH

J. M. J. van Leeuwen

Technological University, Delft, The Netherlands

ABSTRACT

The calculational methods in the position space renormalization approach to phase transitions and critical phenomena are discussed with a special emphasis on the variational method. It is shown that the variational technique leads to accurate free energies, but is of limited value in determining the nature of the phase transition. Results for the correlation functions and new critical phenomena appearing from the search of critical surface are discussed.

1. INTRODUCTION

In past decades critical phenomena have become increasingly intriguing with the discovery of curious power law and scaling behavior for the thermodynamic quantities and the correlation functions. The exponents turn out to be no magic numbers but yet of universal character. The ideas of Wilson[1)] on renormalization, based on Kadanoff's scaling picture[2)], have given a clue to the understanding of scaling and universality and provided a technique of calculating the exponents and scaling functions.

In the position renormalization approach one carries out the renormalization program while staying close to the microscopic description of the system, represented by N degrees of freedom s_i interacting via an hamiltonian $H(s)$. The free energy F is obtained as

$$F = \log \mathrm{Tr} \exp \mathcal{H}(s) \tag{1}$$

Note that the factor $-1/k_B T$ is included in $\mathcal{H}$ and F.

The system can be renormalized by coupling it to $N'(<N)$ degrees of freedom s_i', of the same nature as the s_i through a coupling hamiltonian $\mathcal{H}_I(s',s)$ which has to fulfil the restriction

$$\mathrm{Tr} \exp \mathcal{H}_I(s's) = 1 \tag{2}$$

This condition implies that the combined system has again the free energy F

$$\log \mathrm{Tr} \mathrm{Tr}' \exp[\mathcal{H}_I(s',s) + \mathcal{H}(s)] = F \tag{3}$$

Then a renormalization transformation (RT) is defined as

$$\exp[G + \mathcal{H}'(s')] = \mathrm{Tr} \exp[\mathcal{H}_I(s',s) + \mathcal{H}(s)] \tag{4}$$

The constant G is taken out because it is convenient to work with traceless hamiltonians

$$\mathrm{Tr}\mathcal{H}(s) = \mathrm{Tr}'\mathcal{H}'(s') = 0 \tag{5}$$

$\mathcal{H}'(s')$ is viewed as the hamiltonian for the new (or renormalized) system. The free energy F' of the new system

$$F' = \log \mathrm{Tr}' \exp \mathcal{H}'(s')$$

is related to F by

$$F = G + F' \tag{6}$$

The transition from the condition (2) to (3) assumes classical degrees of freedom. One can extend the definitions to quantum variables by using rather (3) as restriction on $\mathcal{H}_I(s',s)$[3].

The RT(4) is a map in hamiltonian space from $\mathcal{H}$ to $\mathcal{H}'$. We shall think of

$H(s)$ as represented by interaction parameters K and see (4) as a map in K-space from K to K'

$$\begin{cases} K' = K'(K) \\ g = G/N = g(K) \end{cases} \tag{7}$$

Then (6) can be written[4)] as a transformation for the free energy per degree of freedom $f = F/N$

$$f(K) = g(K) + b^{-d} f(K'(K)) \tag{8}$$

with $b^{-d} = N'/N$ the dilution factor (or b the linear contraction factor since d is the dimension of the system).

We denote the sequence of renormalized hamiltonians, which result from K as starting point via repeated application of the RT(7), by K, K', K", ... $K^{(n)}$... .
Then iterating (8) yields the free energy in terms of g(K) as

$$f(K) = \sum_{j=0}^{\infty} b^{-dj} g(K^{(j)}) \tag{9}$$

assuming the boundary condition

$$\lim_{j\to\infty} b^{-dj} f(K^{(j)}) = 0 \tag{10}$$

which is usually fulfilled due to the boundedness of the free energy.

The RT(7) becomes interesting for the description of critical phenomena when it has the following two properties:

I. The RT exhibits a (non-trivial) fixed point i.e. a K^* with the property

$$K'(K^*) = K^*$$

II. Around K^* the functions g(K) and K'(K) are regular.

Neither of the two properties are automatically fulfilled and we will give examples where I or II is violated. In general one may say that it is the purpose of the renormalization approach to explain the singular properties of the free energy f(K) from the regular functions g(K) and K'(K). All is contained in the solution (9) for f(K) in terms of g(K) and K'(K) and indeed all can be deduced from (10) provided I and II hold[4)5)].

2. MODELS

The renormalization scheme (4) or (7) is in practice limited by the ease with which a hamiltonian $H'(s')$ can be retrieved from a partial trace. For discrete variables this is substantially easier and most of the more or less succesful applications of the position space renormalization approach are carried out for the following models, which all consider the degrees of freedom localized on a lattice. It means that also the new degrees of freedom are associated with a lattice of the same structure and that b is the reduction factor of the lattice constant.

a. Ising-like models. The variables are $s_i = \pm 1$ (corresponding to spin $\frac{1}{2}$; higher order spins have also been considered[6)]). The interaction hamiltonian is of the type

$$H(s) = H\sum_i s_i + K\sum_{<ij>} s_i s_j + \ldots \tag{11}$$

with H as magnetic field, K as nearest neighbor interaction etc. The model has been studied most extensively for the ferromagnetic phase transition and critical point.

b. n-component Ising-models[7)]. The variable s has n components s_α ($\alpha = 1\ldots n$) but only 2n states are allowed, given by:

$$s_\alpha = (0,0\ldots,\pm\sqrt{n},\ 0\ldots0) \tag{12}$$

such that the following constraints are fulfilled

$$\sum_\alpha s_\alpha^\rho = 0 \qquad \rho \text{ odd}$$
$$= 2n^{\rho/2} \qquad \rho \text{ even} \tag{13}$$

This model contains in n a parameter on which the critical parameters depend; n = 1 yields the Ising model. The interest in the model is that, in view of (13) the self avoiding random walks are selected in the limit $n\to 0$. The hamiltonian is of the type (11) but also interactions involving s_α^2 do appear even in the $n\to 0$ limit[7].

c. Potts-like models. The variable s runs through q-equivalent states 1, ... q . The energy expression is

$$H(s) = H\sum_i \delta_{s_i 1} + K\sum_{<ij>} \delta_{s_i s_j} + \ldots \tag{14}$$

where $\delta_{s_i s_j}$ = 1 when s_i and s_j are in same state
= 0 otherwise.

The model may also be viewed as a spin vector model in q-1 dimension through the relation

$$\delta_{s_i s_j} = [1 + (q-1)\vec{s}_i . \vec{s}_j]/q \tag{15}$$

where the unit vectors $\vec{s}_i$ point in the corners of a q-1 dimensional tetraheder[8]. Here also the critical properties depend on q. Whereas in the previous model the specific heat exponent α decreased with n it increases with q. For q = 2 the model is Ising-like. For $q\to 1$ the percolation problem is recovered[9].

3. APPROXIMATION STRATEGIES

Given the fact that even the simplest interaction hamiltonian $H(s)$ leads upon renormalization to a more complex $H'(s')$ in any dimension d>1, an exact determination of the RT(7) is out of the question. In the search for a reliable approximation one may distinguish three strategies:

i. Start with a simple coupling $H_I(s',s)$ and design an approximation scheme (e.g. cumulants, clusters, finite lattice approximations) and carry out the approximations up to the required accuracy. In practice most of the calculations are carried out in low order such that the results are only qualitative. The calculation of Niemeyer and Van Leeuwen yields evidence that their transformation converges to the known result[10]. More convincingly Wilson[11] has shown by an extraordinary computational achievement that the method works.

In these approximations the regularity of the RT is automatically fulfilled as one handles only a finite number of degrees of freedom to construct the approximate RT. The confidence that in the exact RT no singularities develop is only based on the stability of successive approximations.

ii. Instead of using an apriori given RT one can built in a number of variational parameters in the coupling $H_I(s',s)$, which can be used to optimalize the RT within a given approximation. In this method, introduced by Kadanoff[12], one constructs upper or lower bounds for the free energy,which is used as criterion for optimalization. We will discuss the method in the next section.

iii. The most desirable approximation is an expansion in a small parameter. Then one should find that the choice of the renormalization scheme is unimportant. The only case where the RT equations can be expanded systematically is the case of weak interactions. Thus around the high temperature fixed point such an expansion exist, but this fixed point is not very interesting from the view point of phase transitions and critical phenomena. In order to combine criticality with weak interactions one has to perform a limit. The best known example is the Kac potential which reads for Ising spins

$$H(s) = (\gamma^d/2)\sum_{ij}\sum K(\gamma(\vec{r}_i-\vec{r}_j))s_i s_j \tag{16}$$

$K(\vec{x})$ is some short ranged function and γ is the small parameter. These potentials have a finite total strength K

$$K = \gamma^d \sum_j K(\gamma(\vec{r}_i - \vec{r}_j)) \tag{17}$$

and a range R

$$R^2 = \sum_j |\vec{r}_i - \vec{r}_j|^2 K(\gamma(\vec{r}_i - \vec{r}_j))/K \tag{18}$$

which diverges as γ^{-1} for $\gamma \to 0$.

The RT equations can be expanded in powers of γ[13]. The analysis is somewhat complicated by the appearance of dangerous irrelevant variables, but when these are handled properly one finds in lowest order a fixed point which is stable for d>4 and unstable for d<4, i.e. for d>4 a critical system transforms to a system with a longer range potential and for d<4 the range R shrinks upon transformation. The whole procedure is very analoguous to the ε-expansion.

4. THE VARIATIONAL METHOD

Starting point is the convexity of the free energy. Consider F_λ

$$F_\lambda = \log \mathrm{Tr} \exp[H_0(s) + V(s)] \tag{19}$$

as function of λ, then $\partial^2 F_\lambda / \partial\lambda^2 > 0$ for any $V(s)$. Integrating $\lambda \partial^2 F/\partial\lambda^2$ and $(1-\lambda)\partial^2 F/\partial\lambda^2$ between 0 and 1 one finds the bounds

$$F_{\lambda=0} + \langle V \rangle_{\lambda=0} < F_{\lambda=1} < F_{\lambda=0} + \langle V \rangle_{\lambda=1} \tag{20}$$

where $\langle V \rangle_\lambda = \partial F_\lambda / \partial\lambda$. For a lower bound one only needs a managable H_0. The upper bound can be used in practice only if one can find a V with $\langle V \rangle_{\lambda=1} = 0$. (Note that lower and upper bounds are in fact upper and lower bounds to the true free energy as a factor $-1/K_0T$ is here included).

Since the determination of the RT(4) is in fact a free energy type of calculation, bounds can be found by a separation

$$\mathcal{H}_I(s',s) + \mathcal{H}(s) = \mathcal{H}_o(s',s) + \mathcal{V}(s) \tag{21}$$

A lower bound RT follows from defining

$$G_1 + \mathcal{H}'_1(s') = \log\mathrm{Tr}\exp\mathcal{H}_o(s',s) + \langle\mathcal{V}(s)\rangle_o \tag{22}$$

where the average $\langle\ \rangle_o$ involves an average over the old degrees of freedom with $\mathcal{H}_o(s',s)$ as weight factor.

According to (19) we have

$$G_1 + \mathcal{H}'_1(s') \lesssim G + \mathcal{H}'(s') \tag{23}$$

and thus

$$G_1(K) + b^{-d}f(K'_1)) \leq g(K) + b^{-d}f(K'(K)) = f(K) \tag{24}$$

Defining the lower bound $f_1(K)$ as the solution of

$$f_1(K) = g_1(K) + b^{-d}f_1(K'_1(K)) \tag{25}$$

this yields an $f_1(K)$ obeying

$$f_1(K) \leq f(K) \tag{26}$$

as (24) holds for all steps in the renormalization sequence.

For an upper bound one usually takes instead of $\langle\mathcal{V}\rangle_{\lambda=1}=0$ the weaker condition

$$\mathrm{Tr}'\mathrm{Tr}\mathcal{V}(s)\exp[\mathcal{H}_o(s',s) + \mathcal{V}(s)] = 0$$

which is equivalent to

$$\mathrm{Tr}\mathcal{V}(s)\exp[\mathcal{H}(s)] = 0 \tag{27}$$

as can be seen by using (21) and (2). Similarly one derives

$$G_a + F'_a = \log\mathrm{Tr}'\exp[G_a + H'_o(s')] = \log\mathrm{Tr}'\mathrm{Tr}\exp H_o(s's)$$
$$> \log\mathrm{Tr}'\mathrm{Tr}\exp[H_o(s's) + V(s)] = F$$

such that

$$G_o(K) + bf^{-d}(K'_o(K)) \geq f(K) \tag{28}$$

and it follows that the RT induced by $H_o(s's)$ is yielding an upper bound for the free energy when (27) is fulfilled.

By the potential moving technique Kadanoff[12] e.a.have constructed $V(s)$ obeying (27) and the resulting free energies are remarkable accurate approximations to what is known exactly or what is believed to be reliable from other sources[14]. As the RT is relatively simple the method has found widespread application[15] for the determination of critical singularities. The following comment shows that the variational method is of limited value for the determination of the nature of the singularities[15].

Let us take the case that K represents a single (nearest neighbor) interaction (a temperature-like variable) and that the coupling $H_I(s's)$ contains a single variational parameter ρ which has to be optimalized with the resulting free energy as criterion. Thus one should find a function $\rho(K)$ on which the RT depends:

$$\begin{cases} K' = K'(K;\rho(K)) \\ g = g(K;\rho(K)) \end{cases} \tag{29}$$

The free energy follows from (8) and $\rho(K)$ from the condition of stationarity

$$\begin{cases} f(K) = g(K;\rho(K) + b^{-d}f(K'(K;\rho(K)) \\ \;\;0 \;\;= g_\rho(K;\rho(K) + b^{-d}f_K(K')K'_\rho(K;\rho(K)) \end{cases} \tag{30}$$

where the subscripts ρ and K denote differentiation with respect to ρ and K. The two equations have to be solved simultaneously since $\rho(K)$ and f(K) mutually determine each other. To obtain a fully consistent set of solution for the entire K range is fairly involved and requires advanced computational optimalization techniques[17]. One is however most interested in the shape of $\rho(K)$ around the fixed point. Schematically the situation is depicted in fig. 1. in the ρ,K plan.

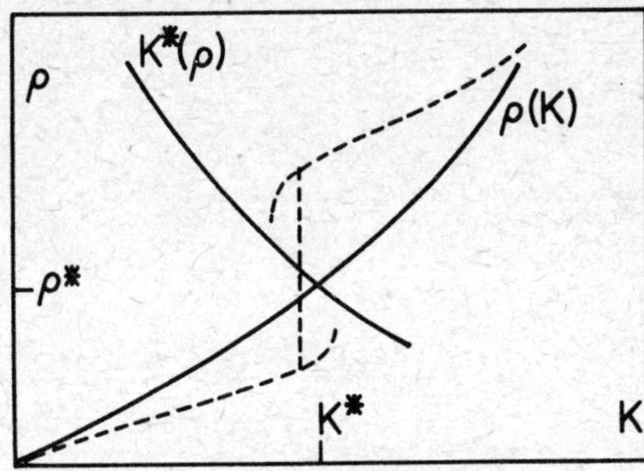

fig. 1. Line of fixed points $K^*(\rho)$ and optimal curve $\rho(K)$.

Firstly the curve $K^*(\rho)$ is determined which is the locus of the fixed K^* for given ρ. Then the intersection (ρ^*,K^*) of this curve with the optimal $\rho(K)$ has to found. It involves[12] the simultaneous determination of ρ^* and f_K^* the free energy derivation at the fixed point. As noted by Knops[18] this is not sufficient for the determination of the singularity exponent for the free energy. When the free energy has a singularity at K^* of the type.

$$f_{sing}^{(K)} \sim |K-K|^{2-\alpha} \tag{31}$$

then the (specific heat) exponent α is given by

$$\alpha = 2-d\log\gamma/\log\lambda$$

$$\lambda = [\partial K'/\partial K]_{K^*} + [\partial K'/\partial\rho]_{K^*}[\partial\rho/\partial K]_{K^*} \tag{32}$$

Thus also the slope ρ_K^* of $\rho(K)$ at K^* has to be determined. This requires a simultaneous determination of ρ_K^* and f_{KK}^* from (30). One may guess from the way the derivatives of $\rho(K)$ and $f(K)$ are connected at K^*, that the singularity (31) in $f(K)$ induces a singularity in $\rho(K)$. From (30) it can be derived that $\rho(K)$ develops a singularity of the type

$$\rho_{sing}(K) \sim |K-K^*|^{1-\alpha} \tag{33}$$

Then one sees that only a consistent determination of α is possible when $\alpha<0$ otherwise ρ_K diverges. We have determined for the square lattice the fully consistent α to be $\alpha = -0.123413$ which should be compared to $\alpha = 0.001756$ found by Kadanoff[12] e.a. omitting the second term in (32) (the exact value is $\alpha = 0$).

Several possiblities do occur in the variational solution. Amongst other we have found a case where the interaction between the curve $\rho(K)$ and the singularity of $f(K)$ is so strong that $\rho(K)$ breaks into two branches one for $K>K^*$ and one for $K<K^*$. The transition then becomes first order skipping the fixed point as is indicated by the dashed line in fig. 1.

The real worry of this variational approach is that if one lets the RT determine itself what is the best RT, by optimalization of the free energy, the RT(29) becomes singular near the fixed point or even excludes the fixed point. Thus the variational approach tends to violate basic assumptions I and II for the proper machinery of the renormalization approach.

5. CORRELATION FUNCTIONS

The RT is set up as a method for calculating the free energy. The correlation functions can be calculated analoguously although the transformation has to be extended[5]. Let the correlation function (for a spin system) be given by

$$G(\vec{r}_i - \vec{r}_j) = \langle s_i s_j \rangle \tag{34}$$

The usual way to deal with correlation functions is to express the correlation function of the renormalized system in terms of the old correlation functions. As a first step one arrives at

$$G'(\vec{r}'_{i'} - \vec{r}'_{j'}) = \mathrm{Tr}'\, s'_{i'} s'_{j'} \langle \exp H_I(s',s) \rangle \tag{35}$$

In general such a relation is of little use as it mixes correlation functions of different type. For the so-called linear weight factors (35) reduces to a relation

$$G'(\vec{r}_{i'} - \vec{r}_{j'}) = \rho^2 \sum_{i\epsilon i'} \sum_{j\epsilon j'} G(\vec{r}_i - \vec{r}_j) \tag{36}$$

where the sums over i and j run over the sites of the cells i' and j'; ρ is the linear coupling parameter. A relation like (36) can be used to determine the scaling behavior of the correlation functions. As a side remark it shows that linear weights only work when ρ is chosen such as to lead to the correct magnetic exponent. The correlation function itself cannot be calculated from (36) as it expresses the new correlation function in the old one and not vice versa. Thus one cannot use the iterative solution (10) as for the free energy.

The calculation of the correlation functions can be set up in close parallel to the free energy calculation[5]. As an illustration we give here a simplified version of the type of equations one arrives at:

$$G(\vec{r}_i - \vec{r}_j) = C(\vec{r}_i - \vec{r}_j) + \sum_{i'j'} \sum T_{i'i} G(\vec{r}'_{i'} - \vec{r}'_{j'}) T_{j'j} \tag{37}$$

Here $C(\vec{r}_i - \vec{r}_j)$ and $T_{i'i}$ are extension of the function G and K' introduced in (7). They can be related to the reponse to a spatially varying magnetic field δH_i.

$$\begin{cases} C(\vec{r}_i - \vec{r}_j) = \delta^2 G/\delta H_i \delta H_j \\ T_{i'i} = \delta H'_{i'}/\delta H_i \end{cases} \tag{38}$$

The functions $C(\vec{r}_i - \vec{r}_j)$ and $T_{i'i}$ are assumed to be short ranged, which is equivalent to the assumption of regularity of the g(K) and K'(K). Equation (37) shows a great similarity to the Ornstein Zernike relation where the long range $G(\vec{r}_i - \vec{r}_j)$ is expressed in the supposedly short range direct correlation. In fourier language (37) becomes

$$\tilde{G}(\vec{K}) = \tilde{C}(\vec{K}) + b^{-d} \tilde{T}(-\vec{K}) \tilde{G}'(b\vec{K}) \tilde{T}(\vec{K}) \tag{39}$$

In this form the equation for the correlation function is perfectly similar to the free energy equation (8) (the thermodynamic dependence on K is suppressed in (39)). Very little has been done to calculate the correlation functions from the renormalization approach along the lines. Fig. 2 shows a preliminary result for the scaling function obtained by Hilhorst[19] based on the first cumulant approximation for a triangular lattice.

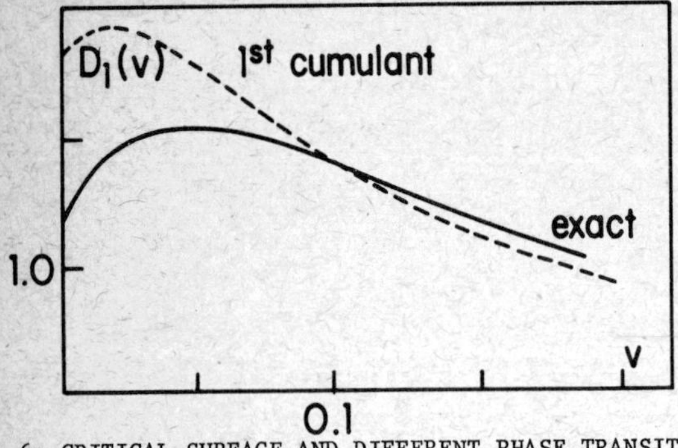

fig. 2. Scaling function $\tilde{G}(k,t) = k^{-2+n} D_1((k\xi)^{-1/\nu})$.

6. CRITICAL SURFACE AND DIFFERENT PHASE TRANSITIONS

Most of the applications are aimed at the understanding of the ferromagnetic phase transition in Ising type systems. A preknowledge of the type of phase transition is necessary for the appropriate choice of the coupling $H_I(s',s)$. In the Ising ferromagnetic case it is sufficient to preserve the up down symmetry to guarantee that a fixed point, if any,appears for field zero. In general a coupling $H_I(s's)$ appropriate for the ferromagnetic transition will not work for the anti-ferromagnetic transition. It is therefore interesting to see what happens when a ferromagnetic coupling $H_I(s's)$ is used in the anti-ferromagnetic regime. In fig. 3 a critical curve is plotted in the plane of nearest neighbor interaction II, K and next nearest neighbor interaction (dashed line), as it was calculated by Nauenberg and

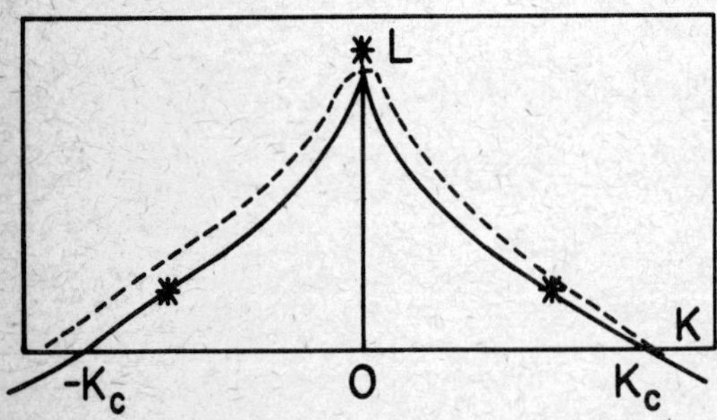

Nienhuis[4] on the basis of their RT for a square lattice. The curve shows a sharp bending near K = 0 and does not display the full symmetry under K↔ -K interchange. In practice a sharp bending in a critical surface is an indication that a different phase transition takes place. In this case it is not hard to design a RT which fully acknowledges the role of the two sublattices both on site and cell level. As a result separate ferromagnetic and antiferromagnetic fixed points appear symmetrically with respect to the L-axis and in addition a new fixed point on the L-axis appears, which causes the critical line to cusp at K = 0. The lesson to be learned from this example is that a RT becomes singular when it fails to preserve the relevant symmetries for the phase transition.

Antiferromagnetic phase transitions are but one example of more complicated arrangements of up and down spins. Fig. 4 shows a triangular lattice where + spins are surrounded by - spins. One can consider such states in coexistence with an all - spin configurations. Transitions of this type do occur in absorbed systems[20]. Domany e.a.[20] identify this transition with the 3-state Potts model transition, because of the 3-fold degeneracy of the state shown in fig. 4. In the lattice gas language one may see this also as a model for the gas-liquid transition provided one introduces repulsive nearest neighbor and attractive next nearest neighbor coupling.

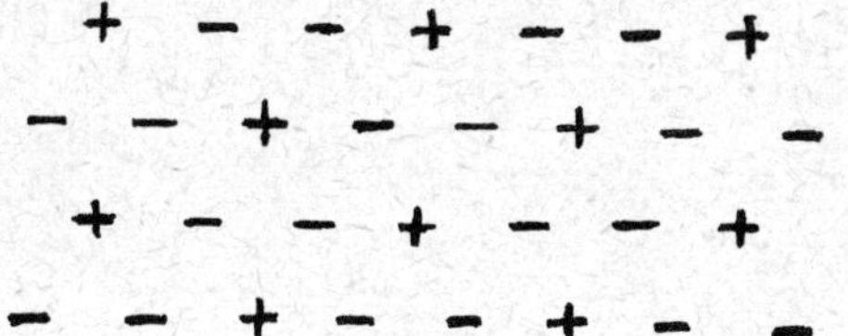

fig. 4. Antiferromagnetic arrangement on a triangular lattice.

This brings in the question whether the gas-liquid transition is of the Ising ferromagnetic type. The experimental evidence[21] points at the moment in the affirmative direction, but theoretically further research is required.The position space renormalization approach lacks a reliable calculational method easy enough to carry out to sufficient high order for the more complex transition as discussed above.

REFERENCES

1. K.G. Wilson, Phys. Rev. B4 (1971) 3174, 3184
   K.G. Wilson and M.E. Fisher, Phys. Rev. Letters 28 (1972) 240.
2. L.P. Kadanoff, Physics 2 (1966) 263.
3. J. Rogiers and R. Dekeyser, Phys. Rev. B13 (1976) 4886
   D.D. Betts and M. Plischke, Can. J. Phys. 54 (1976) 1553
   J. Rogiers and D.D. Betts, Physica 85A (1976) 553
   R.C. Brower, F. Kuttner, M. Nauenberg and K. Subbarao, Phys. Rev. Lett. 38 (1977) 1231
   A. Stella and F. Toigo, preprint, Univ. of Padova, Italy.
4. M. Nauenberg and B. Nienhuis, Phys. Rev. Lett. 33 (1974) 1598.
   J.M.J. van Leeuwen in "Fundamental Problems in Statistical Mechanics III" (1975) (E.G.D. Cohen ed.) North Holland/American Elsevier.
5. Th. Niemeyer and J.M.J. van Leeuwen in "Phase Transitions and Critical Phenomena VI" (C. Domb and M.S. Green ed.)(1976).
6. A. Nihat Berker and M. Wortis, Phys. Rev. B14 (1976) 4946.
   T.W. Burkhardt, Phys. Rev. B 14 (1976) 1196.
7. H.J. Hilhorst, Phys. Lett. 56 A (1976) 153.
   A. Aharony, J. Phys. A 10 (1977) 389.
8. K.P. Zia and D.J. Wallace, J. Phys. A 8 (1975) 1495.
9. P.W. Kasteleyn and C.M. Fortuin, J. Phys. Soc. Japan Suppl. 26 (1969) 11
   A.B. Harris, T.C. Lubensky, W.K. Holcomb and C. Dasgupta, Phys. Rev. Lett. 35 (1975) 327 .

10. Th. Niemeyer and J.M.J. van Leeuwen, Physica 71 (1974) 17.

11. K.G. Wilson, Rev. of Modern Physics 47 (1975) 773.

12. L.P. Kadanoff, Phys. Rev. Lett. 34 (1975) 1005

L.P. Kadanoff, A. Hougton and M.C. Yalabik J. Stat. Phys. 14 (1976) 171.

13. H.J.F. Knops, J.M.J. van Leeuwen and P.C. Hemmer, J. Stat. Phys. 17 (1977) No. 4.

14. S.L. Katz and J.D. Gunton, Phys. Rev. B to appear.

15. Ch. Dasgupta, Phys. Rev. B. 14 (1976) 1221.

Ch. Dasgupta, Phys. Rev. B. 15 (1977) 3460.

T.W. Burkhardt, H.J.F. Knops and M.P.M. den Nijs, J. Phys. A 9 (1976) 1179.

M.P.M. den Nijs and H.J.F. Knops, preprint, Univ. of Nijmegen.

16. W. van Saarloos, J.M.J. van Leeuwen and A.A.M. Pruisken to appear

17. M.N. Barber, preprints, Univ. of New South Wales.

18. H.J.F. Knops, Physica 86A (1977) 448.

19. H.J. Hilhorst this conference.

20. Eytan Domany, M. Schick and J.S. Walker, Phys. Rev. Lett. 38 (1977) 1148.

21. R. Hocken and M.R. Moldover, Phys. Rev. Lett. 37 (1977) 29.

CHAPTER 15

STATISTICAL MECHANICS OF MOLECULAR AND NUCLEAR COLLISIONS*

R. D. Levine

Department of Physical Chemistry, The Hebrew University,
Jerusalem, Israel

*Work supported by the United States – Israel Binational Science Foundation, Jerusalem, Israel.

ABSTRACT

A link is forged between single collision phenomena and non-equilibrium statistical mechanics, based on the principle of maximal entropy. We argue that while phase space is not uniformly filled during a collision, the constraints that are present are simple, can be readily determined from the data by the procedure of 'surprisal analysis' and have an immediate intuitive physical significance. These claims are illustrated by several examples of both experimental and computational origin. A theoretical interpretation of the results is provided. The information theoretic procedure can be derived as a consequence of the dynamics (be they quantal or classical). The constraints (required to obtain an exact solution) are specified via an algebraic procedure. This not only justifies the approach of surprisal analysis but also provides a method of predicting the constraints. The number of coupled equations of motion that need be solved to compute the exact time evolution is typically far smaller than the number of coupled quantum states and equals to the number of constraints.

1. INTRODUCTION

There is a very extensive compilation of experimental and computational (simulation) results for collisions of composite particles (be they nuclei or molecules) leading to both internal excitation and reactions (rearrangements). A notable feature of direct nuclear or molecular collisions is the considerable specificity in the population of accessible final quantum states. In particular, exoergic (energy releasing) processes tend to yield products with high internal excitation. The distribution among the final channels is often counter to that expected on the basis of available volume in phase space. Similarly, and as expected on the basis of microscopic reversibility, there is considerable selectivity in the energy consumption. Cross sections are frequently far higher when the required threshold energy is provided by internal excitation of the projectiles rather than by their relative kinetic energy.

We have addressed ourselves[1-3] to the following questions: (a) optimal means of characterization of the final state distribution in the collision, (b) compaction of the voluminous body of experimental data (or computer simulation thereof) and (c) development of theoretical schemes which will complement the currently available quantum mechanical (or classical) procedures for the analysis and for the dynamical solution of a rearrangement collision of composite particles.

2. SURPRISAL ANALYSIS

Consider as an example the rearrangement collision

$$F + H_2 \rightarrow H + HF(v). \tag{1}$$

Here the stronger (by about 34 Kcal/mol) HF bond replaces the H_2 bond. Some of the excess energy is released as vibrational excitation of the HF molecule. The experimental[4] distribution of HF vibrational states (after the single collision that led to their formation) is shown in the bottom panel of figure 1. The dashed line is the same distribution computed on the basis of the assumption that all products quantum state (that are allowed by conservation of constants of the motion) are equally probable. The computation of this, so called 'prior' distribution[1-3] is quite straightforward. The total energy E can be partitioned between the translational (E_T) rotational (E_J) and vibrational (E_V) degrees of freedom, $E=E_T+E_J+E_V$. The prior probability for product molecules in a specified vibrational and rotational level is:

$$P^o(v,J) \propto (2J+1)E_T^{1/2} = (2J+1)(E-E_V-E_J)^{1/2}. \tag{2}$$

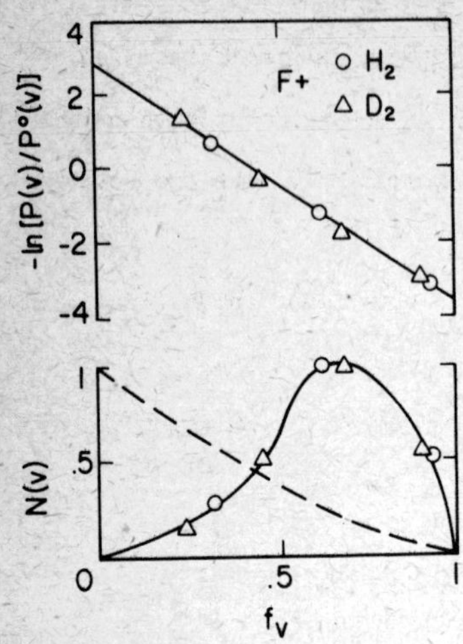

Fig. 1. Products vibrational state distribution[4] in the F + H_2 (circles) and F + D_2 (triangles) reaction vs. f_v=E_v/E, the fraction of the total energy E in the vibration (bottom panel). The continuous line is equation (6) with λ_v = -5.7. The dashed line is the prior distribution $P^o(v)$, equation (3). Top panel: The surprisal vs. f_v.

Here 2J+1 is the degeneracy of the rotational level J and the number of translational states (with energy in the range E_T to E_T+dE_T) per unit volume is $A_T E_T^{1/2} dE_T$. $P^o(v)$ is obtained by summing $P^o(v,J)$ over all rotational levels allowed by the conservation of energy ($0 \leq E_J \leq E-E_v$). In the limit where the rotational energy levels can be approximated as those of a rigid rotor (E_J = Bhc J(J+1)) and summation over J can be replaced by integration over E_J

$$P^o(v) = (E-E_v)^{3/2}/\Sigma_v(E-E_v)^{3/2}. \qquad (3)$$

The prior probability is only a function of f_v=E_v/E, the fraction of available energy in the product vibration. As is evident in the bottom panel of figure 1 the same appears to be true for the actual distribution. (The two distributions shown would be quite different when plotted vs. the vibrational quantum numbers since HF and DF differ in their vibrational frequencies by a ratio of about $2^{1/2}$).

The observed and prior distribution are qualitatively different. On prior grounds the probability is a monotonically decreasing function of the vibrational excitation while the actual results show population inversion. As a measure of their deviance we use the surprisal,

$$I(v) = -\ell n[P(v)/P^{o}(v)], \tag{4}$$

shown vs. f_v in the top panel of figure 1. It is seen that I(v) can be well approximated by a linear dependence on f_v

$$I(v) = \lambda_o + \lambda_v(E_v/E) \tag{5}$$

or

$$P(v) = P^{o}(v)\exp[-\lambda_o-\lambda_v(E_v/E)]. \tag{6}$$

Since P(v) is normalized, the complete distribution can be specified by a single parameter, λ_v. In figure 1, λ_v is negative, reflecting the population inversion.

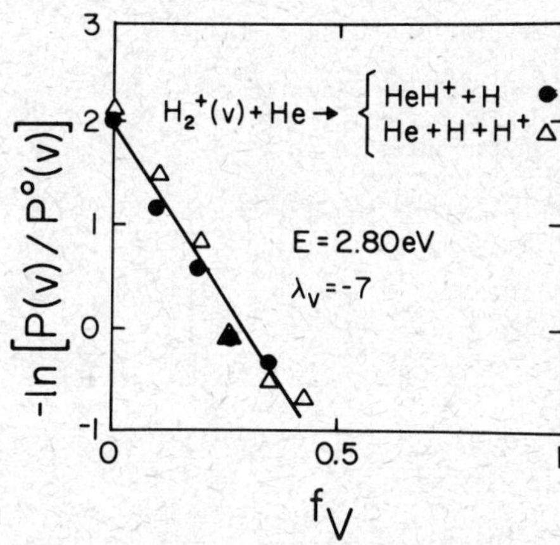

Fig. 2. Surprisal analysis of the influence of reagent vibrational excitation on the rate[6] of reaction (7) at a given total energy E = 2.80 eV.

A complementary aspect of the non-statistical products state distribution is shown in figure 2. If all final states are formed, on prior grounds, at the same rate it follows from microscopic reversibility that all initial states should react with the same rate. The experimental observation is that different partitioning of the same total energy among the degrees of freedom of the reagents can markedly affect the reaction rate. For the ion-molecule reaction

$$H_2^{+}(v) + He \rightarrow \begin{cases} HeH^{+} + H \\ He + H + H^{+} \end{cases} \tag{7}$$

the cross section was measured[6] as a function of v. The surprisal analysis[5], figure 2, again provides a compact representation of the deviance of the observed and the prior distribution.

Inelastic collisions can be similarly analyzed[7,8]. Figure 3 shows the results[8]

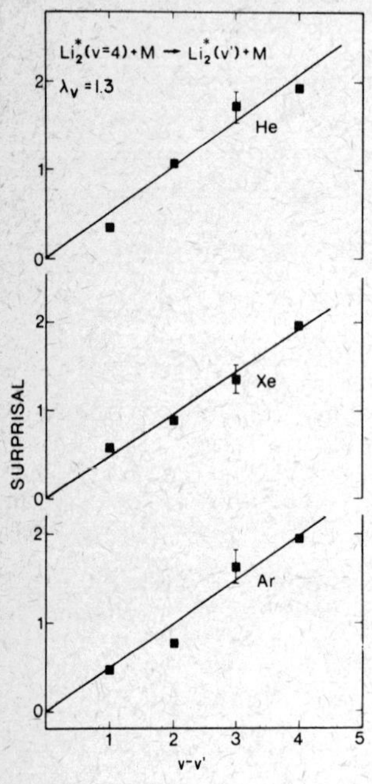

Fig. 3. Surprisal analysis of the rates for vibrational energy transfer. Shown is the surprisal of the measured[9] rates into different final vibrational states v' for the process (8) vs. v-v', for several collision partners. The bars indicate the experimental uncertainty in the rates

for the

$$Li_2^*(v=4) + M \rightarrow Li_2^*(v') + M \qquad (8)$$

collision[9] for various collision partners M. Here, as well as in figure 4 the initial state has a thermal distribution of the energy in the other degrees of freedom and this needs to be taken into consideration in the computation of the prior distribution.

Surprisal analysis has also been carried out for translational or rotational and also for joint distributions. Figure 4 shows the results obtained[10] for low temperature rotational energy transfer

$$HD(J) + He \rightarrow HD(J') + He. \qquad (9)$$

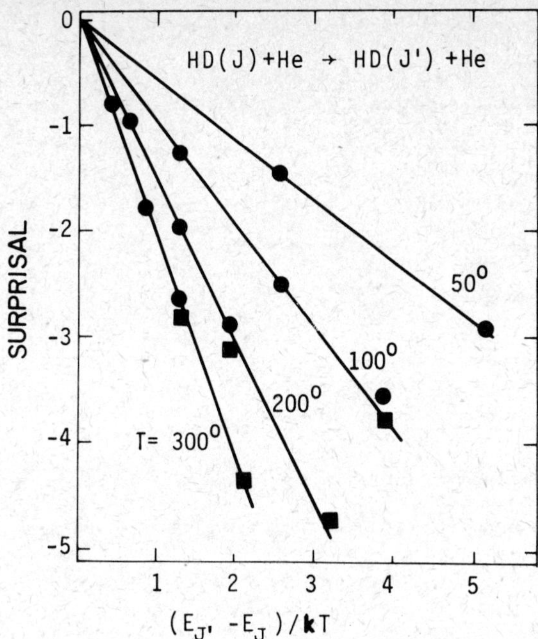

Fig. 4. Surprisal analysis[10] of the final rotational state distribution determined[11] by a numerical (close coupling) solution of the Schrödinger equation. T is the translational temperature. Note that several J → J' transitions with about the same rotational energy gap (E_J-$E_{J'}$) have indeed the same surprisal.

The cross sections were determined as a function of the translational energy by a close coupling solution of the Schrödinger equation and the rates were obtained[11] via a thermal averaging over E_T.

Computational studies of the distribution of states during the collision also often show a simple structure of the surprisal. It should however be noted that a single constraint (cf. (5)) is a fairly frequent but not a universal characteristic. This is particularly so for joint distributions (e.g. in both vibrational and rotational states, P(v,J) or in both energy and angle) where considerable correlations are found. Preliminary applications to direct nuclear reactions have also been reported[12]. Similarly, scattering processes for which closed analytic solutions exist can also be analyzed in this fashion.

3. DYNAMICS

Given a finite set of linearly independent operators A_r, $(r=0,\ldots,N, A_0=I)$ can the representation of the density operator during the collision in the form

$$-\ell n\rho(t) = \sum_r \lambda_r(t) A_r \tag{10}$$

be more than a useful empirical approximation? The exact equation of motion for $\ell n\rho(t)$ is[12]

$$i\hbar\partial \ell n\rho(t)/\partial t = [H, \ell n\rho(t)] \tag{11}$$

where H is the Hamiltonian. Substituting (10) in the equation of motion (11) leads to

$$i\hbar \sum_r A_r \partial\lambda_r/\partial t = \sum_s \lambda_s [H, A_s]. \tag{12}$$

Since the operators A_r are linearly independent, (12) obtains if the set of operators used in the expansion (10) is closed under commutation with the Hamiltonian

$$[H, A_s] = i\hbar \sum_r A_r g_{rs} \,. \tag{13}$$

When such a set of operators can be constructed not only is (10) an exact solution but the substitution of (13) into (12) provides

$$\partial\lambda_r/\partial t = \sum_s g_{rs}\lambda_s. \tag{14}$$

The solution of the complete equation of motion has thus been reduced to solving a set of N+1 linear coupled differential equations for the N+1 functions of time, the $\lambda_r(t)$'s. In classical mechanics the equations of motion are obtained via the correspondence $(i\hbar)^{-1}[\ ,\ ] \rightarrow \{\ ,\ \}$ (the Poisson bracket). Hence equation (14) is unchanged if the phase functions A_r are closed under the operation $\{H, A_r\}$.

To make contact with the scattering problem one still needs to show that the initial state can be written in the form (10). (The initial values of the λ_r's provide the boundary conditions for the solution of (14)). We have been able to do so[12] and to use this approach to provide exact solutions for the collision dynamics in a number of examples. (These agree with the previously known analytic solutions when they are available).

The 'differential' relation (13) can be readily integrated to the form[12]

$$\langle A_r\rangle(t) \equiv \mathrm{Tr}\{A_r\rho(t)\} = \sum_s F_{sr}\langle A_s\rangle(-\infty) \ . \tag{15}$$

The (t-dependent) matrix $\underset{\sim}{F}$ is independent of the initial state and satisfies the equation of motion

$$-\partial\underset{\sim}{F}/\partial t = \underset{\sim}{F}\cdot\underset{\sim}{g} \tag{16}$$

where $\underset{\sim}{g}$ is defined by (13) and the initial value of $\underset{\sim}{F}$ (as $t \to -\infty$) is the identity matrix. The set of initial values $\{\langle A_s\rangle(-\infty)\}$ completely specifies the magnitudes of the $\langle A_r\rangle(t)$'s at any subsequent time and for any initial state.

4. MAXIMAL ENTROPY

The expansion (10) defines a density operator whose entropy

$$S[\rho] = -\mathrm{Tr}\{\rho\ell n\rho\} \tag{17}$$

is maximal[13] among all density operators $\tilde{\rho}$ that are consistent with the given mean values of the 'constraints'. Explicitly, density operators consistent with the magnitude of the constraints means that for any A_r in the closed set

$$\mathrm{Tr}\{A_r\tilde{\rho}(t)\} = \mathrm{Tr}\{A_r\rho(t)\} \equiv \langle A_r\rangle(t) \ . \tag{18}$$

For all $\tilde{\rho}(t)$'s for which (18) holds

$$S[\tilde{\rho}(t)] \leq S[\rho(t)] \tag{19}$$

with equality if and only if $\tilde{\rho}(t) = \rho(t)$. We now show that if (10) is true initially (i.e. for $t \to -\infty$) it remains true throughout the collision.

We choose the initial state ρ_{in} ($\rho(t) \to \rho_{in}$, $t \to -\infty$) to be the state of maximal entropy consistent with the specified initial values $\langle A_r\rangle(-\infty)$,

$$S[\tilde{\rho}_{in}] \leq S[\rho_{in}] \ . \tag{20}$$

Of course, it may well happen that some of the λ_r's will be initially zero. Now let ρ_{in} evolve to $\rho(t)$ while $\tilde{\rho}_{in}$ independently evolves to $\tilde{\rho}(t)$. Because of the closure property (13) which implies (15), both $\tilde{\rho}(t)$ and $\rho(t)$ remain consistent with the magnitude of the constraints (i.e. (18) is satisfied). But for unitary time evolution $S[\rho(t)] = S[\rho_{in}]$ and similarly for $\tilde{\rho}(t)$. Thus (20) implies (19) or an initial state of maximal entropy remains a state of maximal entropy throughout the collision.

5. CONCLUDING REMARKS

A phenomenological analysis of the final state in a collision in terms of a few constraints offers a simple route to the representation and compaction of experimental and computational data. It was shown possible to introduce a corresponding statistical mechanics formulation of the time evolution of systems with a few degrees of freedom. The constraints sufficient to specify the final state after the collision are the set of operators closed under the action of the scattering operator S (cf. (15))

$$SA_rS^{\dagger} = \sum_s F_{sr}A_s \quad \text{or} \tag{21}$$

$$\langle A_r\rangle(\infty) = \sum_s F_{sr}\langle A_s\rangle(-\infty) \ .$$

The matrix $\underset{\sim}{F}$ also specifies the magnitude of the final Lagrange parameters in terms of their initial value

$$\lambda_r^{out} = \sum_s (\underset{\sim}{F}^{-1})_{rs}\lambda_s^{in} \ . \tag{22}$$

Given a set of constraints, the matrix $\underset{\sim}{F}$ can be computed (cf. (16)). For any initial state which is of maximal entropy subject to these constraints (irrespective of the magnitude of the λ_r^{in}'s) the matrix $\underset{\sim}{F}$ specifies (via (21) or (22)) the exact final state.

REFERENCES

1. R.D. Levine in Invited Lectures VIII International Conference Physics of Electronic and Atomic Collisions (Institute of Physics, Beøgrad, 1973) p. 567.
2. R.B. Bernstein and R.D. Levine, Adv. Atom. Mol. Phys. 11, 215 (1975).
3. R.D. Levine and A. Ben-Shaul, Chemical and Biochemical Applications of Lasers (Academic Press, N.Y., 1977) Vol. II p. 145.
4. D.S. Perry and J.C. Polanyi, Chem. Phys. 12, 419 (1976) and references therein.
5. R.D. Levine and J. Manz, J. Chem. Phys. 63, 4280 (1975).
6. W.A. Chupka, J. Berkowitz and M.E. Russel, Abstracts VI International Conference Physics of Electronic and Atomic Collisions (MIT Press, Cambridge, 1969) p. 71.
7. M. Rubinson and J.I. Steinfeld, Chem. Phys. 4, 467 (1974).
8. I. Procaccia and R.D. Levine, J. Chem. Phys. 63, 4261 (1975).
9. G. Ennen and Ch. Ottinger, Chem. Phys. 3, 404 (1974).
10. I. Procaccia and R.D. Levine, Physica 82A, 623 (1976).
11. S. Green, Physica 76, 609 (1974).
12. Y. Alhassid and R.D. Levine, J. Chem. Phys., in press; Phys. Rev., in press.
13. W.M. Elsasser, Phys. Rev. 52, 987 (1937); R.L. Stratonovich, Sov. Phys. J. E.T.P. 1, 426 (1955); E.T. Jaynes, Phys. Rev. 106, 620 (1957); ibid, 108, 171 (1957); E.H. Wichmann, J. Math. Phys. 4, 884 (1963).

CHAPTER 16

HOLOGRAPHIC OBSERVATION OF DIFFUSION WAVES IN LIQUID ^{3}He - ^{4}He MIXTURES *

S. G. Lipson & J. Landau

Physics Department, Technion, Haifa, Israel

*Supported in part by the U.S. – Israel Binational Science Foundation.

ABSTRACT

Holographic Interferometry has been used to study mass diffusion in liquid helium isotopic mixtures. When phase separation is disturbed by sudden heating, a wavefront is seen to propagate through the liquid. A phenomenological explanation is offered, involving critical behaviour of the diffusion constant around the λ-line. Some other studies using the same apparatus are also mentioned.

1. INTRODUCTION

In this paper we shall describe a new technique which we have used to investigate the dynamics of phase-separation in ^3He-^4He liquid mixtures. The method relies on the known relationship between refractive index and isotopic concentration, and uses holographic interferometry to form a concentration map of an experimental cell at various times during the formation or destruction of the phase-separated state[1] which exists below 0.85 K.

The most surprising result of the experiments is the observation of a "diffusion wave" - a wave front propagating through the fluid when the phase-separated state is destroyed - whose behaviour is determined by diffusion, although its appearance is that of a propagating wave. It seems that this unexpected result follows from the completely different dynamic behaviour of the superfluid and normal liquids, which are in contact during the propagation of the wave. In addition, detailed analysis of the interferograms may give us some information on the critical behaviour of the diffusion at concentrations and temperatures around the lambda line.

The helium under observation is contained in a small, essentially two-dimensional cell which is imaged into a 35 mm camera using light from a He-Ne laser. The image of the cell interferes with a holographic reconstruction of the empty cell, and the interference pattern so produced gives visual information on optical density changes in the cell. In the case of isotopic mixtures at temperatures below 1.5 K, there is a simple almost temperature-independent relationship between refractive index and isotopic (^{3}He) density[2], so that the interference patterns are relatively easy to interpret as ^{3}He concentration maps. In our apparatus one interference fringe (a change in optical thickness of one wavelength, λ, corresponds to a concentration change of 2.2 % at the vapour pressure. It is not a high sensitivity compared with what is available from dielectric constant measurements using a capacitance bridge[3]; the advantage of the technique is its high spatial resolution. In this account we shall concentrate mainly on one study which has been initiated using this technique. It investigates the role of diffusion in establishing a new equilibrium when the phase-separated state is disturbed by sudden heating. In addition we shall briefly illustrate some other studies which are currently in progress.

2. THE OPTICAL TECHNIQUE

The technique we have used - real-time holographic interferometry - is carried out as follows. Before the experiment begins (at 77 K) one creates a hologram of the empty cell and the windows in the heat shields. The hologram is processed and replaced in its original position. Reconstruction of the hologram gives a virtual image of the cell which is superimposed on the real cell. When changes occur in the cell, one observes an interference pattern between the cell as it is at that instant and the reconstruction of the empty cell, so that any distortions occuring in both cases cancel one another. In practice one displaces the reconstructed image angularly by a small amount so as to produce a straight-line reference pattern in the absence of any difference between the two cases, because interpretation of the results is thereby simplified.

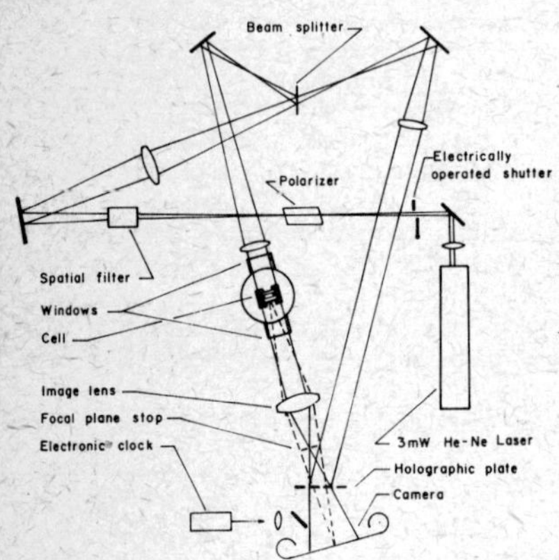

Fig. 1. The optical system

The optical system is shown schematically in figure 1. A full account of the experimental method together with a more detailed discussion of the results has recently been submitted for publication[4].

3. EXPERIMENTS ON THE ESTABLISHMENT OF EQUILIBRIUM

Let us consider the situation in which equilibrium has been established at a temperature T_1, below the tri-critical temperature T_t. The cell contains a fixed quantity of helium of mean concentration $\bar{c}$ at the vapour pressure (or very slightly above, so that no liquid-vapour meniscus is present). The liquid is separated into the lower, He^4 rich, superfluid phase and the upper, He^3 rich, normal phase. In the equilibrium state the values of the chemical potentials of the two isotopes are equal in the two phases. In any equilibrium situation the relative number of moles m of each of the two phases (labelled u-upper, and l-lower) is calculated from the lever rule:

$$m_l c_l + m_u c_u = (m_u + m_l)\bar{c}$$

The experiment which has produced the most exciting results is that in which the mean concentration $\bar{c}$ is slightly below the tricritical concentration (67%) and the phase separated liquid (A and Z) is heated suddenly to a temperature just below the λ-line (J) (fig. 2). We expected that the concentration step at the phase boundary would diffuse away in the usual manner of diffusive processes: getting less and less sharp with time. In fact we observed the formation of a wavefront which propagated slowly up the cell, hardly losing its sharpness. The velocity of propagation was about 5×10^{-3} cm.sec^{-1}. More careful observation of the shape of the wavefront showed it to contain a definite but continuous kink which propagated at the same velocity, having developed out of the phase boundary itself. At the conference, a cine-film of the wavefront propagation was shown, and some "stills" are reproduced here as figure 3.

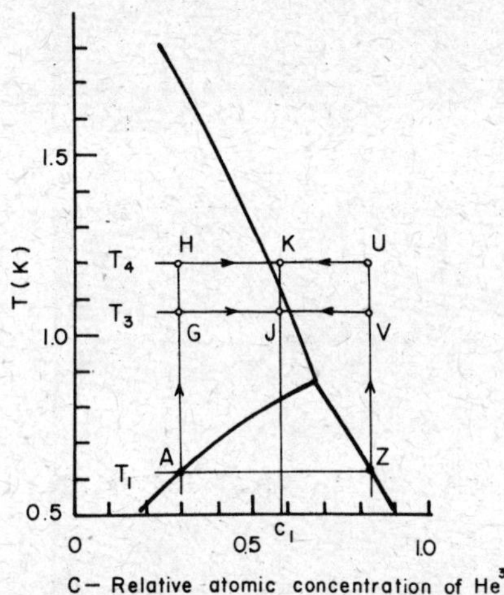

Fig. 2. Phase diagram for the diffusion wave experiment

The explanation of the "wave front" appears to lie in diffusion processes, despite its non-diffusive character. We shall attempt here to illustrate the explanation by dealing with the "wave front" and the "kink" separately.

The "wave-front" appears to be associated with the enormous difference (at least two orders of magnitude) between the large diffusion constant for ^{3}He in superfluid ^{4}He and the small diffusion constant for ^{4}He in normal ^{3}He. The boundary between these two regions, at temperatures above the tricritical, is the λ-line.

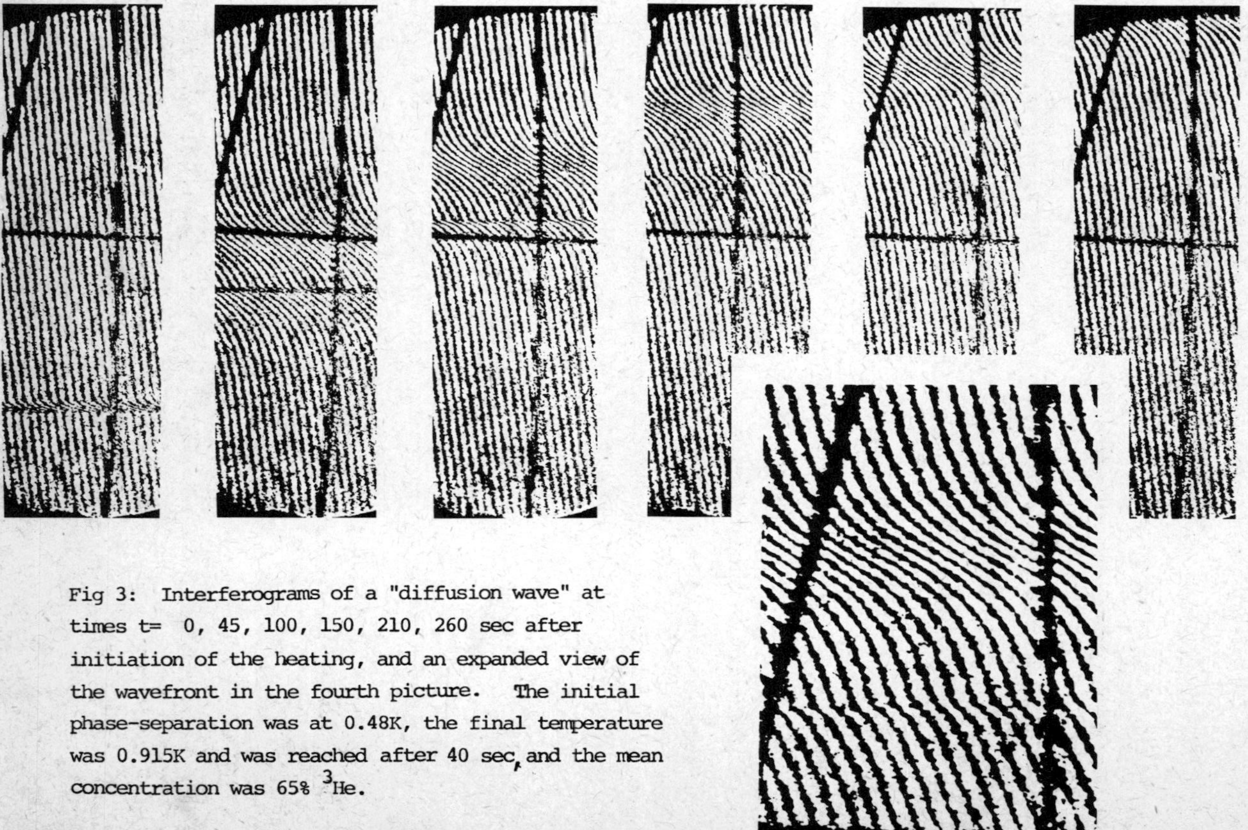

Fig 3: Interferograms of a "diffusion wave" at times t= 0, 45, 100, 150, 210, 260 sec after initiation of the heating, and an expanded view of the wavefront in the fourth picture. The initial phase-separation was at 0.48K, the final temperature was 0.915K and was reached after 40 sec, and the mean concentration was 65% ^{3}He.

We have solved the one-dimensional diffusion equation for this situation in a finite cell, using a simple computer program. The result (fig. 4) does indeed show the propagation of a wave front very similar to that observed, when we start with a step function of concentration at the position of the phase boundary. We have not yet derived the relationship between the front velocity and the diffusion constants, although it is reasonable to guess that the velocity is proportional to the lower diffusion constant.

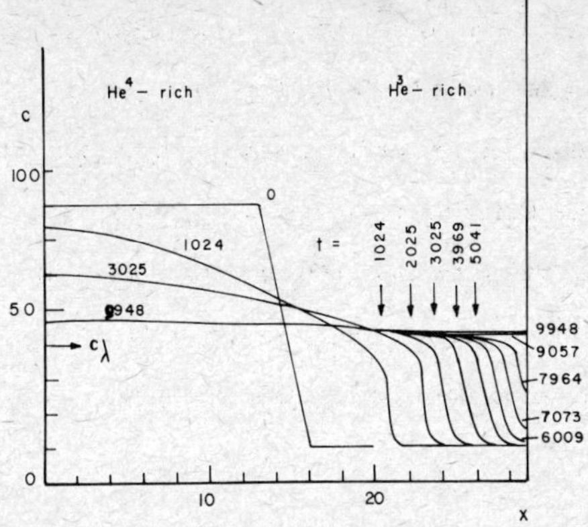

Fig. 4. Simulation of the diffusion wave (the numbers represent the times of the various concentration profiles).

The "kink" does not follow from the simple model described above. It appears to be related to the critical behaviour of the diffusion constant around the λ-line, for which some experimental data does exist[5]. An examination of the shape of the kink leads one to the surprising conclusion - in contrast to what has been described above - that near to the λ-line the diffusion constant is very large on the normal (^3He rich) side and very small on the superfluid (^4He rich) side. A more careful analysis of some of the earlier experiments has been made by fitting pairs of successive interferograms to the diffusion equation

$$\frac{\partial c}{\partial t} = \frac{\partial}{\partial x}\left(D\frac{\partial c}{\partial x}\right)$$

in which D is a function of concentration c. The results of three such fittings are shown in fig. 5. There is a divergence of the diffusion constant as the λ-line is approached on the normal side, and a very low value of the diffusion constant on the superfluid side, which rapidly increases as the λ-line is left behind. An explanation of the latter phenomenon has been given privately by A. Ron, M. Revzen and

L. Trainor, who consider it to be associated with second viscosity of the superfluid in the region of the λ-transition. We should only point out in addition that, whereas the diffusion constants calculated on the normal side all fall approximately on a single curve as a function of $(c-c_\lambda)$, those on the superfluid side do not, and to our thinking this indicates that something other than diffusion is involved in the transport of ^3He in the superfluid. An obvious candidate is second sound.

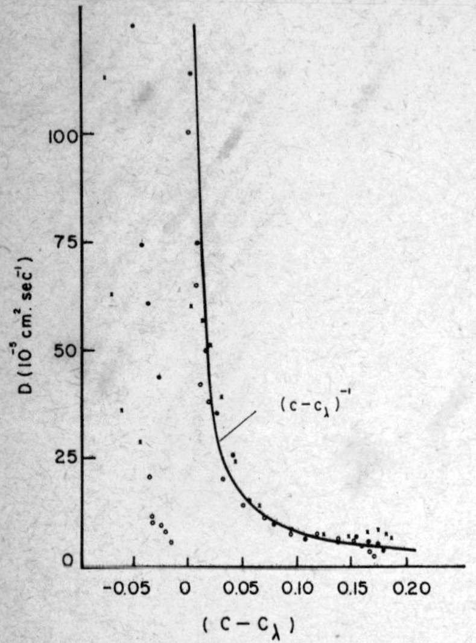

Fig. 5. Experimentally deduced diffusion constant as a function of concentration of ^3He at 0.915 K (data of fig. 3).

3. OTHER STUDIES

Some other experiments which have been started, although they are still in their infancy, involve solidification and crystallization of ^3He-^4He mixtures under pressure. A film of such processes was also shown during the conference. So far, the observations have been little more than qualitative. We eventually hope to study phase separation in solid helium mixtures by this method. Some illustrations are shown in figures 6, 7 and 8.

ACKNOWLEDGEMENTS

The work described in this paper has been mainly carried out by the authors' graduate students: Jacob Pipman[6], Norberto Bochner and Leslie Salem. Many useful discussions on the physics have been held with Charles Kuper, Amiram Ron, Michael

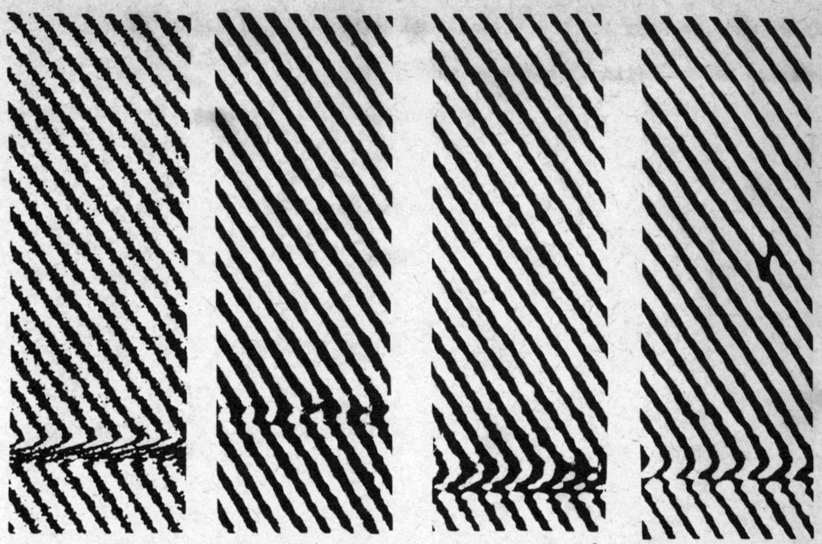

Fig 6: Interferograms of the meniscus between the ^{3}He-rich (upper) and ^{4}He-rich (lower) phases

a) at .425K and the vapour pressure,

b) at .625K and 15 atm. pressure,

c) at .380K and 15 atm. pressure,

d) at .450K and 20 atm. pressure.

Fig 7: Four phases coexist at 22.6 atm. and 0.4K: ^{3}He-rich liquid and solid above the phase boundary, and ^{4}He-rich liquid and solid below it.

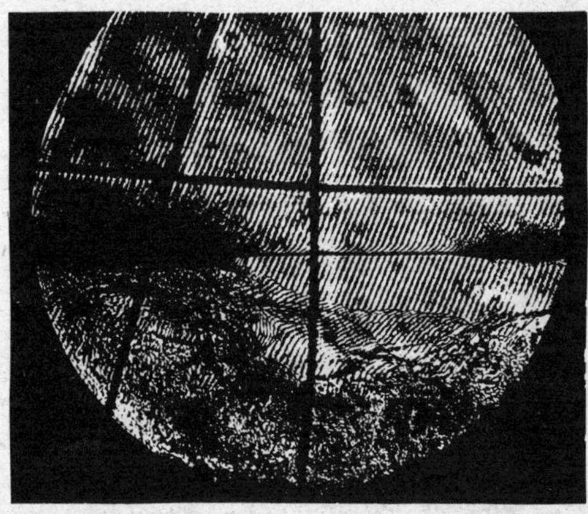

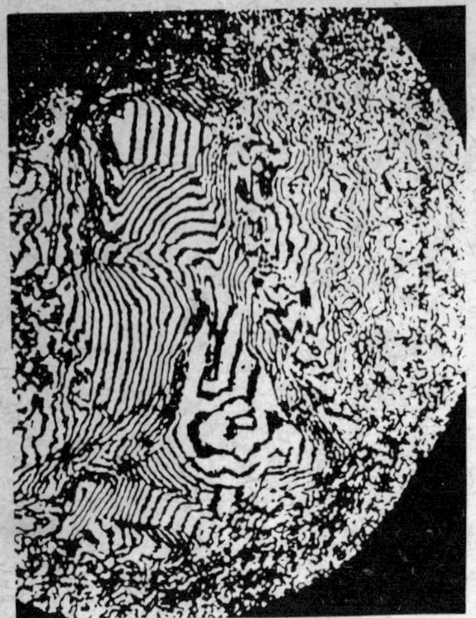

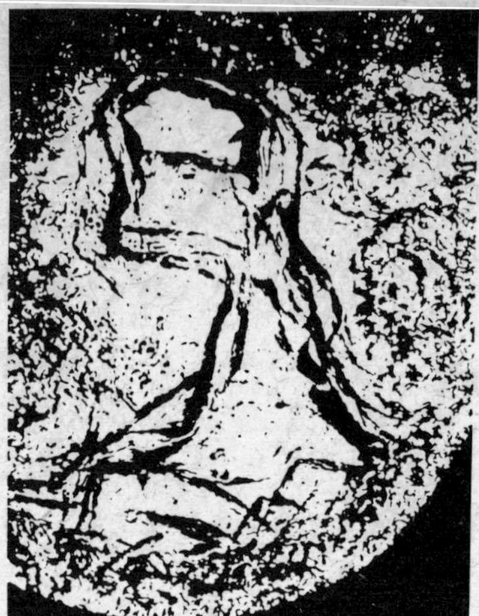

Fig 8: Photographs of annealing solid (above) and solidification (below). The upper photographs are almost simultaneous; the left-hand one uses the interference method, and the right-hand one is photographed directly.

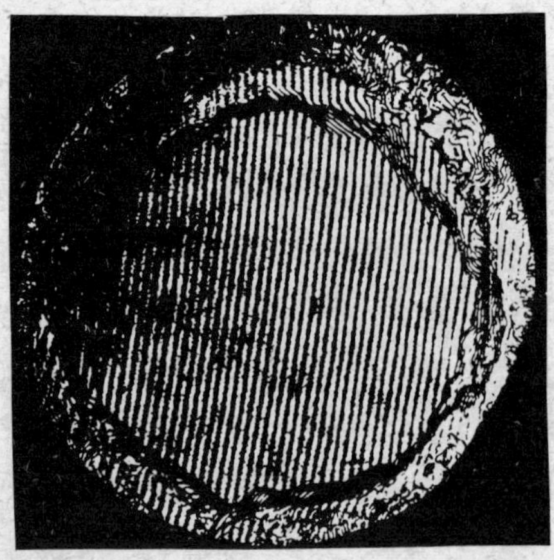

Revzen, Lynn Trainor, and David Bergman. In addition, criticism and encouragement have been received from David Edwards, Gunther Ahlers and Horst Meyer.

1. G. Ahlers in "The Physics of Liquid and Solid Helium, Part I," edited by K.H. Bennemann and J.B. Ketterson (John Wiley and Sons, New York, 1976) Chap. 2.

2. E.C. Kerr and R.H. Sherman, J. Low Temp. Phys. 3, 451 (1970).

3. H.A. Kierstead, J. Low Temp. Phys. 24, 497 (1976).

4. J. Pipman, S.G. Lipson, J. Landau, and N. Bochner, submitted to J. Low Temp. Phys.

5. G. Ahlers and F. Pobell, Phys. Rev. Lett. 32, 144 (1974).

6. Present address: Physics Dept., M.I.T. Cambridge, Mass.

CHAPTER 17

PHASE TRANSITIONS IN STRONGLY RANDOM SYSTEMS: A REVIEW*

T. C. Lubensky†

Department of Physics, University of Pennsylvania,

Philadelphia, Pa. 19104, U.S.A.

*Supported in part by grants from the National Science Foundation and the Office of Naval Research

†Alfred P. Sloan Foundation Fellow.

ABSTRACT

A review of recent progress in the theory of strongly random systems will be presented. Topics to be discussed include 1) Edwards and Anderson Spin Glasses, 2) Percolation, 3) Magnetic phase transitions in diluted systems near the percolation threshold and 4) Conductivity of a randomly diluted resistor network.

This paper will present a review of recent progress in the theory of phase transitions and critical phenomena in strongly random systems. Its primary emphasis will be on work carried out by the group at the University of Pennsylvania which includes A. B. Harris, Jing-huei Chen, Chandan Dasgupta, and Ronald Fisch in addition to myself. No attempt will be made to reference all papers in this field.

In homogeneous (non-random) systems, materials within the same universality class exhibit the same critical behavior. In recent years, largely as a result of the renormalization group, we have learned that a large number of factors participate in determining universality classes. Among these factors are 1) spatial dimensionality d, 2) spin dimensionality n (or more generally the symmetry of the order parameter if it cannot be described by an n-component vector), 3) range of interactions, 4) symmetry of the Hamiltonian (this includes accidental vanishing of certain potentials leading to tricritical points and Lifshitz points in addition to terms which lead, for example, to cubic rather than full O_n symmetry), and 5) interactions with non-critical degrees of freedom such as phonons or the vector potential in superconductors. Randomness opens up the possibility of a whole new set of universality classes. There is the potential for as much richness in probability distributions for random variables as there is in the Hamiltonians of homogeneous systems.

Random systems are usually divided into two broad categories - annealed and quenched. In annealed systems the random variables are in thermal equilibrium with other degrees of freedom. Since they are usually non-critical, they can be removed to yield an effective homogenous Hamiltonian. In quenched systems, randomness is frozen in. Essentially any variable of a homogenous system may be considered as a random variable. Random systems that have received the most attention to date include 1) systems with random nearest neighbor bonds,[1-7] 2) systems with nearest neighbor exchange and sites that can be occupied at random with atoms (or spins) of type A or B,[2,8-11] 3) systems with random external fields at each site[12-14] and 4) systems with orientationally random crystal fields.[15,6] In this paper, we will consider the first two types of random systems.

The most general spin Hamiltonian describing both bond and site nearest neighbor randomness[16] is

$$H = -\sum_{\vec{x},\vec{x}'} J(\vec{x}, \vec{x}') \, \vec{S}(\vec{x}) \cdot \vec{S}(\vec{x}') \tag{1}$$

where

$$J(\vec{x}, \vec{x}') = J_1 + J_2 \, \epsilon(\vec{x}) \, \epsilon(\vec{x}') + J_3 \, (\epsilon(\vec{x}) + \epsilon(\vec{x}')) + J_4 \, \mu(\vec{x},\vec{x}') \tag{2}$$

where $\epsilon(\vec{x})$ is a random site variable taking on the value +1 if $\vec{x}$ is occupied by an A atom and -1 if it is occupied by a B atom,and $\mu(\vec{x}, \vec{x}')$ is a bond random variable with zero average which may be described by an essentially arbitrary probablity distribution $P(\mu)$. If atoms A and B have nearest neighbor exchanges J_{AA}, J_{BB} and J_{AB}, then $J_1 = \frac{1}{4}(J_{AA} + J_{BB} + 2J_{AB})$, $J_2 = \frac{1}{4}(J_{AA} + J_{BB} - 2J_{AB})$ and $J_3 = \frac{1}{4}(J_{AA} - J_{BB})$. Note that the free energy associated with this Hamiltonian has the following symmetry property: $F(J_1, J_2, J_3, J_4) = F(J_2, J_1, J_3, J_4)$. This can be derived by introducing new variables $\vec{\tau}(\vec{x}) = \epsilon(\vec{x})\vec{S}(\vec{x})$ and $\tilde{\mu}(\vec{x}, \vec{x}') = \epsilon(\vec{x})\epsilon(\vec{x}') \, \mu(\vec{x}, \vec{x}')$ and tracing over $\vec{\tau}$ rather than $\vec{S}$. A system described by the Hamiltonian in Eq. (1) is weakly random if the average value of J, $[J]_{av}$ is much greater than fluctuations in J, $[(\delta J)^2]_{av}^{1/2}$. The remarkable thing about such weakly random systems is that their universality class is determined completely by the sign of the specific heat exponent α, and the universality class of the homogeneous system and <u>not</u> by the probability distribution for μ and ϵ.[1-3] If $\alpha < 0$, the random system falls into the same universality class as the homogeneous system. If $\alpha > 0$, the system falls into a new random universality class. When there is strong randomness, i.e. $[(\delta J)^2]_{av} \gtrsim [J]_{av}^2$, then the critical behavior and universality class depends on the values of J_1, J_2, J_3 and J_4 and on the probability distribution for μ.

We will now consider a number of special cases of strong randomness.

1. $\underline{J_1 = J_2 = J_3 = 0}$: Spin Glass[4,5] (SG)

To treat this case, we use the familiar replication trick[2,4]

$$
\begin{aligned}
[\ell n Z]_{av} &= \lim_{n\to 0} 1/n\, \ell n[Z^n]_{av} \\
[Z^n]_{av} &= \mathrm{Tr}\, e^{-\beta \mathcal{H}_{eff}} \\
\beta\mathcal{H}_{eff} &= -\sum_{p=1}^{\infty} \langle \sum_{\vec{x},\vec{x}'} \rangle\, C_p (\beta J \sum_{\alpha=1}^{n} \vec{S}^{\alpha}(\vec{x}) \cdot \vec{S}^{\alpha}(\vec{x}'))^p
\end{aligned}
\tag{3}
$$

where $\vec{S}$ is an m component spin, α is the replication index, and C_p is the pth cumulant of μ: $C_1 = [\mu]_{av} \equiv 0$; $C_2 = \frac{1}{2}([\mu^2]_{av} - [\mu]^2_{av})$, etc. If μ is described by a Gaussian probability distribution, we have the original Edwards and Anderson Model[4] for a spin-glass, with order parameter $Q^{\alpha\beta}_{ij}(\vec{x}) = \langle S^{\alpha}_i(\vec{x}) S^{\beta}_j(\vec{x})\rangle \equiv [\langle S_i(\vec{x})\rangle\langle S_j(\vec{x})\rangle]_{av}$; $\alpha \neq \beta$. Eq. (3) can be converted into a field theory with reduced Hamiltonian[5]

$$
\beta\bar{\mathcal{H}}_{S.G.} = \int d^d x\, \{\tfrac{1}{2} r \mathrm{Tr} Q^2 + \tfrac{1}{2}\mathrm{Tr}(\nabla Q)^2 + W \mathrm{Tr} Q^3 + \dots\} \tag{4}
$$

Normally, the presence of a third order term in a reduced Hamiltonian leads to a first order transition. The first order minimum in the free energy, however, corresponds to a negative value of Q. Q is a positive definite quantity and the first order minimum is inaccessible. The system, therefore, falls into the second order minimum[5] with $\beta = 1$. The critical dimension is therefore six rather than four ($d + 2\beta + \gamma = d_c \nu$ with mean field values $\alpha = 0$, $\beta = 1$, $\gamma = 1$, $\nu = \frac{1}{2}$). That d_c is 6 has been verified by R. Fisch and A. B. Harris[17] using a Padé analysis of the high temperature series for the Ising spin-glass susceptibility $\chi = \sum_{\vec{x}'} [\langle S(\vec{x}) S(\vec{x}')\rangle^2]_{av}$. The exponents obtained from this series are in good agreement with those obtained by the ϵ-expansion near 6 dimensions for $m = 1$: $\nu = \frac{1}{2} + 5m\epsilon/12(2m-1)$; $\eta = -m\epsilon/3(2m-1)$; $\gamma = 1 + m\epsilon/2m-1$. The series for χ become difficult to interpret for $d \leq 4$ indicating the possibility of a breakdown of the Edwards and Anderson spin glass state in low dimension.

2. $\underline{J_2 = J_3 = 0}$: Ferromagnet - Spin-glass[6] (FM-SG)

In this case, there is competition between ferromagnetic and spin-glass ordering for positive J_1. Thus, both $\vec{S}^{\alpha}$ and $Q^{\alpha\beta}_{ij}$ must be included in the reduced Hamiltonian

$$
\beta\bar{\mathcal{H}} = \beta\bar{\mathcal{H}}_{S.G.} + \sum_{\alpha} \int d^d x\, \{\tfrac{1}{2}\, \vec{S}^{\alpha}\cdot\vec{S}^{\alpha} + \tfrac{1}{2}\nabla\vec{S}^{\alpha}\cdot\nabla\vec{S}^{\alpha} + U|\vec{S}^{\alpha}|^4\} + \int d^d x \Big((V \sum_{\alpha} \vec{S}^{\alpha}\cdot\vec{S}^{\alpha})^2 + \int d^d x\, W_1 Q^{\alpha\beta}_{ij}\cdot S^{\alpha}_i \cdot S^{\beta}_j\Big) \tag{5}
$$

This model yields spin glass and random ferromagnetic phases with a phase boundary between them and a multi-critical point where this boundary terminates at the paramagnetic phase. Critical exponents for this point have been calculated in $6 - \epsilon$ dimensions for Ising systems ($m = 1$): $\nu = 1/2 + 2/3\ \epsilon$, $\phi = 1 + 1/2\ \epsilon$, $\eta_s = \eta_Q = -1/2\ \epsilon$. When $m = 2$ or 3, ν and ϕ become complex and things are not understood.

3. $J_1 = J_3 = J_4 = 0$: Alloy Anti-Ferromagnet[8-11] (AAF)

The thermodynamics of this model are identical to that of the homogeneous m component spin system. This can be seen by introducing $\vec{\tau}(\vec{x}) = \epsilon(\vec{x})\ \vec{S}(\vec{x})$ and observing that $Z = \mathrm{Tr}_\tau\{\exp(\sum_{\langle\vec{x},\vec{x}'\rangle} J_2\ \tau(\vec{x}) \cdot \tau(\vec{x}'))\}$ is completely independent of the probability distribution for $\epsilon(\vec{x})$. Thus, the critical dimension for the system is four rather than six. If $[\epsilon]_{av} = 0$, there is no average magnetization: $[\langle\vec{s}(\vec{x})\rangle]_{av} = [\langle\epsilon(\vec{x})\ \tau(\vec{x})\rangle]_{av} = 0$. On the other hand $[|\langle\vec{s}(\vec{x})\rangle|^2]_{av} = [\epsilon^2]_{av}|\langle\vec{\tau}\rangle|^2$ is non-zero. This is precisely the order parameter introduced by Edwards and Anderson to describe the bond random spin glass. The two systems are, however, quite different; and we prefer to distinguish them[16] by referring to the former as an Alloy anti-ferromagnet (AAF) and the latter as a (Edwards and Anderson) spin glass (SG).

4. $J_1 = J_3 = 0$: Alloy Anti-ferromagnet - Spin Glass[16] (AAF-SG)

From the general symmetry property of the free energy for Hamiltonian (1), we have $F(0, J_2, 0, J_4) = F(J_2, 0, 0, J_4)$. Since we have verified that there is a phase boundary separating the ferromagnetic and spin glass states in high enough dimension, there must also be a phase boundary separating the AAF and SG states.

5. $J_1 = J_2 = J_3 = 0$: $P(\mu) = p\delta_{\mu 1} + (1-p)\ \delta_{\mu o}$: Percolation[7,18-21]

In this case bonds are present with probability p and absent with probability $1 - p$. Thus, if $[\mu]_{av} < p_c$, the percolation concentration, there is no infinite cluster of connected spins. This means that there can be no phase transition at any finite temperature. If $[\mu]_{av} > p_c$, there is an infinite cluster of connected spins and a transition to a random ferromagnetic state will occur at a temperature $T_c(p)$ that goes to zero at $p = p_c$. The point $p = p_c$, $T = 0$ may be viewed as a multi-critical point. Along the axis $T = 0$, the transition is described entirely in terms of the geometrically driven percolation transition with

$$[\langle s\rangle]_{av} \sim P(p) \sim (p-p_c)^{\beta p}$$
$$\chi T \sim S(p) \sim (p-p_c)^{-\gamma_p} \tag{6}$$

$[\langle s(x)\ s(x')\rangle]_{av} \sim \langle w(\vec{x},\vec{x}')\rangle \sim |\vec{x}-\vec{x}'|^{-d+2-\eta_p} f(|\vec{x}-\vec{x}'|/\xi_p)$ where $P(p)$ is the probability that a site is in the infinite cluster, $S(p)$ is the mean square cluster size, $w(\vec{x},\vec{x}')$ is a function that is unity if $\vec{x}$ and $\vec{x}'$ are in the same cluster,

$\xi_p \sim |p-p_c|^{-\nu_p}$ is the percolation correlation length and β_p, γ_p, η_p, and ν_p are exponents for the percolation transition.

To describe heuristically scaling behavior in the vicinity of the point $p = p_c$, $T = 0$, it is convenient to consider an idealization of the infinite cluster for $p > p_c$ introduced by deGennes and by Skal and Shklovskii[22]. Points in the infinite cluster can be classified by the number of bonds that must be broken in order that no path exists between that point and infinity. The "backbone" depicted schematically in the adjoining figure consists of those points for which at least two bonds must be broken to isolate it from infinity. Dangling ends consisting of points for which only one bound must be broken do not play any role in the propagation of spin correlations near $T_c(p)$. Nodes (points for which at least three bonds must be broken to isolate them from infinity) are separated on the average by the percolation correlation length, ξ_p. The number of bonds $L \sim |p-p_c|^{-\zeta}$ between nodes is larger than ξ_p. Furthermore, we expect L to be less than or equal to the number of steps in a self-avoiding walk between nodes. Thus, we expect $\nu_p \leq \zeta \leq \nu_p/\nu_{saw}$ where ν_{saw} is the self avoiding walk exponent. For $d > 6$, ζ should be equal to its upper bound of $\nu_p/\nu_{saw} = 1$. The "backbone" between nodes is essentially one-dimensional in character. It is, therefore, relevant to compare the temperature dependent correlation length $\xi_1(T)$ of a one dimensional spin system with L. The true d-dimensional structure of the random lattice is felt only when $\xi_1(T) > L$. We expect, therefore, that

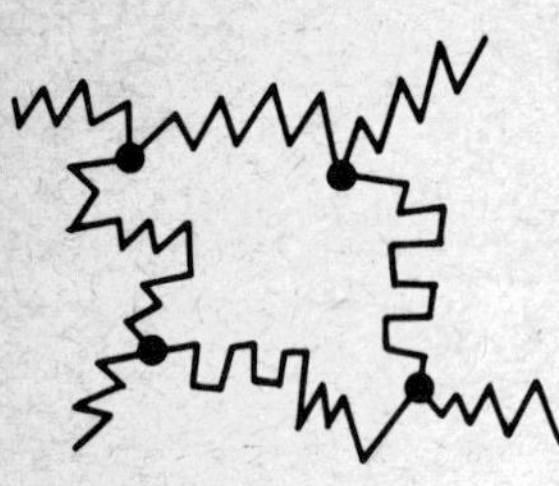

Fig. 1: Backbone of the infinite cluster

$$\chi^T \sim (p-p_c)^{-\gamma_p} f(\xi_1(T)/L) \tag{7}$$

Note that this implies that $\xi_1(T_c(p)) \sim |p-p_c|^{-\zeta}$ for $\zeta \leq 1$. Since $\xi_1(T) \sim \beta J$ for Heisenberg, one would expect substantial differences in the behavior of $T_c(p)$ near $p = p_c$ for the two systems. Unfortunately, most systems have some anisotropy which becomes important near $T = 0$. It may turn out, therefore, that only Ising systems can be investigated experimentally.

6. Random Resistor Networks

This problem has a great deal in common with the problem just considered. Here, we are interested in the macroscopic conductivity Σ of a network in which each bond carries a conductance $\sigma \equiv R_o^{-1}$ with probability p and a conductance zero with probability 1 - p. Σ goes to zero at p_c as $(p - p_c)^\mu$. By dimensional analysis one expects Σ to be proportional to $R_{eff}^{-1}\ a^{-(d-2)}$ where R_{eff} is the effective resistance between nodes and a measures the separation of nodes. Using $R_{eff} \sim LR_o$ and $a \sim \xi_p$ for the infinite cluster above p_c, we obtain $\Sigma \sim R_o^{-1}\ L^{-1}\ \xi_p^{-(d-2)}$ or

$$\mu = \zeta + (d-2)\ p \tag{8}$$

In order to discuss this problem more fully in the language of phase transition, it is convenient to introduce a resistive susceptibility[23] that exists both below and above p_c: $\chi^{(r)} = \sum_{\vec{x}'} \langle w(\vec{x},\vec{x}')\ R(\vec{x},\vec{x}')\rangle$ where $R(\vec{x},\vec{x}')$ is the resistance between points $\vec{x}$ and $\vec{x}'$. We expect $\chi^{(r)}$ to obey a scaling relation of the form

$$\chi^{(r)} \sim (p - p_c)^{-\gamma(r)} g(q\xi_p) \tag{9}$$

where q is the wave number. For $p > p_c$, $\chi^{(r)} \sim P^2(p)/\Sigma q^2$ and for $p < p_c$, $\chi^{(r)} \sim (p - p_c)^{-\gamma(r)}$. Thus, using Egs. (6), (8) and (9), we obtain

$$\gamma^{(r)} = \mu - (d-2)\ \nu_p + \gamma_p = \zeta + \gamma_p \tag{10}$$

7. <u>ϵ-Expansion for the Dilute Magnet and Random Resistor Network Near Pc</u>

The statistics of the percolation problem can be described by the therodynamics of the one state limit of the s-state Potts model described by the Hamiltonian.[18]

$$H = -sJ \sum_{\langle\vec{x},\vec{x}'\rangle} (\delta_{\sigma(\vec{x})\sigma(\vec{x}')} - 1) - s \sum_{\vec{x}} h(\vec{x})\ \delta_{\sigma(\vec{x})1} \tag{11}$$

where $\sigma(\vec{x})$ is a variable that can be in any of s different states. Note that $(\delta_{\sigma\sigma'}-1)$ is just the projection operator $P_{\sigma\sigma'}$ onto the states for which $\sigma = \sigma'$. It is straightforward to show that the non-local susceptibility $\chi(\vec{x},\vec{x}') = (s-1)^{-1}\ \delta^2 \ell nZ/\delta h(\vec{x})\,\delta h(\vec{x}')$ satisfies

$$\chi_s\ (\vec{x}, \vec{x}') = (\Sigma p^{N_b}\ s^{N_c})^{-1}\quad (\Sigma\ \ p^{N_b}\ q^{N-N_b}\ s^{N_c}\ w(\vec{x},\vec{x}')) \tag{12}$$

where $q = (1-p) = e^{-sJ}$, N_b is the number of occupied bonds, N the total number of bonds and N_c the number of clusters. Thus, $\lim_{s\to 1} \chi_s(\vec{x},\vec{x}') = \langle w(\vec{x},\vec{x}')\rangle$ describes the percolation problem. On the other hand, if $s \to 0$, one obtains for a homogenous network

$$\lim_{s\to 0} \chi_s\ (\vec{x},\vec{x}') = 1 - 1/J'\ \ R\ (\vec{x},\vec{x}'),\quad J' \to \infty \tag{13}$$

where $R(\vec{x}, \vec{x}')$ is the resistance between points $\vec{x}$ and $\vec{x}'$ and $J' = J/\sigma_o$ where σ_o is the fundamental conductance of a bond. This result follows because $R(\vec{x}, \vec{x}')$ is proportional to the number of spanning trees with a bond between $\vec{x}$ and $\vec{x}'$ divided by the number of spanning trees with no bond between $\vec{x}$ and $\vec{x}'$. If the network is random rather than homogeneous, one may use Eq. (13) to obtain

$$\lim_{s\to 0} \sum_{\vec{x}'} \langle w(\vec{x}, \vec{x}')\rangle \chi_s(\vec{x}, \vec{x}')\rangle = S(p) - 1/J'\chi^{(r)}(p) \equiv \tilde{\chi}(p) \tag{14}$$

Stephan and Grest[24] have calculated critical exponents near $p = p_c$, $T = 0$ in $6-\epsilon$ dimensions by demonstrating that the replicated spin Hamiltonian reduces to the one state Potts model near $T = 0$:

$$[Z^n] = \prod_{\langle \vec{x},\vec{x}'\rangle} (1 + ve^{\beta J \sum_\alpha \vec{s}^\alpha(\vec{x})\cdot\vec{s}^\alpha(\vec{x}')}) \equiv \mathrm{Tr}\, e^{-\beta \mathcal{H}_{eff}} \tag{15}$$

where $v = p/1-p$. For Ising spins, this can be expressed in terms of a closed set of projection operators

$$-\beta\mathcal{H}_{eff} = \sum_{\vec{x},\vec{x}',k} A_k P_k(\vec{x},\vec{x}') \tag{16}$$

Where $A_k = \ln(1 + ve^{-2\beta kJ})$ and $P_k(\vec{x},\vec{x}')$ is the projection operator onto the state where the $\vec{x}$ and $\vec{x}'$ have relative spin flips in k replicas of $\vec{s}^\alpha$. When $T = 0$, $A_0 = \ln(1/1-p)$ and $A_k = 0$ for $k \geq 1$, and Eq. (16) becomes equivalent to the Potts model Hamiltonian of Eq. (11) with a 2^n state rather than an s state variable at each site. Thus Eq. (16) describes percolation when $T \to 0$ as $n \to 0$ ($2^n \to 1$). To calculate the crossover exponent for turning on $e^{-2\beta J}$, it is convenient to use an alternate representation of Eq. (16)

$$-\beta\mathcal{H}_{eff} = \sum_{\vec{x},\vec{x}'} K_k\, S^{(k)}(\vec{x}) \cdot S^{(k)}(\vec{x}') \tag{17}$$

where $\vec{S}^{(1)}(\vec{x}) = s^\alpha(\vec{x})$, $\vec{S}^{(2)} = s^\alpha(\vec{x})\, s^\beta(\vec{x})$, $\alpha \neq \beta$, etc., where the dot product is understood to include summations over the replication indices and where $2^n K_k = \log(1/1-p) + (r-2k)\, ve^{-2\beta J}$. Thus at $T = 0$, $p = p_c$, all possible replica order parameters including the spin $\langle s^\alpha\rangle$, the spin glass order parameter $\langle s^\alpha s^\beta\rangle$, and others $\langle s^\alpha s^\beta s^\gamma \ldots\rangle$ order simultaneously. Stephen and Grest[24] have calculated the crossover exponent $\phi' = \zeta$ for turning on $e^{-2\beta J}$ in Ising systems and βJ in xy and Heisenberg systems to second order in ϵ. They find $\zeta = 1 + O(\epsilon^3)$.

The random resistor network near six dimensions can be treated in a fashion exactly analogous to the spin problem just discussed[25]. Using the representation $(s\,\delta_{\sigma\sigma'} - 1) = (s - 1)\,\vec{m}\cdot\vec{m}'$ where $\vec{m}$ is a vector that points to the vertices of an s-dimensional simplex, one can replicate the s-component Potts model to obtain a model with bonds (conductances) randomly removed. Following the steps just outlined, we obtain

$$\tilde{\chi}(p) = (p - p_c)^{-\gamma_p}\, \tilde{g}\, (J(p - p_c)^{-\zeta}) \tag{18}$$

where $\zeta = 1 + O(\epsilon^3)$. Thus, using Eqs. (18), (14), and (9) we obtain $\gamma^{(r)} = \zeta + \gamma_p = 1 + \gamma_p$ in agreement with the heuristic picture. It is uncertain whether the resistive and spin ζ remain equivalent in lower dimension. In particular, a $1 + \epsilon$ dimensional calculation for the random resistor problem since the zero state homogeneous Potts model has no transition below d = 2 (a 2 + ϵ dimensional expansion is possible for this case).

References

1. A. B. Harris and T. C. Lubensky, Phys. Rev. Lett. 33, 1540 (1974); T. C. Lubensky, Phys. Rev. B9, 3573 (1975).
2. G. Grinstein and A. Luther, Phys. Rev. B 13, 1329 (1976).
3. D. E. Khmelnilsky, Soviet Physics - JETP (to be published).
4. S. F. Edwards and Anderson, J. Phys. F 5, 965 (1975).
5. A. B. Harris, T. C. Lubensky and J. H. Chen, Phys. Rev. Lett. 36, 415 (1976).
6. J. H. Chen and T. C. Lubensky, Phys. Rev. B, May (1977).
7. A. B. Harris, T. C. Lubensky, W. K. Holcomb and C. Dasgupta, Phys. Rev. Lett. A 48, 103 (1974).
8. D. C. Mattis, Phys. Lett. 56A, 421 (1976).
9. J. M. Luttinger, Phys. Rev. Lett. 37, 778 (1976).
10. A. Aharony and Y. Imry, Solid State Commun. 20, 889 (1976).
11. A. Aharony, (preprints)
12. L. J. Sham and Bruce R. Patton, Phys. Rev. B 13, 3151 (1976).
13. Y. Imry and S. Ma, Phys. Rev. Lett. 35, 1399 (1975); G. Grinstein, Phys. Rev. Lett. 37, 944 (1976); A. Aharony, Y. Imry and S. Ma, Phys. Rev. Lett. 37, 1364 (1976).
14. A. Aharony (preprint).
15. A. Harris, M. Plischke and M. J. Zuckermann, Phys. Rev. Lett. 31, 160 (1973); j. de Phys. (Paris) 35, C4-265 (1973).
16. A. B. Harris and T. C. Lubensky (to be published).
17. R. Fisch and A. B. Harris, Phys. Rev. Lett., 38, 785 (1977).
18. P. W. Kasteleyn and C. M. Fortuin, J. Phys. Soc. Jpn. Suppl. 16, 11 (1969); Physics (utr.) 57, 536 (1972).
19. D. Stauffer, Z. Phys. B22, 161 (1975).
20. T. C. Lubensky, Phys. Rev. B15, 311 (1977).
21. H. E. Stanley, R. J. Birgeneau, P. J. Reynolds, J. Nicoll, J. Phys. C9, L553 (1976).
22. P. G. de Gennes, J. Phys. Lett. 37, L1 (1976); A. S. Skal and B. I. Shklovskii, Fiz. Tekh. Polup. 8, 1582 (1974) [Sov. Phys. - Semicond. 8, 1029 (1975)].
23. A. B. Harris and R. Fisch, Phys. Rev. Lett. 38, 796 (1977).
24. M. J. Stephen and G. S. Grest, Phys. Rev. Lett. 38, 567 (1977).
25. C. Dasgupta, A. B. Harris and T. C. Lubensky (to be published).

CHAPTER 18

GEOMETRIC FACETS OF STATISTICAL PHYSICS:
SCALING AND FRACTALS

B. B. Mandelbrot

IBM Watson Research Center, Yorktown Heights, N.Y. 10598 U.S.A.

ABSTRACT

The geometric facets of statistical physical systems have not received anywhere near the attention they deserve. This paper examines some examples relative to systems that satisfy a condition of scaling, including some systems near a critical point. Percolation and intermittent turbulence, among others, are shown to involve nonstandard sets which the author has termed fractals. Their principal parameter, which need not be an integer, is a scaling exponent, sometimes an old and sometimes a new one, the fractal (Hausdorff Besicovitch) dimension. Fractals are also encountered in other very different investigations in physics; in particular, the strange attractors of dynamics are fractals.

1. INTRODUCTION

The study of scale invariant systems has focused almost exclusively on the fact that scaling implies that all functional relationships between physical quantities are given by power laws. Such systems involve a host of exponents, for example (but not limited to) the critical exponents. However, the consequences of scaling are not limited to the functional relationships: indeed, whenever a physical system involves actual visible geometric shapes, their form should also be expected to be scale-invariant and cries out for investigation. Geometry may serve either to reinterpret some old scaling exponents, or, more notably, to introduce fresh ones. It is clear that the standard shapes such as lines or planes are scale-invariant, but they cannot suffice for the present purposes, because the shapes involved in statistical physics are ordinarily extremely convoluted, irregular and fragmented. In other words, the geometric study of scaling statistical systems must be expected to involve other, highly nonstandard sets. Examples of such scaling physical systems will be given, then the corresponding nonstandard sets will be described briefly. They are fractals[1]. Other roles of fractals in physics will be sketched.

2. EXAMPLES OF GEOMETRIC PROBLEMS RAISED BY STATISTICAL PHYSICS

A first example is classical Hammersley percolation in the plane when the basic probability takes the critical value. Because of the property of scaling, the asymptotic relative area of the largest critical cluster could only be either 0 or 1, and it is obvious that the correct alternative is the former one: the relative area is small and tends to zero as the overall lattice size L divided by the lattice step ζ tends to infinity. However, a rescaled critical cluster is clearly not a standard curve; it branches out and loops extensively. As $L/\zeta \to \infty$, this curve's propensity to branch out increases without bound. Hence, in intuitive terms, the critical cluster must be viewed as a "shape in between a curve and a plane domain". It would be most desirable to characterize its form in precise mathematical terms, and to extend the study of all the classical problems, such as the issue of universality, to the cluster's geometric properties.

A second example is SARW: the self avoiding random walk. Computer experiments, including those of C. Domb, suggest that if R designates the walk's overall linear size (e.g., the distance between its ends, or the radius of the smallest circumscribing circle), then the walk's "bulk" as measured by the number of sites is proportional to R^D (using Domb's notation, $D=2/\gamma$). In the plane, one finds $D \sim 4/3$; in 3-dimensional space, $D \sim 5/3$. The formula R^D is formally the same as the formula for the contents of a ball in Euclidean D dimensional shapes, but in the present instance the exponent D is known to be a fraction. Hence an SARW is also an in-between shape.

A third broad class of examples is encountered in the study of turbulence, which of course involves both the earliest applications of scaling ideas (L. F. Richardson 1926; A. N. Kolmogorov 1941, 1962) and some of the most recent ones (many are described in my book[1]). Here again the standard shapes fail to suffice and the call for "in-between shapes" was heard most insistently.

A fourth class of examples is involved in the study of scaling random fields. This class overlaps the third class above, since the earliest example is provided by Kolmogorov's "self similar" turbulent velocity field. Furthermore, the study of critical phenomena, e.g., of the Ising model, involves scaling random measures (as opposed to fields). It is natural to inquire about the shape of the clusters within which the spin is up or down.

It should be noted that the theory of turbulence uses the term "similar" in a peculiar generic sense that also includes geometric "affinity". This usage may simply be traceable to conflict between Russian and English. A different and even more obvious mistranslation is involved in the term automodel, which Sinai[2] and others apply to probability distributions. There is evidence that this term does not refer to a new and distinct concept. In all instances it seems best to keep to the physicists' term, "scaling".

In other areas of statistical physics the formula R^D for the volume, or its equivalent for the surface, is postulated a priori. One example is relative to the Fisher droplet model[3]. The geometric task in such instances is to develop an actual construct having the desired properties.

Additional examples will be mentioned in Section 7.

3. FRACTAL SETS AND THE FRACTAL (HAUSDORFF BESICOVITCH) DIMENSION

In order to model the geometry of the various "in-between" shapes exemplified in the above list, and in particular to extend the notion of dimension so as to allow its value to be a fraction, I have recently[1] singled out a certain very broad class of classical but nonstandard geometric shapes, and advanced the notion of fractal. Each fractal is, from the most concrete viewpoint, defined through a sequence of instructions which actually allow it to be drawn, at least approximately. Many examples of various sorts are given in my book[1]. And at the most abstract end there is an underlying mathematical theory satisfying every standard of rigor. Most notably, the notion of fractal dimension coincides with a notion that is classical in mathematics but, so to speak, obscure: Hausdorff Besicovitch dimension.

Fractal is a new term and concept, but specific examples of fractal sets are widely known. One recent example familiar to many in this conference has been made up to provide a model of the Fisher droplet[4]. Furthermore, many fractals have been familiar to pure mathematicians since circa 1900, but used to be viewed as extremely pathological and totally counter, or at best non-, intuitive. Examples are Cantor sets and Peano curves.

A different fractal is, on the contrary, absolutely familiar to all physicists: it is Brownian motion. In a space of Euclidean dimension $E \geq 2$, its simplest and most important property from the present viewpoint is the following: Consider a cubic box of side L conditioned to be intersected by a Brownian motion, and pave said big box by small cubes of side ζ. Out of the $(L/\zeta)^E$ small cubes defined in this manner, mark those which our Brownian motion intersects. It turns out that there are $N \sim (L/\zeta)^2$ of them. Because Brownian motion has the property of scaling invariance, each nonempty cube of side ζ contains a piece of Brownian motion that is identical to the whole except for a similarity of ratio $r = \zeta/L$. Hence one could have expected to end up with some relationship of the form $N \sim (1/r)^D$, D being a constant. The specific properties of this process yield the universal value $D=2$, independent of E.

The same construction applied to a vector manifold (line or plane) would have yielded $N \sim (1/r)^D$ with D the manifold's dimension. More generally, elementary geometry has accustomed us to finding that all the facets of the loose and intuitive notion of dimension, including the vectorial and topological facets, involve different mathematical definitions, but yield identical values. For Brownian motion, however, the situation is entirely different. Its topological dimension is $D_T=1$ because Brownian motion is topologically a curve. On the other hand, we just observed that $D=2$. The fact that D exceeds D_T expresses that Brownian motion, when viewed from a certain nontopological vantage point, is "more than linelike".

Said vantage point has indeed been investigated by mathematicians, starting with Hausdorff (1919). To be more accurate, they did not stress the D as defined above, to be called the similarity dimension, because its definition is overly informal and is restricted to special sets. While from the viewpoint of the physical applications I propose, especially in the study of scaling systems, the similarity dimension is usually sufficient, from the mathematical viewpoint it raises difficulties and lacks generality. Therefore, it was modified by Hausdorff and Besicovitch so as to yield a dimension that is defined for every set in a metric space (however pathological). See Section 9 below. Nevertheless, the result coincides with the similarity dimension sufficiently often to allow the physicist to view the two notions as identical in a first approximation. Furthermore, it can be shown that the topological and fractal dimensions, and the embedding Euclidean dimension, are related by the inequalities

$$D_T \leq D \leq E$$

For standard figures, like those in Euclid, one has $D_T = D$. By contrast, the "anomalous" double inequality $D_T < D \leq E$ was used in my book[1] to define the notion of fractal set.

Thus, the prototypical fractal is Brownian motion, but it is atypical: while topological dimension is necessarily an integer, fractal dimension need not be an integer. A ready example is provided by the self avoiding random walk. For it, $D_T=1$ while D is 2/Domb's γ (see Section 2).

An unavoidable question is whether or not the Hausdorff Besicovitch dimension can also serve to give precise mathematical meaning to a powerful heuristic tool of theoretical physics, namely to spaces of fractional dimension (say, $4-\varepsilon$) as opposed to sets of fractional dimension imbedded in integer dimensional spaces. The answer, as I know it, is ambiguous. Since the Hausdorff Besicovitch definition applies to any set in a metric space, it applies to a whole ball, and it is conceivable that the space be such that said D is fractional. Since, on the other hand, the topological dimension is always an integer, it would have to be smaller than D, meaning that such a space is fractal. Unfortunately all the examples I know are readily imbedded in larger integer dimensional spaces.

In addition to the fractal and topological dimensions, the rough notion of dimension has a multiplicity of other facets. Among those mentioned in Fractals[1] as being susceptible of rigorous mathematical study, a notable one refers to spectra and is called the Fourier dimension. It seems that Fourier dimension is involved in physics[5]. This dimension's value need not coincide with D or D_T, and controversy about which is the better one is fruitless.

4. PERCOLATION IN THE PLANE: CRITICAL CLUSTERS OF DIMENSION D=1.78

In studying the fractal properties of diverse natural configurations, I was naturally led to evaluate the above defined quantities r and N in the case of critical clusters in Hammersley's percolation on a square lattice. Brute force simulations yielded $D \sim 1.78$. It is easy to verify that in this particular case one has $D=2-\eta$, η being a standard exponent in percolation theory. (This is a generalization of an equality of Novikov & Stewart, as interpreted in my book[1] and in my papers on turbulence.) Hence $D \sim 1.78$ gives an indirect empirical estimate of η. A different kind of computer simulation (independent but nearly simultaneous) was carried out by Leath[6], whose simulations are compatible with $D \sim 1.78$.

The classic equality $2-\eta=\gamma/\nu$ shows that $D=\gamma/\nu$. The observation that γ/ν is dimensionlike was reached independently in a different and interesting way by Stanley et al.[7] – but without the backing of a rigorous definition of D. The value of the ratio γ/ν had been evaluated through series expansions by Dunn et al.[8], who obtain essentially the same value. They also compare the percolation problem with the Ising problem and observe the following: while the values of γ and ν are very different in the two problems, the value 1.78 they obtain for γ/ν is nearly the same as the value 1.75 known to hold for the Ising model. They recall furthermore that Suzuki[9] had proposed a weak form of universality which asserts that absolute exponents such as γ and ν may depend on the details of interaction, but reduced exponents such as γ/ν must be universal: independent of the system's Hamiltonian. The exponent η is by definition a reduced exponent. In the case of percolation, my arguments and those of Stanley et al. underline the importance of one reduced exponent by showing that the familiar ratio $\gamma/\nu=2-\eta$ has a direct new geometric interpretation as a fractal dimension. Suzuki universality goes one step further to suggest that the "true" value of D is 1.75. This value is not excluded by my simulations and is so "simple" as to suggest the possibility of a direct analytic derivation.

Stanley et al.[7] contribute a further intriguing thought to this topic by sketching a possible relationship between the estimate $\gamma/\nu=1.78$, written as 17/9 (hence, $D \neq 1.75$), and some geometric features of the self avoiding random walk.

Kirkpatrick[15] presents further developments and in particular focuses upon the critical clusters' "skeletons".

5. THE ISING MODEL

Despite the above-mentioned closeness of the corresponding critical values of η (and their identity assuming Suzuki universality), percolation and Ising clusters seem to involve totally different geometric structures.

Following the usual Kadanoff scaling argument, average the spins within boxes of side $\sigma\zeta$ and rescale the results by dividing them by their typical values. Well below the critical point the spins are near independent. Hence, when $\sigma \to \infty$ while $\sigma\zeta \equiv 1$, the limit of rescaled average spins is a white Gaussian measure. A measure, not being a function of points but a function of domains such as the above boxes, does not involve an actual zeroline or zeroset separating positive and negative domains; the best one can do is to draw a line between boxes within which the averages are positive and negative. As $\zeta \to 0$, the basic box grid becomes tighter, and in addition the boundary fills space increasingly tightly, so that the simple algorithm used to calculate the dimension yields the degenerate result D=E, E being the Euclidean dimension of the embedding space.

At the critical point, the principal difference is that the above limit measure is nonwhite (pink), again with no defined zeroset. Hence, again, the estimated D is the Euclidean dimension of the embedding spaces.

On the line, E=1, and the distinction between function and measure is less vital. The boxes are intervals, and the measure of any box can be defined through the measure of a varying interval (0,t), which is a function. In the case of white Gaussian measure, said function is Brownian motion. But the Brownian motion and its increments process have separate and distinct generalizations to multidimensional variable space: the present white Gaussian measure generalizes the increments and the Brownian function studied extensively in Fractals[1] generalizes the Brownian motion. Similarly, except for E=1, the above defined pink measure is distinct from the increments of the fractional Brownian function[1].) [Erratum: on p. 279 of said book[1], line 10 of the entry Automodel should begin by "On the line (E=1) ...".]

6. THE EXPONENTS OF INTERMITTENT TURBULENCE: D AND INEQUALITIES INVOLVING IT

The different physical properties of a system near a critical point are well known (as we saw in the case of percolation) to be distinct exponents related by inequalities that become equalities under the scaling assumption. I have established that a different but somewhat analogous situation holds true of intermittent turbulence. A first difference is that in turbulence dimension is a new exponent, which does not in general reduce to a combination of old ones. Also, scaling is compatible with inequality between exponents, while equality demands an specially strong form of scaling. Furthermore, my argument is not based upon any a priori analogy presumed to exist between turbulence and critical point behavior or any other chapter of physics. An earlier theory of intermittency, which is due to Kolmogorov and Yaglom and can be characterized by the use of the lognormal Ansatz, had not recognized that there is a multiplicity of distinct exponents, because it felt able to deduce all the basic properties from one constant, say Yaglom's μ.

The present section is a very sketchy overview of results given elsewhere[10]. First, recall that the classical picture of intermittent turbulence, as it emerged in 1962, involves the following exponents.

B: The spectral density of turbulent velocity takes the form $k^{-5/3-B}$. The term B is a correction to the classic Kolmogorov 5/3.

Q: The covariance of local dissipation takes the form R^{-Q} and the variance of the dissipation contained within a ball of radius R is R^{2-Q}.

M_h: The h-th moment of the dissipation contained within a ball of radius R is R^{M_h}. (In particular, $M_1=1$ and $M_2=2-Q$.)

Expressed in terms of B and Q alone, the main result[10] resides in the basic inequality

$$0 \leq 3B \leq Q$$

In the Kolmogorov-Yaglom model, 3B is much smaller than Q. However, in a different model which I proposed to call <u>absolute curdling</u> (in the case of turbulence, it was pioneered by Novikov & Stewart), 3B is identical to Q.

For the above inequality to become fully meaningful, one needs to add a further exponent D, which is nonclassical and can be interpreted as a similarity dimension.

D: Let a basic spatial cube of side L be paved by small cubes of side ζ, with $\zeta/L << 1$. Intermittent turbulence is characterized by an exponent D such that dissipation is either wholly or overwhelmingly concentrated in a very small set that can be covered by $(L/\zeta)^D$ of these small cubes. This new exponent D clearly satisfies $2<D\leq3$. Taking account of this D, the basic inequality becomes enriched to read

$$0 \leq 3B \leq 3-D \leq Q$$

To summarize these inequalities, and the classical probabilistic inequalities between moments, we reach the following result: there must exist a cup convex function f(h) passing through the following points:

$$(2/3,B),\ (1,0),\ (2,2-Q=M_1),\ \dots\ (h,M_h)$$

Equality of 3B, 3–D and Q corresponds to the case where the graph of this f(h) reduces to a straight line. A consequence is that all exponents can be deduced from any one of them. The exponent D should be viewed as first among equals, because it relates to the problem's basic geometric substrate. Geometrically speaking, a special and unique virtue of a straight f(h) is that it requires the dissipation to be uniform or <u>homogeneous</u> over a <u>closed</u> subset of space to which it is restricted. The carrier of such dissipation is akin to a cluster in Hammersley percolation in the sense that it has a well-defined boundary and is "all skin" with no significant "inside".

Every other f(h) implies that the bulk of dissipation is homogeneous over a small set, but some dissipation must be considered separately, thus involving two remainders: one remainder is spread around everywhere else and another remainder is concentrated in sharp peaks. These terms are unexpectedly significant, in the sense that different moments of the dissipation turn out to be very much affected by one or by the other. For example, if the graph of f(h) is not straight, its next simplest form is a parabola. This f(h) happens to be implicit in the lognormal <u>Ansatz</u> and indeed the approximations that underlie the <u>Ansatz</u> turn out to be equivalent to curve fitting of the above f(h) by a parabola. The corresponding values of B and Q are inordinately affected, respectively, by the first and by the second of the remainder measures mentioned above. It is difficult to avoid postulating that this influence is of limited physical significance, and that the basic physics is mostly dependent upon D.

It may be observed that the original derivation of the above inequalities[10] begins with a specific model I call <u>curdling</u>. In this model, a cup convex function f(h) is part of the basic assumptions. A linear f(h) corresponds to <u>absolute curdling</u>. However, the preceding argument suggests that the basic distinction between closed and open carriers of dissipation can be established without invoking curdling; it is natural therefore to conjecture that the validity of the basic irregularities is not dependent on any specific model.

As a useful restatement of absolute curdling, Frisch et al.[11] devised the "β-model". It is more specific on some points and less demanding on others, but yields identical results. See also Siggia[12].

Universality. The D of turbulent dissipation may or may not be universal. The question is open and deserves attention.

Some other aspects of turbulence involving fractal dimensions also raise the issue of universality. For example, the boundary of a domain resulting from turbulent dispersion is a fractal (see my contribution to a recent seminar[13]); is its D a universal exponent?

7. FRACTALS AND "STRANGE" ATTRACTORS

The geometric shapes listed in Section 2 are, so to speak, drawn by Nature itself, but physics involves many other geometric shapes drawn by Man to illustrate Nature. Many seem to be fractal. Take turbulence as an example: quite independently of my work based upon fractals (which started in 1963), Cantor sets and other fractals are involved in the so-called strange attractors. Ancient investigations in dynamics (going back to Poincaré) having been revived, this field has become very active[13]. Using data from Ibanez and Pomeau[14], Velarde (private communication) estimates the fractal dimension of a Lorenz attractor to be 2.09.

8. ASIDE

The principle underlying the Kadanoff scaling argument has been used (quite independently, and in fact somewhat earlier) in my own work on speculative price changes. Their statistical distribution turns out to be "stable" in the sense of Paul Lévy; stable processes are now finding their way into physics. The topic is summarized in Fractals[1] – the bulk of which is devoted to applications of fractal sets in astronomy and in diverse contexts.

9. DEFINITIONS OF HAUSDORFF MEASURE AND DIMENSION

The definition of similarity dimension, as given in Section 3, may seem uncomfortably informal. Therefore it is good to put on record the underlying rigorous notion of fractal dimension.

The first step is Hausdorff measure. Consider a metric space and a positive gauge function $h(\rho)$ that tends to 0 with ρ. Given a set S and a covering of S by balls of radii ρ_m, one defines the approximate measure as $\Sigma h(\rho_m)$. Next, for every $\rho>0$, one defines the best approximate measure as being $\inf \Sigma h(\rho_m)$, the infimum being carried over all the coverings that satisfy $\sup \rho_m \leq \rho$. Definition. The limit best approximate $\lim_{\rho\downarrow 0} \inf \Sigma h(\rho_m)$ is called the Hausdorff measure of S for the gauge $h(\rho)$.

The next step is to focus attention on gauge functions of the form $h(\rho) = \rho^d$. Theorem: For every set S, there exists a value of d, call it D, such that when $d < D$, the Hausdorff measure of S for the gauge ρ^d is infinite and when $d > D$, this measure is zero. Definition. The above D is called the Hausdorff Besicovitch dimension, or fractal dimension.

10. REFERENCES

1. Mandelbrot, B.B. Fractals: Form, Chance, and Dimension. San Francisco & London: W.H. Freeman. 1977. Also, Les objets fractals: forme, hasard et dimension. Paris & Montreal: Flammarion. 1975.
2. Sinai, Ja.G. Statistical Physics (IUPAP, 1975, Budapest), 139-149. Ed. L. Pál & P. Szépfalusy.
3. Fisher, M.E. Physics **3,** 1967, 255-283.
4. Fisher, M.E. & Nelson, D.R. Ann. Phys. (NY) **91,** 1975, 226-275.
5. Dhar, D. J. Math. Phys. **18,** 1977, 577.
6. Leath, P.L. Phys. Rev. **B 14,** 1976, 5046-5055.
7. Stanley, H.E., Birgeneau, R.J., Reynolds, P.J. & Nicoll, J.F. J. Phys. C (Solid St. Phys.) **9,** 1976, L553-L560.
8. Dunn, A.G., Essam, J.W., & Ritchie, D.S. J. Phys. C (Solid St. Phys.) **8,** 1975, 4219-4235.
9. Suzuki, M. Prog. Theor. Phys. Japan **51,** 1974, 1992-1993.
10. Mandelbrot, B.B. Turbulence and Navier Stokes Equations, R. Temam, ed. New York: Springer. 1976.
11. Frisch, U., Sulem, P.L., & Nelkin, M. Submitted to J. Fl. Mech.
12. Siggia, E.D. Phys. Rev. **A15,** 1977, 1730-1750. Also, "A Model of Intermittency in Three Dimensional Turbulence" (preprint, July 1977).
13. Chorin, A., Marsden, J. & Smale, S. (eds.) Seminar in Turbulence: Berkeley 1976. New York: Springer.
14. Ibanez, J.L. & Pomeau, Y. Submitted to J. Non-equilibrium Thermodynamics.
15. Kirkpatrick, S. Electrical Transport and Optical Properties of Inhomogeneous Media, D.B. Tanner, ed. A.I.P. Conference Proceedings (Ohio State U., September 1977.)

CHAPTER 19

UNIVERSALITY AND SCALING IN FULLY-DEVELOPED TURBULENCE

Mark Nelkin

Cornell University, Ithaca, N.Y. 14853, U. S. A.

ABSTRACT

Turbulent fluid flow exhibits many simplifying features in the limit of high Reynolds number. The large scale structure of the flow is independent of Reynolds number, but is strongly dependent on the boundary conditions. The small scale fluctuations are isotropic and universal in that they depend on the large scale flow only through a small number of parameters. Scale similarity is manifest through the power law behavior of equal time correlation functions, and through the power law dependence on viscosity of averages at a single spatial point. A variety of scaling exponents can be defined, and several of these have been reasonably well measured. In this paper phenomenological scaling arguments relating measurable exponents are reviewed and critically analyzed. The 1941 Kolmogorov theory is discussed as a mean field theory obtained by neglecting fluctuations in the energy cascade. Attempts to calculate these fluctuations from the Navier-Stokes equation are briefly described, and the role of spatial dimensionality is discussed.

INTRODUCTION

Turbulent fluid flow has by now become a regular feature of these statistical physics conferences. At Statphys 10 in Chicago, Bob Kraichnan[1] gave an excellent review of the statistical theory of turbulence. I strongly recommend that you read the introduction to that paper before plunging into the details of the present more specialized review. At Statphys 12 in Budapest, Paul Martin[2] discussed the exciting recent developments relating to the transition from laminar to turbulent flow.

Here at Statphys 13, I want to discuss the opposite limit in which the Reynolds number becomes very large, and turbulent fluctuations are excited over a very wide range of length scales.

Fully developed turbulence is still an unsolved theoretical problem. There are a bewildering variety of observed phenomena which are in principle contained within the Navier-Stokes equations of hydrodynamics. There is no reason to doubt the validity or applicability of the underlying equations, but we have just not been clever enough to solve them in the domain of greatest physical interest. What understanding we have is based on phenomenological theory strongly guided by experiment. There is no lack of experimental information, but much of it is directed more at applications than at fundamental understanding, and the level of precision is not very high. Despite these limitations it seems to me that there is a valid basic physical picture, and that there are challenging theoretical problems posed by the data. I will try to indicate to you what some of these problems are, and tell you a little about the very preliminary attempts at their solution. Perhaps at Statphys N ($N \geq 14$) someone will report more systematic theoretical progress, but for now I believe that the most important thing is to critically analyze what the observed phenomena suggest about the underlying physics.

We restrict our attention to an incompressible fluid, and neglect all thermal effects. The governing equation for the velocity field is the Navier-Stokes equation

$$\frac{\partial \vec{v}}{\partial t} + (\vec{v}\cdot\vec{\nabla})\vec{v} = -\frac{1}{\rho}\vec{\nabla}p + \nu\nabla^2\vec{v} \qquad (1)$$

Taking the divergence of Eq. (1), and using the incompressibility condition

$$\vec{\nabla}\cdot\vec{v} = 0 \qquad (2)$$

the pressure field can be eliminated by solving the Poisson equation

$$\nabla^2 p = -\rho\,\vec{\nabla}\cdot(\vec{v}\cdot\vec{\nabla})\vec{v} \qquad (3)$$

The spatially non-local quadratic nonlinearity expressed by substituting the formal solution of Eq. (3) back into Eq. (1) presents one of the essential difficulties of the problem.

Suppose we have a fixed flow geometry characterized by a largest length scale L, and a characteristic velocity difference U across a distance of the order of L. A natural dimensionless parameter for the problem is the Reynolds number $R=UL/\nu$, where ν is the kinematic viscosity of the fluid. Consider what happens as R increases. For convenience take U and L fixed, and let the viscosity of the fluid decrease. At first for small R the flow is laminar. The nonlinear terms in Eq. (1) are unimportant, and the problem is well understood. For moderate R there may be a complicated sequence of instabilities leading eventually to a turbulent flow in which velocity correlations decay to zero in the limit of long times. This transition to turbulence is different in different geometries. In certain simple geometries it has recently become reasonably well understood, as has been reviewed by Paul Martin in Budapest. More recent work supports the picture he described there, but that story is for another speaker and another day.

As the Reynolds number increases further the flow becomes fully turbulent in a very complicated manner, exhibiting for example reasonably well defined large scale structures. The overall large scale properties of the flow are of great practical interest, but are understood in only a rough semi-empirical way. As the Reynolds number becomes very large, the large scale properties of the flow become nearly independent of Reynolds number, but still depend greatly on the flow geometry. Their understanding is of both practical and basic interest, but this remains too hard a problem for even an optimistic theoretical physcist.

Consider a spatial Fourier analysis of the flow in terms of wave numbers k, or scales of motion $r \sim 2\pi/k$. The viscosity is a singular perturbation which will be important at small enough scales no matter how small its numerical value. In the limit of very large Reynolds number (very small viscosity) the nonlinear terms in the equation of motion will excite a very broad range of scales of motion, limited at the upper end by the geometrically defined length scale L, and at the lower end by a dynamically determined cutoff η below which viscous dissipation takes over. A plausible assumption is that for scales of motion $r \gg \eta$, viscosity has no effect. If we look only at scales of motion large compared to η, the fluid could just as well have zero viscosity. In particular the rate of energy dissipation per unit mass will be determined entirely by the large scale properties of the flow and be independent of viscosity in the limit of zero viscosity. This assumption, which is generally made but unproved, lies at the heart of our physical picture of high Reynolds number turbulence.

UNIVERSALITY AND SCALING - THE ENERGY SPECTRUM

If we examine the Fourier transformed Navier-Stokes equations there surely is a direct coupling between wave numbers $k \sim L^{-1}$ and wave numbers $k \sim \eta^{-1}$. Why then is it reasonable to neglect the direct effects of the smallest scales back on the large scale motions? The reason is that these effects are not expected to be dynamically important. The large scale motions will primarily convect the small scales without distortion. The effective dynamical interactions which transfer energy among scales of motion should be between comparable scales. This locality in k-space of the important dynamical couplings is a central idea, again not proven, but supported by a great deal of approximate theory and experiment. It goes back to Richardson's famous 1922 poem, of which I will spare you one more repetition.

If the dynamical couplings are local we might expect a certain degree of universality in the small scales. Details of the large scale motions will be lost as the energy is transferred to smaller scales. In addition dynamical theory suggests that the pressure forces will lead towards local isotropy of the small scale motions. Thus the small scale motions might be expected to exhibit statistical properties which are independent of the geometry of the large scale flow. Furthermore an estimate of relevant time scales indicates that characteristic times decrease rapidly as scale size decreases. Thus the small scales of motion should be in a statistically steady state, and the important statistical information should be contained in equal time correlation functions. For a review of the underlying physics see Batchelor[3] or Orszag[4].

Since there is no internally determined length scale intermediate between the external length scale L and the viscous dissipation scale η, it is natural to expect that correlation functions for scales of motion satisfying

$$\eta << r << L \quad , \qquad (4)$$

should be of power law form, and should define universal scaling exponents. The range defined in Eq. (4) is called the inertial subrange.

The simplest equal time correlation function is the velocity correlation

$$\langle v_1(\vec{x}) \; v_1(\vec{x} + \vec{r})\rangle \quad ,$$

or its Fourier transform the energy spectrum E(k). Defining E(k) dk as the turbulent kinetic energy per unit mass between k and k + dk, and introducing ε, the rate of energy dissipation per unit mass, the universal scaling form for the energy spectrum in the region $kL>>1$ is expected to be

$$E(k,\nu) = \varepsilon^{2/3} \; k^{-5/3} \; (kL)^{-B} \; f(k\eta) \quad . \qquad (5)$$

The dependence on the viscosity ν is entirely in the dissipation length scale η. The cutoff function f(x) is expected to have a finite limit for small x and to show a sharp cutoff for $x \geq 1$. Approximate dynamical theories indicate that the cutoff is probably exponential.

The parameter ε plays a central dynamical role. It is the rate of energy transfer through the inertial subrange, and it is also the rate at which energy is dissipated by viscosity. The former role is the more fundamental, but to calculate ε for a given flow is extremely difficult. The energy dissipation rate is, however, directly measurable. In terms of energy spectrum E(k) it is given by

$$\varepsilon = 2\nu \int_0^\infty E(k,\nu)\, k^2 \, dk = \nu \langle\omega^2\rangle \quad , \qquad (6)$$

where $\langle\omega^2\rangle$ is the mean square vorticity. In order for ε to be constant in the limit of zero viscosity, $\langle\omega^2\rangle$ must diverge as ν^{-1}.

Substituting Eq. (5) into Eq. (6) allows us to calculate the dependence of the dissipation length scale η on viscosity. The result is

$$\eta = \text{const.} \quad \varepsilon^{-x/3} \quad \nu^{x} \quad L^{-Bx} \tag{7}$$

where

$$x = (\frac{4}{3} - B)^{-1} \tag{8}$$

The original 1941 Kolmogorov theory is recovered with the additional assumption that the external length scale L is dynamically irrelevant. This corresponds to B=0 in Eqs. (5), (7) and (8), and leads to the familiar 5/3 law for the inertial range energy spectrum, and to the Kolmogorov expression

$$\eta_0 = \text{const.} \quad \varepsilon^{-1/4} \quad \nu^{3/4} \tag{9}$$

for the dissipation length scale. With this additional assumption the scaling form of Eq. (5) and the values of the scaling exponents follow from dimensional analysis. If we allow a dependence on L there is no guarantee that the scaling form of Eq. (5) is preserved, and even if it is the exponent B is not determined by dimensional arguments. Our point of view is that the universal scaling form of Eq. (5) is more general than the 1941 Kolmogorov theory from which it first came. To someone familiar with modern statistical physics this is a very natural assumption, but there is no theoretical support for its validity.

Having written down Eq. (5) there are many questions we can ask: What is the experimental evidence for the scaling form of Eq. (5)? What is the numerical value of the exponent B expressing deviations from the Kolmogorov 5/3 law? What other universal scaling exponents can be defined by other correlation functions? Can we write down a reasonable phenomenological scaling theory relating measurable exponents to each other? Is an experimental test of such a scaling theory feasible? Is there a well defined approximate "mean field theory" for the problem? Can we calculate the universal exponents from the Navier-Stokes equation? How do exponents depend on spatial dimensionality? These are the questions we will address in the remainder of this paper. We can give reasonable answers to most of them, but it should be emphasized once more that nothing is solidly established.

There are many complicating factors that I have chosen not to discuss. For example the $k^{-5/3}$ energy spectrum is seen for the component of the velocity difference along the mean wind speed even for wave numbers small enough that the fluctuations are not statistically isotropic. Also almost all the measurements are at one spatial point as a function of time, and use Taylor's frozen turbulence assumption to convert measured time correlations into spatial correlations. This assumes that the large scales convect the small scales past the point of observation without distortion. This is consistent with our scaling picture, but should be separately evaluated. Despite these complications there seems little doubt that the basic Kolmogorov scaling picture is a valid first approximation to the physics.

FLUCTUATION CORRECTIONS TO 1941 KOLMOGOROV

What does it mean physically to have the external length scale L appear in Eq. (5) as it will if the exponent $B \neq 0$? Assume that the energy cascade takes place in steps, each of which is a factor of about two in wave number. The quantity kL is then just a measure of the number of cascade steps needed to reach wave number k. If the energy spectrum E(k) depends on the number of cascade steps as well as on the amount of energy cascading this indicates that there is a buildup of fluctuations in the cascade process. (This leads to the qualitative idea of the 1941 Kolmogorov theory as a mean field theory in which fluctuations are neglected. This will be made explicit later in this paper. It is consistent with this interpretation that the 5/3 law deduced from the dimensional arguments of Kolmogorov is the same independent of the spatial dimensionality of the fluid.)

Given that the exponent B is probably too small to measure directly, how would we look for fluctuation effects experimentally? Compared to usual statistical mechanical systems we have much more freedom to study the statistical properties of the random field of interest. Experimentally we essentially have a tape recording of the random velocity field $v_1(\vec{x})$ (the 1-component being along the wind direction), and we can form any correlation function we want to study. The best choice would be that most directly related to fluctuations in the energy transfer process.

Historically it was realized in the early 1960's that fluctuation effects were important. This was first pointed out in two famous 1962 papers by Obukhov[6] and Kolmogorov[7]. Subsequent work focussed on the statistical properties of the local energy dissipation, and is reviewed in Chapter 25 of Monin and Yaglom[5]. More recently[8,9] it has been emphasized that a more appropriate emphasis would be on fluctuations in the energy transfer rate. The literature in this area is extremely confusing, and it has taken me about five years to sort it out. I think I now have the essential physics straight, and I will present it from my current point of view rather than historically.

Consider an interval of length r extending from the point $\vec{x}$ to the point $\vec{x} + \vec{r}$, and take as a basic random variable defined on this interval the velocity difference

$$\Delta v(r) = v_1(\vec{x} + \vec{r}) - v_1(\vec{x}) \tag{10}$$

A random variable with the dimensions of energy transfer defined on this interval is

$$\varepsilon_t(r) = \Delta v^3(r)/r \qquad . \qquad (11)$$

The moments of this random variable are simply related to the so-called longitudinal structure functions

$$D_n(r) = \langle[\Delta v(r)]^n\rangle \qquad , \qquad (12)$$

which are the subject of considerable attention in the turbulence literature[5]. In the inertial subrange $\eta \ll r \ll L$, it is reasonable to assume the power law form

$$D_n(r) \approx C_n \quad \varepsilon^{n/3} \quad r^{n/3} \quad (r/L)^{\zeta_n} \qquad . \qquad (13)$$

The physics of fluctuations in the cascade process is largely contained in the "anomalous" exponents ζ_n. In particular $D_2(r)$ is simply related by Fourier transformation to the energy spectrum E(k), and we can make the identification

$$\zeta_2 = B \qquad . \qquad (14)$$

In the 1941 Kolmogorov theory we have $\zeta_n = 0$ for all n.

Equation (11) suggests that we pay special attention to the cases n=3 and n=6. If we can interpret $D_3(r)/r$ as an average energy transfer we would expect it to be independent of r. This can be shown from the Navier-Stokes equation[10] so that our expectation that $\zeta_3 = 0$ is in fact correct. The quantity $D_6(r)/r^2$ is the most direct measure we can find of fluctuations in energy transfer. The phenomenological picture is that these fluctuations are spatially intermittent; that is they concentrate the activity on smaller and smaller portions of the available space as the scale of motion becomes smaller. The most direct scale similar expression of this tendency would be a negative value of the exponent ζ_6. We would of course like to calculate this exponent from the Navier-Stokes equation, but for now we consider the more modest objective of relating it to other measurable quantities. Here we can make contact with the historical development of the subject, and consider the fluctuations in the energy dissipation.

As a one dimensional surrogate for the local viscous dissipation we introduce the variable

$$\tilde{\varepsilon}(\vec{x}) = \nu\, \psi^2(\vec{x}) \qquad , \qquad (15)$$

where

$$\psi(\vec{x}) = \partial v_1 / \partial x_1 \qquad . \qquad (16)$$

Now consider the average of $\tilde{\varepsilon}(\vec{x})$ over a linear interval of length r

$$\varepsilon^*(r) = r^{-1} \int_0^r \tilde{\varepsilon}(\vec{x})\, dx_1 \qquad . \qquad (17)$$

Within a numerical constant the average of the random variable $\varepsilon^*(r)$ is just ε, and is independent of r. Since the 1962 suggestion of Kolmogorov[7], a great deal of attention has focussed on the statistical properties of $\varepsilon^*(r)$. Since the local dissipation is not of any great dynamical importance, this attention seems misplaced. It is worthwhile, however, to consider the simplest measure of dissipation fluctuations which is $\langle[\varepsilon^*(r)]^2\rangle$. Again scale similarity suggests that for $\eta \ll r \ll L$,

$$\langle[\varepsilon^*(r)]^2\rangle = \text{const. } \varepsilon^2 \quad (L/r)^\mu \quad , \tag{18}$$

with the exponent μ positive. For homogeneous turbulence Eq. (18) is equivalent to

$$\langle\tilde{\varepsilon}(\vec{x})\ \tilde{\varepsilon}(\vec{x}+\vec{r})\rangle = \text{const. } \varepsilon^2 \quad (L/r)^\mu \quad . \tag{19}$$

Equation (19) is the usual experimental definition of the exponent μ. It has been measured by several workers[5,11] and has a value of about 0.5. It is suggestive but far from proven by the experiments that this exponent is universal in the sense that it is the same for all high Reynolds number incompressible flows.

In 1962 Obukhov[6] suggested the relation

$$D_n(r) = C_n \ \langle\left[\varepsilon^*(r)\right]^{n/3}\rangle \quad r^{n/3} \quad , \tag{20}$$

for the inertial subrange. At first sight Eq. (20) seems to be no more than an ad hoc generalization of the 1941 Kolmogorov theory, but in the physical context constructed here it takes on considerably more plausibility. Equation (20) is in fact equivalent to the statement that the energy transfer variable $\varepsilon_t(r)$ defined by Eq. (11) has the same statistical properties as the energy dissipation variable $\varepsilon^*(r)$ defined in Eq. (17). This is just a statement that the viscous dissipation will reflect the spatial intermittency developed by the nonlinear energy transfer without modifying it.

Also in 1962 Kolmogorov[7] suggested that the statistics of the dissipation variable $\varepsilon^*(r)$ should be lognormal. When combined with Eq. (20) and Eq. (18) this allows the exponents ζ_n to be calculated with the result

$$\zeta_n^{(K)} = \mu\, n(3-n)/18 \tag{21}$$

We now know that the lognormality of $\varepsilon^*(r)$ is at best approximate, and that approximate lognormality is irrelevant to the calculation of the exponents ζ_n. Thus Eq. (21) has no particular merit. Equation (20) is probably correct, but it does not appreciably simplify things.

Historically, however, Kolmogorov's 1962 suggestion has led to considerable experimental work on the statistics of the dissipation variable $\varepsilon^*(r)$. In particular the exponent μ is reasonably well known. This does not help us with the exponents ζ_n for arbitrary n, but if Eq. (20) is valid, it does tell us that

$$\zeta_6 = -\mu \tag{22}$$

The one experiment[12] on $D_6(r)$ is consistent with Eq. (22), but the precision is not very great. We will assume Eq. (22) to be correct, noting that it does not depend on any assumptions about the statistics of the dissipation $\varepsilon^*(r)$ such as Kolmogorov's lognormality assumption.

ONE EXPONENT SCALING THEORY

We have introduced the exponents ζ_n and suggested that the first simple attempt to relate them is not correct. All we have so far is

$$\zeta_2 = B, \quad \zeta_3 = 0, \quad \zeta_6 = -\mu \quad .$$

The first of these statements is just renaming an exponent, the second follows from the Navier-Stokes equation, and the third is a physical assumption relating two measurable exponents. In order to simplify things further we must first complicate them by introducing another family of measurable exponents.

Consider the structure functions $D_n(r)$ for r small compared to the dissipation length scale η. It is reasonable to assume that the $D_n(r)$ are analytic in this regime so that

$$D_n(r) \simeq \langle\psi^n(\vec{x})\rangle \; r^n, \quad n\geq 2, \quad r\ll\eta \tag{23}$$

where $\psi(\vec{x})$ is defined by Eq. (16). The quantities $\langle\psi^n(\vec{x})\rangle$ will diverge as the viscosity goes to zero, and we will assume this divergence to be of power law form

$$\langle\psi^n(\vec{x})\rangle \sim \nu^{-y_n} \quad . \tag{24}$$

From the assumption that the dissipation ε is independent of ν for small ν we have $y_2=1$, but we have no a priori information about y_n for $n\geq 3$.

Of more direct physical interest are the skewness and kurtosis for the probability distribution of the velocity derivative $\psi(\vec{x})$. These are given by

$$S = -\langle\psi^3\rangle \,/\, \langle\psi^2\rangle^{3/2} \sim \nu^{-\theta} \quad , \tag{25}$$

and

$$F = \langle\psi^4\rangle \,/\, \langle\psi^2\rangle^2 \sim \nu^{-\alpha} \quad , \tag{26}$$

where

$$\theta = y_3 - 3/2 \quad , \tag{27}$$

and

$$\alpha = y_4 - 2 \quad . \tag{28}$$

The 1941 Kolmogorov theory gives $y_n = n/2$, and the skewness and kurtosis become constant in the limit of zero viscosity. Fluctuation corrections to the 1941 theory are manifest in an increasingly non-Gaussian probability distribution for $\psi(\vec{x})$ as the viscosity decreases. The most natural scale similar expression of this effect is in positive values for the exponents θ and α. Experiments are not sufficiently precise to give numerical values for these exponents, but the measured kurtosis increases by a factor of about 10 and the skewness by a factor of about 3 over the available range of data.

There is a length scale η_n at which the small r behavior of the structure function $D_n(r)$ given by Eq. (23) joins on to the inertial range behavior given by Eq. (13). The dissipation length scale η given by Eq. (7) is just η_2. We now make the strong but plausible additional assumption that there is only a single dissipation length. More precisely we assume that the η_n for all n have the same power law dependence on viscosity. This gives the exponent relation

$$y_n = \left(\frac{2n}{3} - \zeta_n \right) x \quad , \tag{29}$$

where x is given by Eq. (8) and we recall that $\zeta_2 = B$ by definition. Thus the exponents y_n and ζ_n are not independent of each other, but we still have no relation among the ζ_n for different values of n.

We can easily get some exponent inequalities by using the Schwarz inequalities

$$\langle \psi^3 \rangle^2 \le \langle \psi^2 \rangle \langle \psi^4 \rangle$$

$$\langle \psi^4 \rangle^2 \le \langle \psi^2 \rangle \langle \psi^6 \rangle \quad . \tag{31}$$

When combined with Eq. (29) these give

$$\zeta_2 + \zeta_4 \le 0 \quad , \tag{32}$$

and

$$\zeta_2 + \zeta_6 \le 2 \zeta_4 \quad . \tag{33}$$

Now we use the assumption of a single dissipation length scale once more in a slightly different way. Consider

$$\nu^2 \langle \psi^4(\vec{x}) \rangle = \lim_{r \to 0} \langle \tilde{\varepsilon}(\vec{x}) \; \tilde{\varepsilon}(\vec{x} + \vec{r}) \rangle \quad , \tag{34}$$

and assume that this limit can be calculated by extrapolating the inertial range form of Eq. (19) back to $r=\eta$. This gives us an additional exponent relation

$$y_4 = 2 + \mu x \tag{35}$$

If we combine Eq. (29) for n=4 with Eq. (35) we have

$$2\zeta_2 - \zeta_4 = \mu \quad . \tag{36}$$

If we now use the assumption that $\zeta_6 = -\mu$, we find that Eq. (36) and the inequalities of Eqs. (32) and (33) are only consistent for

$$B \equiv \zeta_2 = -\zeta_4 = \mu/3 \quad . \tag{37}$$

The argument can be extended to higher n and to the behavior of other correlation functions. The details are given elsewhere[13]. The results are

$$y_n = \frac{n}{2} + \frac{3\mu}{4-\mu}\left(\frac{n}{2} - 1\right) \quad , \tag{38}$$

and

$$\zeta_n = (3-n)\,\mu/3 \tag{39}$$

The exponent relations of Eq. (37-39) express all the measurable exponents in terms of the dissipation exponent μ. In particular they suggest that the correction B to the 5/3 law is $\mu/3$ rather that the 1962 Kolmogorov value of $\mu/9$ (see Eq. 21). Since the experimental value of μ is about 0.5 this suggests a value of $B \approx 0.15$. This is large enough that a modest improvement in experimental accuracy could significantly test this prediction.

We have derived one exponent scaling essentially from the assumption that there is a single well defined dissipation length scale. This is a simple and plausible assumption, and gives much stronger results than one would have expected. Just because these results are so strong, however, we should critically analyze the underlying assumption. There is certainly no a priori dynamical reason why the relations among scaling exponents should be as simple as derived here.

Although the derivation of one exponent scaling given here is new, the results are not. They have been obtained earlier in a variety of ways, most of which seemed much more model dependent than what we have presented. Similar results were first obtained by Novikov and Stewart in 1964 (see chapter 25 of ref. 5 for discussion). A description of intermittency in terms of statistical geometry has been given by Mandelbrot[14], as he will describe for you at this conference. The scaling relations given here correspond to a special case that he calls "fractally homogeneous turbulence".

He also argues that this special case is a more plausible candidate for the correct physics than it might first seem. In a recent paper Frisch, Sulem, and Nelkin[15] have tried to put Mandelbrot's ideas into a simpler physical cascade model, and have explicitly obtained the exponent relations of Eqs. (37-39). From a different point of view, working explicitly by analogy to equilibrium critical phenomena, Nelkin[16] obtained some of the results given here. It turns out with further work that this analogy is essentially identical to the one exponent scaling presented here.

This further work is described in a recent preprint by Nelkin and Bell[13]. This work corrects and extends ref. 16. It overlaps substantially with the discussion presented here, but it has a more abstract formulation which constitutes a kind of assumed operator algebra for turbulence. Consider the dynamical variable

$$A_n(\vec{x}) = \nu^{y_n} \psi^n(\vec{x}), \quad n\geq 2 \quad . \tag{40}$$

This variable is "regularized" in the sense that

$$\lim_{\nu\to 0} \langle A_n(\vec{x})\rangle = \text{const.} \tag{41}$$

A natural generalization of our scaling results is that

$$\lim_{\nu\to 0} \langle A_p(\vec{x}) \; A_q(\vec{x} + \vec{r})\rangle \sim (L/r)^{\mu} \quad . \tag{42}$$

In other words all of the $A_n(\vec{x})$ for $n\geq 2$ are regularized scaling operators with scaling dimension $\mu/2$. In particular the quantity

$$T(\vec{x}) = \eta^2 \; \psi^3(\vec{x}) \quad , \tag{43}$$

which is proportional to $A_3(\vec{x})$, has the same statistical properties as the local dissipation $\tilde{\varepsilon}(\vec{x})$. We can think of $T(\vec{x})$ as a locally defined energy transfer variable which plays the role of local dissipation in the limit of zero viscosity. This statement may be right or wrong, and may be trivial or deep. It is too early to tell. Fortunately an experimental test of Eq. (42) seems feasible.

DYNAMICS OF THE ENERGY CASCADE

Phenomenological nonlinear equations for the time evolution of the energy spectrum E(k,t) have reached a very sophisticated state of development. These are reviewed by Kraichnan[1], and Orszag[3] and more recently by Rose and Sulem[17]. Among the phenomena which have been studied are the approach of E(k,t) to the scaling form of Eq. (5), the change in direction of the energy cascade as dimensionality is varied continuously between d=2 and d=3, and the special phenomena associated with two dimensional turbulence. We will return to the effects of spatial dimensionality, but for now we consider a very simple model of the energy cascade, first introduced by Desnyansky and Novikov[18] and further developed by Bell and Nelkin[19,20].

In this model a discrete set of variables $u_n(t)$, labelled by integers n, represents the energy spectrum. They are defined so that

$$\frac{1}{2}\, u_n^{\,2}(t) = \int_{k_n/\sqrt{2}}^{\sqrt{2}k_n} E(k,t)\, dk \quad , \qquad (44)$$

is the energy contained in a wave number band centered around the logarithmically spaced wave numbers

$$k_n = 2^n\, L^{-1} \quad . \qquad (45)$$

Collapsing all the spectral information in a wave number band into a single dynamical variable is of course very crude, but it still allows a non-trivial nonlinear evolution of the (discretized) energy spectrum. As with the more sophisticated spectral models there are no spatial degrees of freedom built in to the description so that intermittent fluctuations in the cascade process are left out from the beginning. Thus both the sophisticated spectral models and the crude discretized models can be thought of as a kind of mean field theory for the turbulent cascade.

Desnyansky and Novikov proposed nonlinear equations of motion for the variables $u_n(t)$ subject to the constraints that the equations have only quadratically nonlinear terms, introduce no intrinsic length scales via the coupling coefficients, couple nearest neighbor shells only and conserve the total energy of the system in the absence of viscous dissipation or an external force. The most general set of equations satisfying these conditions is

$$\frac{du_n}{dt} = \alpha\, k_n \left[u_{n-1}^{\,2} - 2\, u_n\, u_{n+1} - 2^{1/3}\, C\,(u_{n-1}\, u_n - 2\, u_{n+1}^{\,2}) \right] - \nu\, k_n^{\,2}\, u_n + f_n(t) \qquad (46)$$

The last two terms in Eq. (46) represent the effects of viscosity and an externally imposed driving force. The parameters α and C are not fixed by the constraints.

Considering its simplicity the model behaves in a remarkably sensible way, and contains a surprising degree of richness in its solutions. Explicit time dependent scaling solutions have been found which combine an analytic scaling form with a simple numerical calculation of the scaling function. Details are given elsewhere[20].

The case of greatest physical interest is freely evolving three dimensional turbulence. This seems well described by Eq. (46) with $0<C<1$ and no external forcing. After an initial transient the solutions approach the scaling form of Eq. (5) with $B=0$. The time dependence for large k is accounted for by a power law decay of the total kinetic energy

$$E(t) \sim t^{-w} \quad , \quad \varepsilon(t) \sim t^{-1-w}$$

The exponent w and the detailed shape of the scaling function $f(x)$ in Eq. (5) depend on the parameter C, but the inertial range spectral exponent does not. The parameter C can be fitted to the experimental decay of turbulence behind a grid in a wind tunnel and falls in the range $0.3 \leq C \leq 0.6$. For values of $C>1$ the model exhibits a change in direction of the energy cascade to which we will return later.

DYNAMICS OF FLUCTUATIONS IN THE CASCADE PROCESS

We have considered an energy cascade in which scale size changes by a factor of the order of two in each step. We have also discussed spatial intermittency associated with the non-uniform distribution of energy among "eddies" of size ℓ arising from the disintegration of "eddies" of size 2ℓ. For a quantitative dynamical description the phenomenology suggests a representation which contains elements of spatial locality and at the same time elements of a Fourier decomposition. Some kind of wave packets or phase space cells of the type used by Wilson[21] in critical phenomena seem natural for the problem. There is no well defined orthonormal set of such phase space cells available for a systematic expansion of the velocity field, but there are some semi-quantitative considerations which can shed some light on the physics.

Consider a set of wave number bands as in Eq. (45), but now take into account that the n'th wave number band will contain $\sim 2^{nd}$ modes each of which can be localized in a spatial box of linear dimension 2^{-n} L. Thus the velocity field is expanded in the form

$$v_\alpha(\vec{x},t) = \sum_{n,m} A^\alpha_{n,m}(t) \; u_{n,m}(\vec{x}) \quad ,$$

where n indicates the wave number band, and m indicates the spatial box. For now we neglect all but the crudest features. Complications associated with the incompressibility condition, number of modes per box, orthogonality and completeness are not considered, and in many cases have not been resolved.

The $(\vec{v}\cdot\vec{\nabla})\vec{v}$ terms in the Navier-Stokes equation will couple primarily nearest neighbor wave number bands if our ideas of a local cascade and scaling are correct. They will also be spatially local so that they will typically couple a box of size $2^{-n}L$ in the n'th wave number band to the 2^d boxes in the (n+1) wave number band which are imbedded in it, and to the box in the (n-1) wave number band in which it is imbedded. The pressure forces in the Navier-Stokes equation do not contribute directly to the energy cascade, but they tend to isotropize the velocity field, and they are spatially non-local. The predominant couplings due to these forces will be among boxes in the same wave number band.

Qualitatively we expect the $(\vec{v}\cdot\vec{\nabla})\vec{v}$ terms to contribute to the buildup of intermittency, and the pressure forces to resist this buildup. This shows up in its most dramatic form in the one dimensional Burgers equation. Here the pressure forces are missing and the $v(\partial v/\partial x)$ term leads to the buildup of shock fronts which are an extreme form of spatial intermittency. In the 1941 Kolmogorov theory the pressure forces do not explicitly appear. Only the rate of energy cascade ε due to the $(\vec{v}\cdot\vec{\nabla})\vec{v}$ forces is considered. If the 1941 Kolmogorov theory is approximate, however, it is not because it neglects the effects of pressure forces, but because it overestimates them. If the spatial couplings among boxes in the same wave number band are strong enough then an averaging procedure in which only the total kinetic energy in a wave number band is dynamically relevant becomes reasonable. With such an averaging procedure the "mean field" dynamical equations could plausibly take the form of Eq. (46).

Recently there has been a very interesting attempt by Siggia[22] to take this phase space cell expansion more seriously, estimate the important dynamical couplings, and calculate the buildup of intermittency. In order to make the calculation computationally feasible he had to throw away most of the spatial couplings, but he did make a serious effort at the difficult problem of quantitatively estimating the coupling coefficients. The resulting set of coupled differential equations does show an approximately scale similar buildup of fluctuations in time which Siggia associates with the observed spatial fluctuations in a turbulent cascade. The quantity associated with the exponent μ comes out to be about 0.8 with considerable numerical uncertainty.

The couplings between wave number bands in Siggia's equations are similar to those in Eq. (46), but his dynamical variables, being spatially localized, are not to be directly compared with the u_n in Eq. (46). He also has terms in u_{n-1}^2 and in $u_{n-1}u_n$. In the spatially local picture the former term has no simple physical interpretation, but the latter is naturally associated with the shear of motions of scale

$2^{-n}L$ by larger scale motions. The physics of vortex stretching suggests that this term is dominantly responsible for the cascade of energy to smaller scales. This is to be contrasted with Eq. (46) where the u_n^2 represents the turbulent kinetic energy in the n'th wave number band integrated over all space. Phenomenological considerations there suggested that the u_{n-1}^2 terms dominated the energy cascade, and the term in $u_n u_{n-1}$ contributed to an inverse transfer of energy to larger scales. This reversal of sign in the couplings when going from a spatially local to a global description is surprising and is not understood.

There is some progress in the qualitative physics of the dynamics of intermittency, but a reliable and well understood calculation for three dimensional turbulence is still a challenge for the future. In the spirit of modern statistical physics we might hope to find an easier problem by allowing the spatial dimensionality of the fluid to vary.

VARIABLE SPATIAL DIMENSIONALITY

In two dimensions the vorticity is perpendicular to the velocity. This eliminates the possibility of vortex stretching by the velocity field, which is the dominant dynamical mechanism for three dimensional turbulence. Turbulent flows in two dimensions of a somewhat different character do, however, exist. The most important new feature is that the nonlinear terms now conserve the mean square vorticity (known as enstrophy) as well as the energy. Since the enstrophy can not grow in time a cascade of energy to large wavenumbers is not possible. What does happen in a randomly forced two dimensional fluid was suggested by Kraichnan[23] in a seminal paper ten years ago. If the fluid is forced at wave number k_0, there will be a cascade of enstrophy to large k terminated by viscosity, and a cascade of energy to small k, terminated by the length of time the fluid has been stirred. Dimensional arguments of the Kolmogorov type suggest that the enstrophy cascade will give a k^{-3} energy spectrum at large k, and the energy cascade will give a $k^{-5/3}$ energy spectrum at small k. Numerical solutions of phenomenological integral equations for the energy spectrum E(k,t) support these ideas[24], and there is also limited support from direct numerical simulation of the two dimensional Navier-Stokes equations[25].

The question of intermittency corrections in two dimensions has received limited study. Kraichnan[26] has argued that these corrections, although present, will not change the k^{-3} spectrum. The $k^{-5/3}$ spectrum for the energy cascade to small k could, however, be changed, Phenomonological arguments similar to those in three dimensions suggest that in two dimensions any such change would make the spectral exponent smaller than 5/3, in contrast to what is expected in three dimensions. Whether such corrections to the 5/3 law in two dimensions exist is an open question on which there has been very little theoretical work. It is a somewhat artificial problem, but it is well defined and probably not quite as difficult as the corresponding question in three dimensions. Direct numerical simulation can also be useful for this problem which is not feasible in three dimensions.

For a statistical physicist it is natural to ask what happens as the dimensionality is continuously varied between two and three. (The fluid mechanics community still finds such questions quite bizarre.) Within the "mean field theory" of phenomenological integral equations for E(k,t) this question has been answered recently by Frisch, Lesieur and Sulem[27]. The coefficients in the integral equation for the eddy damped quasi-normal Markovian (EDQNM) approximation are easily continued analytically for continuous dimensionality d. For each d the equation is solved numerically. For any $d>2$ energy is the only exactly conserved quantity, but for d close to 2 enstrophy is approximately conserved. For $2<d<d_c$, with $d_c \approx 2.05$, energy cascades towards small k. For $d>d_c$, energy cascades towards large k. For all $d>2$ the energy cascade has a $k^{-5/3}$ energy spectrum. For $2<d<d_c$ there is also a k^{-m} regime for large k, but this scale similar regime is not associated with the cascade of any conserved quantity. The exponent m starts at 3 for $d=2$ and decreases with increasing d.

Similar results to those found as a function of dimensionality in the EDQNM were found by Bell and Nelkin[19] for the cascade model of Eq. (46) as a function of the parameter C in that equation. For $C>1$ the energy cascade is to small k and has an associated $k^{-5/3}$ spectrum. There is a k^{-m} regime for large k, where the exponent m depends on the parameter C, and there is no conserved quantity cascading through this k^{-m} regime. For $0<C<1$ the energy cascade is to large k, and is associated with a $k^{-5/3}$ spectrum. Comparison with the EDQNM as well as other phenomenological considerations suggest that $C>1$ for two dimensions, and that C decreases monotonically with increasing dimensionality, crossing $C=1$ for $d=d_c \approx 2.05$ and probably approaching zero in the limit of infinite dimensionality.

The effects of variable dimensionality within "mean field theory" are thus reasonably well understood. A more interesting question is the effect of variable dimensionality on the fluctuation corrections to the 1941 Kolmogorov theory. It was suggested[16,28], using a kind of Ginsburg argument on "dangerous" fluctuations, that fluctuation corrections might vanish for $d<8/3$. This argument was based, however, on the intrinsic dynamical importance of dissipation fluctuations. It was pointed out by Frisch, Lesieur and Sulem[27] that using a dynamically more relevant energy transfer variable would completely change the relevant crossover dimension. There is in fact no evidence that fluctuation corrections to scaling exponents vanish in any finite dimensionality.

There has been interest recently in the limit of infinite dimensionality. Nothing has been published, but there is some discussion in a paper for this conference[29].

It has been conjectured that intermittency corrections to the 5/3 exponent vanish in the limit of infinite dimensionality, but the validity of this conjecture is completely open. In fact we do not even understand the physics well enough to be sure whether corrections to the 5/3 exponent increase or decrease with increasing dimensionality.

Recent interest in scaling exponents for turbulence has been largely motivated by the hope that these exponents could be calculated by renormalization group techniques. For certain infra-red problems in a randomly stirred fluid there has in fact been one successful application of these techniques[30]. So far, however, the application of these methods in the non-perturbative situation characterizing an energy cascade has not yet been possible. We understand the phenomenology better, and have every reason to believe that the renormalization group is applicable to the problem, but we are still far from having developed the technical theoretical tools to handle the problems of greatest physical interest.

ACKNOWLEDGEMENTS

This work was supported by the U.S. National Science Foundation under Grant Number ENG 74-08501 A02. During the past three years most of my work on turbulence has been in close collaboration with Thomas L. Bell. His contributions to the ideas presented here have been substantial. During the recent Aspen workshop on turbulence I had extensive discussions with Uriel Frisch, Bob Kraichnan, Paul Martin, David Nelson and Eric Siggia. Their ideas and recent published work have affected what I have presented here, but it should not be assumed that they agree with the conclusions that I have reached.

REFERENCES

1. Robert H. Kraichnan, in "Statistical Mechanics, New Concepts, New Problems, New Applications", University of Chicago Press, Chicago, 1972 p. 201-227.
2. P.C. Martin, in "Statistical Physics, Proceedings of the International Conference, Akademiai Kiado, Budapest, 1976, p. 69-96.
3. G.K. Batchelor, "Theory of Homogeneous Turbulence", Cambridge University Press, 1953.
4. S. Orszag, "Les Houches Summer School of Theoretical Physics, 1973", Gordon and Breach, 1976.
5. A.S. Monin and A.M. Yaglom, "Statistical Fluid Mechanics" Vol 2, English edition updated by the authors, edited by J. Lumley, MIT Press, 1975.
6. A.M. Obukhov, J. Fluid Mech. 13, 77 (1962).
7. A.N. Kolmogorov, J. Fluid Mech. 13, 82 (1962).
8. R.H. Kraichnan, J. Fluid Mech. 62, 305 (1974).
9. M. Nelkin, Phys. Rev. A 9, 388 (1974).
10. L.D. Landau and E.M. Lifshitz, "Fluid Mechanics" Addison Wesley (1959), Eq. (33.19).
11. C.H. Gibson, G.R. Stegen and S. McConnell, Phys. Fluids 13, 2488 (1970) see also chapter 25 of ref. 5.
12. C.W. Van Atta and J. Park, in "Statistical Models and Turbulence" Springer Verlag, 1972, p. 402-426
13. M. Nelkin and T.L. Bell, "One Exponent Scaling for Very High Reynolds Number Turbulence", preprint, 1977.

14. B. Mandelbrot, "Fractals: Form, Chance and Dimension" W.H. Freeman, 1977.
15. U. Frisch, P.L. Sulem and M. Nelkin, "A Simple Dynamical Model for Intermittent Fully Developed Turbulence", preprint, 1977.
16. M. Nelkin, Phys. Rev. A 11, 1737 (1975).
17. H. Rose and P.L. Sulem, "Fully Developed Turbulence and Statistical Mechanics". preprint, 1977.
18. V.N. Desynansky and E.A. Novikov, Prikl. Mat. Mekh. 38, 507 (1974). (J. Appl. Math. Mech.). See also Izv, Akad. Nauk. SSSR, Fiz. Atmos. Okeana (Atmos. Oceanic Phys. 10, 127 (1974).
19. T.L. Bell and M. Nelkin, Phys. Fluids 20, 345 (1977).
20. T.L. Bell and M. Nelkin, "Time-dependent Scaling Relations and a Cascade Model of Turbulence", preprint, 1977.
21. K.G. Wilson, Phys. Rev. B 4, 3184 (1971).
22. E.D. Siggia, Phys. Rev A 15, 1730 (1977).
23. R.H. Kraichnan, Phys. Fluids 10, 1417 (1967).
24. A. Pouquet, M. Lesieur, J.C. Andre and C. Basdevant, J. Fluid Mech. 72, 305 (1975).
25. David Montgomery, private communication, 1977.
26. R.H. Kraichnan, J. Fluid Mech. 67, 155 (1975).
27. U. Frisch, M. Lesieur and P.L. Sulem, Phys. Rev. Lett. 37, 395 (1976).
28. P.G. deGennes, in "Fluctuation, Instabilities and Phase Transistions" edited by T. Riste, Plenum Press, 1975.
29. P.L. Sulem, J.D. Fournier, U. Frisch and H. Rose, "Crossover Dimensions for Fully Developed Turbulence". This conference.
30. D. Forster, D.R. Nelson, and M.J. Stephen, Phys. Rev. A, July, 1977.

CHAPTER 20

STATISTICAL CONCEPTS IN HADRON PHYSICS

H. Satz

Department of Theoretical Physics, University of Bielefeld, Germany

Summary

Concepts and notions from statistical mechanics have played a significant role in various areas of elementary particle theory. We discuss here two specific cases which characterize or extend the limits of conventional statistical physics. In section II, we present the ideal gas approach to multihadron systems (FERMI model) and the empirical reason for its failure. In section III, we consider the thermodynamics of a relativistic hadron gas with resonance interaction (HAGEDORN model). The abundance of new resonance degrees of freedom leads to a finite ultimate temperature T_H of hadronic matter, with $T \rightarrow T_H$ for infinite energy density. We then discuss, in section IV, high energy pion condensation as a relativistic extension of the concept of phase transition: the new phase does not arise from a change of state of the given constituents, but is rather due to particle creation in the interaction between the constituents of the old phase.

I. INTRODUCTION

The collision of two energetic hadrons leads four out of five times to the production of additional hadrons, and the number of such "secondaries" increases appearently indefinitely with increasing collision energy. This empirical result constitutes the main distinguishing feature of strong interactions and makes the solution of the many-body problem essential for the determination of the underlying dynamics.

To help fix the frame for our considerations, let us note a few experimental facts. In a proton-proton collision at the highest presently available machine energy (60 GeV in the center of mass), a typical final state contains about twenty particles, predominantly pions; rougly a third of these are neutral. Measurement and identification problems for neutrals, for fast particles, or simply for that many particles have ensured that at high energies multiparticle data consist more and more of inclusive distributions: one or several of the secondaries are detected, identified and measured, but without any knowledge of the further structure of the event. Consequently, the information provided by this data is compatible with a tremendously large number of possible and realistic final states. One of the principal tasks of multihadron theory is therefore the construction of a scheme suitable for such situations of highly partial or incomplete information.

This task, the formulation of statistical hadron physics, can be approached in classical fashion along two different lines. On one hand, we can formulate a phenomenological hadron thermodynamics; on the other, we can study the limit of particle number $N \to \infty$ for dynamical models of interest. Both problems, as well as their connection (the "derivation of hadron thermodynamics") are at present far from completed.

A further aspect of statistical hadron physics, and one which we shall here particularly emphasize, is of a more structural nature. The interaction dynamics of multihadron systems differs significantly from that conventionally encountered in statistical physics. Hence our subject can lead to new statistical concepts, or shed some light on the limits of existing statistical pictures.

II. THE IDEAL GAS MODEL

Since mutlihadron physics and statistical mechanics are faced by rather similar problems, it is only natural that one of the first attempts to explain hadronic particle production invoked the ideal gas (FERMI model[1]). In contrast to statistical mechanics, however, it became quickly evident that in hadron physics the ideal gas picture was quite inadequate even as the roughest first approximation. Nevertheless, the approach is still of interest today, since it is used to define dynamical multihadron features (deviations from ideal gas behavior) and to illustrate the concepts and methods to be generalized in more realistic "interacting" gas approaches.

Quite generally, we shall speak of a statistical description of multihadron production, if the relative probability of two inclusive ("macroscopic") final state constellations is determined by the ratio of the corresponding level densities, i.e., by the number of compatible exclusive ("microscopic") final states in each case. The simplest level density is that of an ideal gas in a fixed volume V_o and subject to energy conservation. FERMI[1] conjectured that of all the features of multihadron production dynamics, only conservation laws and the short range of the strong interaction would survive the many-body averaging effects, with the interaction range determining V_o. Hence the level density for the FERMI model is

$$\rho_N(Q^2) = \frac{V_o^{N-1}}{N!} \int \prod_1^N d^3p_i \delta^{(4)}(\sum_1^N p_i - Q) \ , \tag{1}$$

where we consider for simplicity a system of N identical scalar particles ("pions") with four-momenta p_i, $p_i^2 = m^2$; Q denotes the incident four-momentum. From this, the average number of particles $\bar{N}(Q^2)$ produced in a hadron-hadron collision at squared CMS energy Q^2 is given by

$$\bar{N}(Q^2) = V_o(\partial/\partial V_o) \ln V_o\rho(Q^2) \ , \tag{2}$$

in terms of the "grand canonical" level density

$$\rho(Q^2) = \sum_N \rho_N(Q^2) \ . \tag{3}$$

The inclusive momentum spectrum of a secondary is

$$F(Q,p) = V_o\rho((Q-p)^2)/\rho(Q^2); \tag{4}$$

it satisfies the sum rule

$$\int d^3p F(Q,p) = \bar{N}(Q^2). \tag{5}$$

The evaluation of the phase space integral (1) and the corresponding sum (3) can be carried out by slight extensions[2,3] of standard methods from statistical mechanics[4,5], in the limit $Q^2 \to \infty$; at low finite values of the energy, numerical techniques are often needed[6]. In the ultra-relativistic limit $m/M \to 0$, with $M \equiv \sqrt{Q^2}$, we have

$$\rho_N(M^2) \simeq \frac{V_o^{N-1}}{N!} M^{3N-4} \left[\frac{(4N-4)!(2N-1)}{((2N-1)!)^2(3N-4)!}\right] \tag{6}$$

and

$$\rho(M^2) \simeq \text{const. } M^{-4}(V_o M)^{-3/2}\exp(\text{const.}(V_o M^3)^{1/4}). \tag{7}$$

Equations (6) and (7) allow us to obtain all asymptotic distributions of the FERMI model.

The secondary spectrum (4) gives with

$$F(M,p) \simeq V_o \exp - (\text{const.}(V_o/M)^{1/4} p_o);\ p_o^2 = \vec{p}^{\,2} + m^2 , \tag{8}$$

an isotropic distribution; it damps the energy p_o of the secondary exponentially, but weaker for larger incident Q^2. The resulting multiplicity

$$\bar{N}(M^2) \simeq \text{const.}(V_o M^3)^{1/4} , \tag{9}$$

increases powerlike with M, as does consequently the average energy of any secondary

$$\bar{p}_o \equiv \int d^3p\, p_o\, F(M,p)/\int d^3p\, F(M,p) \simeq \text{const.}(M/V_o)^{1/4}. \tag{10}$$

Finally, let us consider the production of a two-particle subsystem in an N-body final state; take for simplicity N=3. The distribution in the invariant mass $\bar{M}^2 = (p_1+p_2)^2$ is then given by

$$d\rho_3(M,\bar{M})/d\bar{M} \sim g(M)\bar{x}(1-\bar{x}^2);\ \bar{x} = \bar{M}/M; \tag{11}$$

i.e., it depends on $\bar{M}$ in a smooth, scale invariant fashion.

The basis of the FERMI model, as of any ideal gas model, is an equal a priori probability for all admissable final states. The resulting equidistribution (1) in momentum space is, however, experimentally not observed; the main deviations are of three distinct types.

1) Resonance Structure: Instead of the smooth, scale invariant distribution (11) in the invariant mass $\bar{M}$ of subsystems, one observes sharp resonance peaks. These are distributed in $\bar{M}$ according to the REGGE pattern

$$\alpha(\bar{M}^2) = \alpha_o + \alpha'\bar{M}^2 = n\ ;\ n = 0,1,2,\ \ldots \tag{12}$$

involving a fundamental scale α', the universal slope of all REGGE trajectories.

2) Jet Structure: The momentum spectrum, instead of being isotropic, is strongly collimated around the beam axis. The distribution in momentum p_T transverse to the collision axis is for all secondaries so far rather well described by the form $\exp-\lambda\sqrt{p_T^2+m^2}$, where m is the mass of the secondary and $\lambda \simeq 6\ \text{GeV}^{-1}$ a

universal scale. In the longitudinal direction, momentum distributions become roughly scale invariant, depending only on $x_L = 2p_L/M$.

3) Constituent Structure: The total available energy is asymptotically not equidistributed among the final state particles. An incident proton, for example, appears to retain in the collision some memory of its identity - it generally leads to a final state baryon carrying a fixed fraction of the total incident energy. This partial conversion of Q^2 into production is most likely a reflection of the constituent (quark) structure of hadrons[7] and will not be treated in this review[8].

Instead, we want to study here how the presence of many different resonant states affects the statistical description, and how such a description can relate the two fundamental scale parameters α' and λ.

III. THE ULTIMATE TEMPERATURE OF HADRONIC MATTER

The presence of a resonance interaction between secondaries clearly effects the level density to be used in a statistical description of multihadron production. If the interaction Hamiltonian were known, it would determine the new "real gas" level density. In hadron physics, this is not the case; hence we must find a different way to modify the ideal gas form (1), and that is the aim of HAGEDORN's bootstrap hypothesis[9].

Intermediate resonances allow two distinct ways of arriving at N final state pions: directly, as in the FERMI model, or as decay products of resonances. The latter situation can also be treated in a non-interacting gas approach[10,11,12]: we consider the level density for a system of N-N' free pions and R resonances, which subsequently decay into N' pions. Knowing the relative weights and the decay distributions of the resonances would then complete the statistical picture of pion production in the presence of resonance interaction. Let us make this more precise.

Instead of eq. (1/3), we now consider

$$\tilde{\rho}(Q^2) = \sum_N \frac{V_o^{N-1}}{N!} \int \prod_1^N \{d^4p_i \tau(p_i^2)\} \delta^{(4)}(\sum_1^N p_i - Q) \ , \tag{13}$$

the grand canonical level density for a system of non-interacting pions and resonances. Here $\tau(p^2)$ describes the hadronic mass spectrum, i.e., the mass distribution of the resonances and their relative weights. If we write

$$\tau(p^2) = \sum_{r=o}^{\infty} c_r \, 2p_o \delta(p^2 - m_r^2) \Theta(p_o) ; \ p_o^2 = \vec{p}^{\,2} + m_r^2 \ , \tag{14}$$

then (with c_o=1, m_o=m) the contribution from r=o for all integrations in eq. (13)

gives back the ideal gas (1/3), while the $r\neq o$ terms account for the presence of resonances. The spectral function $\tau(p^2)$ thus contains the relevant information on resonance structure.

In the HAGEDORN model[9], we now extend the statistical description to determine these resonance features: we assume that the level density (13) governs both the decays and the relative weights of heavy resonant systems ("fireballs"). More precisely, we require with

$$\lim_{Q^2\to\infty} \tilde{\rho}(Q^2)/\tilde{\tau}(Q^2) = 1 \tag{15}$$

the so-called strong bootstrap condition[13], where

$$\tilde{\tau}(p^2) = \tau(p^2) - 2p_o\delta(p^2-m^2)\Theta(p_o) \tag{16}$$

denotes the resonant part of the hadronic mass spectrum. Inserting $\tilde{\rho}(Q^2) = \tilde{\tau}(Q^2)$ in eq. (13), we obtain the statistical bootstrap equation[9,13]

$$\tilde{\rho}(Q^2) = 2Q_o\delta(Q^2-m^2)\Theta(Q_o) + \sum_{N=2}^{\infty} \frac{V_o^{N-1}}{N!} \int\prod_1^N\{d^4p_i\tilde{\rho}(p_i^2)\}\, \delta^{(4)}(\sum_1^N p_i-Q) \tag{17}$$

which determines the level density for pions with resonance interaction as well as the hadronic spectrum governing the resonance distribution.

To solve eq. (17), we resubstitute $\tilde{\rho}(Q^2)$ under the integral and obtain by iteration[14]

$$\tilde{\rho}(Q^2) = \sum_N g_N V_o^{N-1} \int\prod_1^N d^3p_i\delta(\sum_1^N p_i-Q) \ , \tag{18}$$

a purely pionic level density. The coefficients g_N are through eq. (17) subject to the recursion relation[15]

$$g_N = \frac{1}{N}\left[2\sum_{k=1}^{N-1} kg_k g_{N-k} - (N-1)g_{N-1}\right] \ , \qquad g_1 = 1 \ , \tag{19}$$

with the asymptotic solution[14,16]

$$g_N \simeq (4\pi)^{-1/2}\, N^{-3/2}(2\ \ln\ 2-1)^{-N+1/2}. \tag{20}$$

The critical feature to note here is the very weak dependence of g_N on N, to be contrasted with the 1/N! of the ideal gas (1/3): as expected, the introduction of resonances provides a level density increasing much more rapidly with energy. The evaluation of the sum (18) with the coefficients (20) yields, again with $M^2 \equiv Q^2$,

$$\tilde{\rho}(M^2) \simeq \text{const.}\ M^{-3} \exp(\beta_H M)\ , \tag{21}$$

where β_H is a constant to be determined shortly. Here we want to compare the level density (21) with the form $\exp(cM^{3/4})$ of the ideal gas (cf. eq. (7)): Resonance interactions in statistical bootstrap form lead to a linearly exponential energy rise of the level density.

The faster growth of $\tilde{\rho}(Q^2)$ has remarkable thermodynamical consequences. The partition function

$$Z(\beta) = \int d^4Q\ e^{-\beta_\mu Q^\mu}\ \tilde{\rho}(Q^2);\ \beta_o>0,\ \beta_o^2-\vec{\beta}^2>0\ , \tag{22}$$

must, because of eq. (17), satisfy

$$\varphi(\beta) = 2Z(\beta) + (1-e^{Z(\beta)})/V_o\ , \tag{23}$$

where

$$\varphi(\beta) \equiv \int d^3p\ e^{-\beta_\mu p^\mu} = (4\pi m^2/\beta)K_2(m\beta)\ ; \tag{24}$$

here $K_2(x)$ is the Hankel function of pure imaginary argument. As a consequence of eq. (23), $Z(\beta)$ has a square root branch point[17]

$$Z(\beta) = Zo + \text{const.}\sqrt{\beta-\beta_H} \tag{25}$$

at $\beta=\beta_H$, where β_H is the solution of

$$\varphi(\beta_H) = (2\ \ln 2 - 1)/V_o\ . \tag{26}$$

Inversion of the Laplace transform (22) using eq. (25) gives the result (21), with β_H determined from eq. (26). The linearly exponential growth of $\tilde{\rho}(M^2)$ thus allows a convergence of eq. (22) only if $\beta>\beta_H$; for the temperature $kT\equiv\beta^{-1}$ this means with

$$T < T_H \equiv (k\beta_H)^{-1} \tag{27}$$

the existence of a finite upper bound: the ultimate temperature of hadronic matter.

Let us apply these results to the inclusive momentum distribution of hadronic secondaries. From eq. (18), we have

$$F(M,p) = V_o^2((\partial/\partial V_o)\tilde{\rho}((Q-p)^2))/\tilde{\rho}(Q^2) \tag{28}$$

and hence asymptotically from eq. (21)

$$F(M,p) \simeq \text{const.} M \exp - (p_o/T_H) . \tag{29}$$

Consequently, the average energy per secondary attains in the limit of high incident energy the constant value

$$\bar{p}_o \simeq -(\partial \ln \varphi(\beta)/\partial\beta_o)_{\beta=\beta_H} , \tag{30}$$

while the average number of secondaries

$$\bar{N}(M) \simeq \text{const.} T_H^3 M , \tag{31}$$

grows linearly with M. We thus see that asymptotically an increase in incident energy is not converted into kinetic energy of secondaries, but just into more secondaries. This in turn gives rise to the upper bound on the temperature, as a measure of kinetic energy.

If we now restrict ourselves to multihadron production orthogonal to the collision axis (so that $p_L=0$ for the observed secondary), then the inclusive distribution (29) gives with

$$F(M,p_L=0, p_T) \sim \exp - \sqrt{p_T^2+m^2} /T_H \tag{32}$$

precisely the transverse mass bound so far observed for all types of hadronic secondaries, from m = .3 GeV for pions to m = 3.1 GeV for 5/4. The universal scale parameter λ of hadronic jet structure thus finds its explanation in terms of the universal ultimate temperature T_H of hadronic matter. Its numerical value is by eq. (26) determined from the range of strong interactions, as this enters in V_o.

So far, our hadron thermodynamics has been of a phenomenological nature: we have incorporated some aspects of resonance structure through the bootstrap hypothesis (15), without reference to details of resonance dynamics. Let us now see to what extent such a picture is "derivable" from what we know of strong interaction dynamics.

The main thing to note is that the REGGE pattern (12) gives in fact a very specific modification of the ideal gas[18]. Consider for simplicity a system of N non-interacting, identical, one-dimensional particles, taking on discrete energy values q_i:

$$q_i = n \; ; \quad i = 1,2, \ldots, N \; ; \quad n = 0,1,2, \ldots \; . \tag{33}$$

For the "cluster" masses

$$\omega_i = \sum_{l=1}^{i} q_L \quad ; \quad i = 1,2, \ldots, N \quad , \tag{34}$$

we then also have

$$\omega_i = n \quad ; \quad i = 1,2, \ldots, N \quad ; \quad n = 0,1,2, \ldots \quad . \tag{35}$$

The level density of the system at given overall energy M can be written

$$\rho(M) = \sum_{n=2}^{\infty} \frac{1}{N!} \{ \sum_{\omega_1=o}^{M} \sum_{\omega_2=\omega_1}^{M} \sum_{\omega_3=\omega_2}^{M} \ldots \sum_{\omega_{N-1}=\omega_{N-2}}^{M} \} = \frac{I_1(2\sqrt{M})-1}{\sqrt{M}} \quad , \tag{36}$$

which asymptotically yields ($I_1(x)$ is a Bessel function of pure imaginary argument)

$$\rho(M) \simeq \text{const.} M^{-3/4} \exp 2\sqrt{M} \tag{37}$$

in analogy to eq. (7). The quantization (35) is not in accord with the REGGE pattern (18); to have that, we must require instead

$$\alpha'\omega_i^2 = n \quad ; \quad i = 1,2, \ldots, N \quad ; \quad n = 0,1,2, \ldots \tag{38}$$

with α' denoting the REGGE slope. The level density resulting from the quantization (38) becomes asymptotically

$$\tilde{\rho}(M) \simeq \text{const.} M^{-3/2} \exp 2\sqrt{\alpha'}\, M \quad , \tag{39}$$

to be compared with the bootstrap form (21). Hence a REGGE distribution of resonances, an ideal gas in the REGGE space (38), leads to a linearly exponential rise of the level density, and the ultimate temperature of hadron thermodynamics is directly related to the universal scale parameter α' governing the distribution of the observed resonances: $T_H \sim (\alpha')^{-1/2}$. - The generalization of our arguments to three-dimensional particles is straightforward[18].

To go one step further, we must consider a dynamical scheme containing the pattern (18). The most complete description up to now is given by the dual resonance model (DRM)[19], which constructs arbitrary N-particle amplitudes from (narrow width) resonance poles in all variables, the resonances lying on trajectories of the form (18). The level density in the DRM is, in accord with the above, linearly exponential in the energy[20]. Moreover, we can calculate the transition from a given initial state to any given final state. If we now consider this transition in a statistical average[21], where we sum without interferences over all compatible final states and average with equal weights over all compatible initial states, then we obtain the momentum distribution (29) with

$$T_H = (3/8\pi^2\alpha^1)^{1/2} \tag{40}$$

again providing the connection between hadron thermodynamics and hadron dynamics.

What remains completely open at present, is why in production from hadronic collisions such statistical arguments can be successfully applied in the transverse direction (universal jet diameter λ), but not for distributions in the longitudinal momenta or in energies of the observed secondaries.

On a more general level, it is of course also still open if all hadronic systems indeed obey the temperature bound (27), making that relation something of a "fourth law of thermodynamics", or if at sufficiently high energy density a phase transition sets in, from a hadron gas to one of hadronic constituents ("quark gas")[22,23,24], whose interaction is not governed by the dynamics we have considered here.

IV. PHASE TRANSITIONS IN HADRONIC SYSTEMS

The transformation of a hadron gas into a quark gas would in many ways correspond to a conventional phase transition. Since hadrons and their resonances are presumably bound states of confined quarks in definite orbital configurations, a new phase is conceivable, in which these configurations are broken. Particle production at large transverse momentum (very central collisions and hence high energy density) does show features of this nature.

In addition to such transitions, however, hadronic systems allow the creation of a new phase in a way not possible on a non-relativistic or weakly coupled relativistic level, and in the present section we want to consider these "new" phase transitions. When ice melts or water evaporates, a given assembly of interacting H_2O molecules changes its collective structure: the new phase is formed at the expense, by transformation, of the old. Relativistic hadronic systems can, upon increase of energy, create new particles through the interaction of the constituents, and these secondaries could provide a new phase <u>in addition to the old</u>.

A first step in this direction is the phenomenon of pion condensation as conjectured in nuclear physics[25,26,27]. Consider dense nuclear matter consisting of neutrons, e.g. in a stellar interior. Fermi statistics prevents all neutrons from being in the lowest possible energy state. Upon increase of energy density, the transition $n \rightarrow p + \pi^-$ becomes allowed; since the proton thus formed can occupy a lower energy level, this state may be energetically favorable and therefore lead to the formation of a pion condensate in addition to the Fermion gas. Given a suitable nucleon-nucleon interaction, similar effects can also occur without change of nucleon charge: $p(n) \to p(n) + \pi^o$. For more details, cf. ref.27.

A rather transparent illustration of new phase transitions can be obtained from a study of clustering effects in strongly interacting boson systems[28]. The partition function for a relativistic ideal Bose-Einstein (B-E) gas[29,30] can be written in a form[30] analogous to the cluster expansion for an interacting system

$$Z(\beta,V,A) = \exp V \sum_{k=1}^{\infty} \frac{A^k}{k^4} \varphi(km,\beta). \tag{41}$$

Here V denotes the volume of the system, and $\varphi(km,\beta)$ is given by eq. (24) with $m \to km$:

$$\varphi(km,\beta) = (4\pi k^2 m^2/\beta) K_2(km\beta). \tag{42}$$

Eq. (41) holds for the continuum of scattering states: the discrete sum over momenta has been replaced by the integration $\int d^3p$ [+)]. The quantity A in eq. (41) is the relativistic activity , related by

$$A = A_{NR}\, e^{m\beta} \tag{43}$$

to the non-relativistic activity A_{NR}. If we define the latter as $A_{NR} = \exp \mu_{NR}\beta$ in terms of the non-relativistic chemical potential μ_{NR}, then eq. (43) can be taken to define with $\mu = \mu_{NR} + m$ the relativistic chemical potential satisfying $A = \exp \mu\beta$. - The physical content of eq. (41) is quite clear: the term k=1 in the sum gives the contribution of a Boltzmann gas of pions, the k=2 term that of a Boltzmann gas with two pion systems of zero relative kinetic energy as constituents (single constituent mass 2m), the k=3 term correspondingly for three pion clusters, and so on.

For a relativistic ideal B-E gas, the temperature range is $0 < T < \infty$, and that of the activity (43) is[29]

$$0 < A < e^{m\beta} \tag{44}$$

The presence of an interaction in general modifies the range of the activity[32]; it can, as we saw in section III, also effect the temperature range. We shall here consider the class of interactions obtained by treating the activity A as

+) This means that the ground state contribution to $Z(\beta,V,A)$, $-\ln(1-A\exp\beta m)$ has been dropped, corresponding to a large volume limit[22,31].

an open parameter characterizing the system ("quasi-ideal gas"[28]), while leaving each fixed k contribution to eq. (41) that of a free system.

Using the asymptotic expansion

$$K_2(x) \simeq \sqrt{\pi/2x}\, e^{-x}(1 + O(1/x)) \,, \tag{45}$$

it is immediately evident that the partition function (41) diverges for $A > \exp m\beta$, in accord with relation (44). In other words,

$$T_* = m/\ln A \tag{46}$$

constitutes for any given $A > 1$ a finite critical temperature of the system; for $A \to 1$ from above, it moves to infinity. In general[22], this T_* could be either an ultimate temperature, as in the statistical bootstrap model, or it could correspond to a phase transition. To distinguish the two, we consider the energy density

$$\varepsilon(V,\beta,A) \equiv V^{-1}(\partial/\partial\beta) \ln Z(V,\beta,A) \tag{47}$$

in the thermodynamic limit $V \to \infty$:

$$\varepsilon(\beta,A) \equiv \lim_{V\to\infty} \varepsilon(V,\beta,A). \tag{48}$$

If $\varepsilon(\beta,A)$ exists at $\beta=\beta_*=(kT_*)^{-1}$, we have a phase transition; if it becomes singular ($T\to T_*$ only as $\varepsilon(\beta,A) \to \infty$),we have an ultimate temperature[22].

For eq. (41) with T_* given by eq. (46), it is readily shown[28] that both the energy density ε and the particle number density

$$\nu(V,\beta,A) \equiv V^{-1}(A\partial/\partial A) \ln Z(V,\beta,A) \tag{49}$$

in the limit $V \to \infty$ exist at $\beta=\beta_*$,while all order derivatives of ε and ν diverge there. The critical temperature T_* thus corresponds to an anomalous first order phase transition[33]. The behaviour of the average energy $\omega(T,A)$ per particle is illustrated in fig. 1: it grows from $\omega = m$ at $T \simeq 0$ to $\omega = \omega_* \equiv \varepsilon(T_*,A)$ at $T = T_*$.

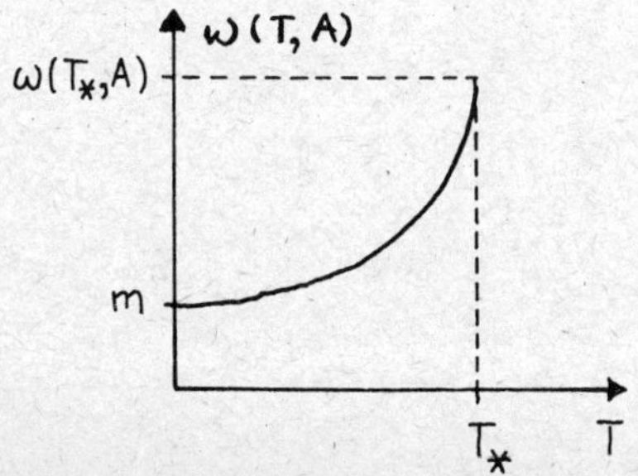

Fig. 1: The average energy per particle in the quasi-ideal B-E gas, as function of the temperature $T < T_*$

To study the behaviour of the system of energy densities $\varepsilon > \varepsilon(\beta_*, A)$, and to investigate the nature of the phases involved, we must use a (grand) microcanonical description, i.e., calculate at fixed energy density $\varepsilon = M/V$ rather than at fixed temperature. For a quasi-ideal gas this is technically possible:

$$r(\varepsilon,A) \equiv \lim_{V\to\infty} (V^{-1} \ln \rho(V,M,A)) \tag{50}$$

can be calculated and in turn gives with

$$\nu(\varepsilon,A) \equiv A(\partial r(\varepsilon,A)/\partial A)_{\varepsilon} \tag{51}$$

the particle number density $\nu(\varepsilon,A)$. For $\varepsilon \geq \varepsilon_*$, the result is

$$\nu(\varepsilon,A) = \frac{1}{m}(\varepsilon - \varepsilon_*) + \nu(\varepsilon_*,A) \tag{52}$$

where $\nu(\varepsilon_*,A)$ agrees with the value $\nu(\beta_*,A)$ of the canonical approach. Correspondingly, we have for the average energy per particle

$$\omega(\varepsilon,A) = \begin{cases} \varepsilon/\nu(\varepsilon,A) & \varepsilon \leq \varepsilon \\ \varepsilon/[(\varepsilon - \varepsilon_*)/m + \nu(\varepsilon_*,A)] & \varepsilon \geq \varepsilon \end{cases} \tag{53}$$

For $\varepsilon < \varepsilon_*$, this gives back our previous result (as function of ε instead of β), while for $\varepsilon \to \infty$ we find

$$\lim_{\varepsilon\to\infty} \omega(\varepsilon,A) = m \tag{54}$$

The functional form of $\omega(\varepsilon,A)$ is shown in fig. 2.

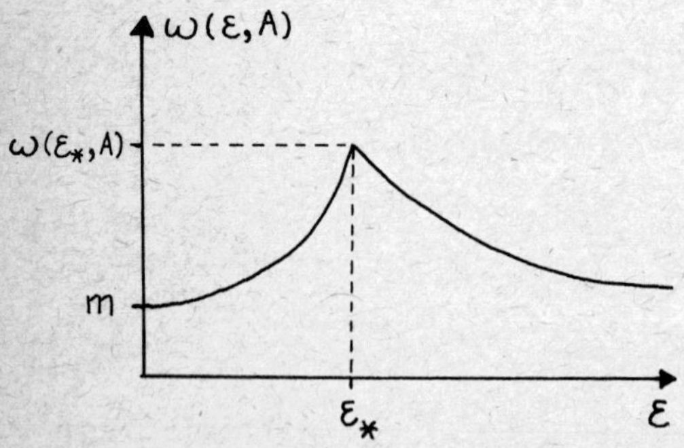

Fig. 2: The average energy per particle in the quasi-ideal B-E- gas, as function of the energy density ε.

The behaviour of $\nu(\varepsilon,A)$ and $\omega(\varepsilon,A)$ provide a clear picture of the underlying physics. Up to $\varepsilon = \varepsilon_*$, an increase in energy density leads to faster constituents as well as to heavier ones - the role of higher pion number clusters increases with ε. Above $\varepsilon = \varepsilon_*$, all further increase of energy density goes into the formation of pions at rest (cf. eq. (52)), i.e., into an infinitely heavy system of condensed bosons: the condensate phase. Note that the gas phase continues to exist undiminished at all $\varepsilon > \varepsilon_*$; but as an increase of ε goes into condensate formation, it becomes negligible as $\varepsilon \to \infty$, and hence $\omega \to m$.

This point is also illustrated by consideration of the temperature $T = (k\beta)^{-1}$ as defined through

$$\beta \equiv (\partial r/\partial \varepsilon)_A \quad ; \tag{55}$$

it grows from zero at low ε to T_* at $\varepsilon = \varepsilon_*$ and then remains at T_* for all $\varepsilon > \varepsilon_*$ Similarly the pressure P remains constant above ε_*. As both T and P measure the kinetic energy of the system, it is evident that above ε_* this neither increases nor decreases. The same behaviour is observed in the coexistence region of usual phase transitions (ice to water, water to water vapor); but here the transition transforms the old phase into the new, and hence T and P change again when this is completed. In the "new" hadronic phase transition, the condensate is created in addition, by the interaction of a gas which does not disappear. Hence the transition is "complete" only for $\varepsilon \to \infty$.

The structure of the high energy B-E condensation just studied bears many similarities to that of the statistical bootstrap. In particular, both provide so many new degrees of freedom (resonances, "condensed" pion clusters) that adding energy to the system does not produce, as dominant mode, an increase of kinetic energy of the constituents. The bootstrap condition, however, forces a structural identity in the composition of light and heavy objects; hence there can be only one phase, and $T \to T_H$ as $\varepsilon \to \infty$. In the quasi-ideal B-E system, gas and condensate differ, and we have a phase transition at finite energy density.

The interplay of pion condensate creation and specific features of hadron dynamics has recently attracted interest[34,35,36]; such condensation (as one instance of soft pion divergence) can decisively alter the asymptotic properties of multihadron systems.

References

1) E. Fermi, Progr. Theoret. Phys. (Japan) 1, 570 (1950)

2) F. Lurçat and P. Mazur, Nuovo Cim. 31, 140 (1964)

3) H. Satz, Nuovo Cim. 37, 1407 (1965)

4) A.I. Khinchin, The Mathematical Foundations of Statistical Mechanics, Dover, New York, 1949

5) P. Mazur and J. Van der Linden, J. Math. Phys. 4, 271 (1963)

6) See e.g. E. Byckling and K. Kajantie, Particle Kinematics, John Wiley, New York 1974

7) H. Satz, Phys. Lett. 25B, 220 (1967)

8) For a recent survey, see e.g. H. Satz in Proceedings of the 1977 European Conference on Particle Physics, Budapest, Hungary, July 1977 (Bielefeld Preprint BI-TP 77/26, 1977)

9) R. Hagedorn, Nuovo Cim. Suppl. 3, 147 (1965)

10) E. Beth and G.E. Uhlenbeck, Physica 4, 915 (1937)

11) S.Z. Belenkij, Nucl. Phys. 2, 259 (1956)

12) R. Dashen, S. Ma and H.J. Bernstein, Phys. Rev. 187, 345 (1969)

13) S. Frautschi, Phys. Rev. D3, 2821 (1971)

14) J. Yellin, Nucl. Phys. B52, 583 (1973)

15) J. Engels, H. Satz and K. Schilling, Nuovo Cim. 17A, 535 (1973)

16) R. Hagedorn and I. Montvay, Nucl. Phys. B59, 45 (1973)

17) W. Nahm, Nucl. Phys. B45, 525 (1972)

18) H. Satz, Phys. Lett. 44B, 373 (1973)

19) G. Veneziano, Nuovo Cim. 57A, 190 (1968)
See e.g. J. Schwarz, Phys. Rep. 8, 269 (1973) or
G. Veneziano, Phys. Rep. 9, 199 (1974)
for further references

20) S. Fubini and G. Veneziano, Nuovo Cim. 64A, 811 (1969)
K. Bardakçi and S. Mandelstam, Phys. Rev. 184, 1640 (1969)

21) M.I. Gorenstein, V.A. Miransky, V.P.Shelest, B.V.Struminsky and G.M. Zinovjev, Phys. Lett. 43B, 73 (1973)

22) N. Cabibbo and G. Parisi, Phys. Lett. 59B, 67 (1974)

23) J.C. Collins and M.J. Perry, Phys. Rev. Lett. 34, 1353 (1975)

24) B.A. Freedman and L.P. McLerran, MIT Preprint 541 (1976)

25) A.B. Migdal, ZhETF 61, 2210 (1971)
A.B. Migdal, O.A. Markin and I.N. Mishustin , ZhETF 66, 443 (1974)

26) R.F. Sawyer, Phys. Rev. Lett. 29, 362 (1972)
D.J. Scalapino, Phys. Rev. Lett. 29, 386 (1972)
R.F. Sawyer and D.J. Scalapino, Phys. Rev. D7, 953 (1973)

27) For a recent survey, see G.E. Brown and W. Weise, Phys. Rep. 27, 2 (1976)

28) I. Montvay and H. Satz, Nuovo Cim. 39A, 425 (1977)

29) M.M. Nieto, J. Math. Phys. 11, 1346 (1970)

30) M. Chaichian, R. Hagedorn and M. Hayashi, Nucl. Phys. B92, 445 (1975)

31) I. Montvay, Nuovo Cim. in press (Bielefeld Preprint BI-TP 76/02)

32) D.E. Miller, Phys. Lett. 61A, 291 (1977) and contribution to this conference (Bielefeld Preprint BI-TP 77/09, 1977)

33) A. Münster, Statistical Thermodynamics, Springer Verlag, Berlin 1969

34) A.I. Bugrij, Kiev Preprint ITP 77-48E

35) J. Letessier, A. Tounsi and H. Satz, Paris Preprint IPNO/TH 77/16 (1977)

36) E.H. de Groot and H. Satz, Bielefeld Preprint BI-TP 77/15 (1977)

CHAPTER 21

METASTABILITY IN THE ISING MODEL

L. S. Schulman

Physics Department, Technion - Israel Institute of Technology, Haifa, Israel

and Physics Department, University of Indiana, Bloomington, Indiana, U.S.A.

ABSTRACT

For the theoretical study of metastable systems there is first the question of defining what it is you are talking about and then the problem of calculating the physical properties of those systems (model or real) which in fact exhibit metastability, oblivious to the theoretical difficulties just alluded to. Furthermore, there is the related open question of whether the free energy has an essential singularity at a first order phase transition.

Our approach has been to attempt analytic continuation of the free energy. For finite volume - where one hopes for evidence of metastability, even should the thermodynamic limit spoil this evidence - we do not study the partition function which is necessarily single valued. Rather we look (for the two-dimensional Ising model) at eigenvalues of the transfer matrix which, on the one hand, give the free energy in the thermodynamic limit, but on the other hand, as roots of a polynomial, are multivalued functions and can show both stable and metastable properties for a given value of the external magnetic field, H.

This approach yields predictions for the magnetization and other properties of the metastable system. It also suggests a certain singularity structure for the free energy in the metastable region. By numerical diagonalization of the $2^N \times 2^N$ transfer matrix for $N = 5,7,9$ and 11 we ascertained first what these predictions were.

The relevance of these predictions to metastable systems was checked with a stochastic dynamical Ising model. For small and moderate H (and low temperature) magnetization agreed fairly well. A surprising result is an apparent relation between the lifetime of the metastable dynamical state and the distance of the branch point in the free energy from the real H axis.

Lifetime dependence on H and temperature for the dynamical system also shows strong support for the critical droplet model of condensation. Even the "mountain pass" picture, used in more general contexts for the prediction of lifetimes, appears here in a simple and explicit way.

Introduction

The practical problems of metastable states are concerned with determining whether they have reproducible properties and with the measurement and prediction of these properties. More fundamentally there is the difficulty of putting metastability in a theoretical framework in the face of various arguments, physical and mathematical, which deny it firm definition.

In this paper I shall propose a way of defining properties of metastable systems, namely through an analytic continuation of the free energy. Having provided this particular framework for metastability in the two dimensional Ising model, I shall examine the predictions obtained and compare them to a naive continuation of the low temperature expansion and to computer experiments on a dynamical Ising model.

Conceptual Problems

Decay from the metastable state occurs through the formation of a droplet of stable phase large enough so that it is energetically favorable for this droplet to grow rather than shrink. The size of the critical droplet is determined by balancing volume energy against surface energy. The lifetime is then assumed to be inversely proportional to the probability of formation of a single critical droplet which in turn is proportional to volume; one therefore expects the lifetime to go to zero in the thermodynamic limit.

The avoidance of this dilemma in mean field theories and the appearance of metastability in such models arises because with infinite range forces there is no distinction between surface and volume energy and the nucleation energy is infinite.

Analytic Continuation of Eigenvalues[1]

In the two dimensional Ising model a phase transition occurs (for $T < T_c$) as H (external magnetic field) goes through zero. One could imagine that the free energy

F(H), which is analytic for ReH > 0, could be analytically continued to ReH < 0, such a continuation being identified with the metastable system. However, even if F can be continued in the thermodynamic limit, one could not discover this from the finite volume partition function, whose logarithm is F. This is because one expects metastability to be expressed through a multivaluedness of the free energy, and the finite volume partition function as a polynomial in exp(±H/kT) must be single valued.

For this reason, C.M. Newman and I sought a function that exhibits multivaluedness for finite volume (where physical arguments favoring metastability are strongest) but which also has as its limit the thermodynamic free energy. For $M \to \infty$ (N x M lattice) the partition function is dominated by the largest eigenvalue of the $2^N \times 2^N$ transfer matrix. For finite N we proposed the analytic continuation of this eigenvalue to ReH < 0. I shall present some motivation for this proposal later.

Spectrum of the Transfer Matrix[2]

Over the last year R.J. McCraw and I have gathered evidence bearing on the proposal through study of the spectrum of the transfer matrix as a function of H. Although the transfer matrix operates on a 2^N dimensional space, translational invariance (periodic boundary conditions) reduces the dimension so that we could do numerical studies for N as large as eleven. Let the eigenvalues of the $2^N \times 2^N$ transfer matrix be $\lambda_j^{(N)}$ $(\lambda_o^{(N)} > \lambda_1^{(N)} > \lambda_2^{(N)} > \ldots)$ and let a "free energy" be defined as

$$F_j^{(N)} = - T \log \lambda_j^{(N)}/N \tag{1}$$

(T = temperature); thus $F_o = \lim_{N\to\infty} F_o^{(N)}$ is the stable free energy. Fig. 1 illustrates transfer matrix "free energies" for N = 7 as a function of H for T = 1.0 ($T_c \cong 2.27$). At H = 0 there is a near crossing of levels. In fact the levels do not cross, and it is just the "near discontinuity" in the slope of $F_o^{(N)}$ that gives

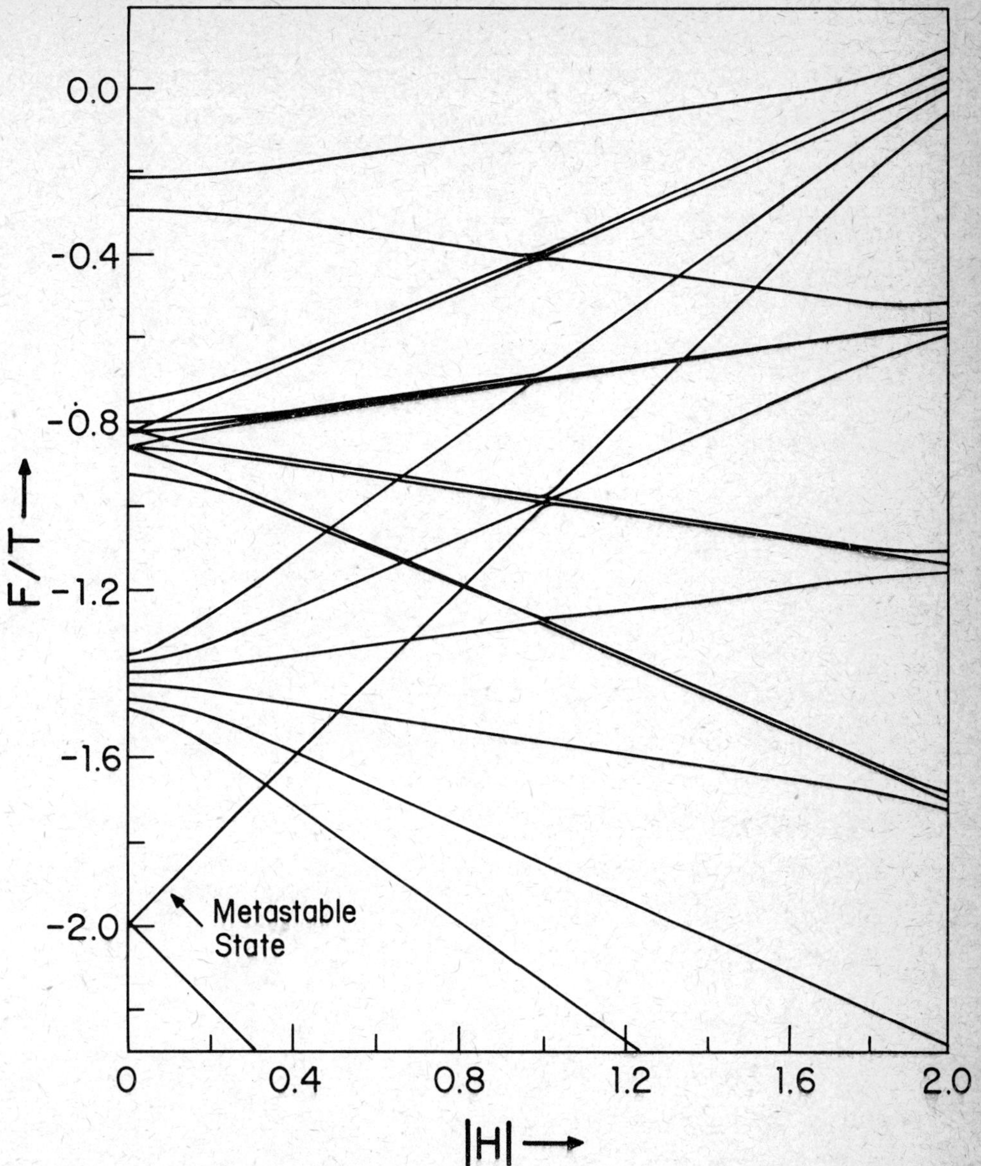

Fig. 1. (Free energy)/T = -log(eigenvalue)/N for $2^7 \times 2^7$ transfer matrix for various eigenstates as a funtion of H for T = 1. Crossings are not accurately depicted.

rise to spontaneous magnetization at the first order phase transition (as $N \to \infty$). Although the levels are not perfectly degenerate for real H, one need not go far into the complex H plane for an exact crossing to occur. This gives rise to a square root branch point in the function $F_o^{(N)}(H)$. An analytic continuation of $F_o^{(N)}(H)$ (from $\mathrm{Re}H > 0$ where it is analytic) to $\mathrm{Re}H < 0$ must go over this branch point in order to follow the metastable state. Fig. 2 shows the cut structures and paths in the complex H plane associated with the metastable continuation.

In the thermodynamic limit the branch points at $H = 0$ coalesce and it might appear that the analytic continuation could be carried through without difficulty. (S. Katsura has recently informed me that he made a proposal[3] to this effect in 1953.) In fact, the situation is far more complicated. As H becomes more negative the energies of other levels shift, and there are many more "near crossings", evident in Fig. 1. Fig. 3 shows details of a crossing at $H = 1$ for $N = 7$. Note the large number of levels involved. Numerical evidence and crude calculations suggest that as N increases a value of H at which a near crossing occurs will retain that property, but that more levels will be involved and that in addition near crossings will occur at new values of H. It seems likely that in the thermodynamic limit the region in which $F_o(H)$ can be continued does not include the negative real H axis and branch points remain a finite distance off this axis, a distance which goes strongly to zero as H approaches zero. Our guess is that the clustering of branch points around $H = 0$ gives rise to an essential singularity.

Despite the pessimism implicit in our foregoing conjecture, it seems nevertheless true that in several ways the picture works. Examination of Fig. 1 shows that even with the crossings, there is a clearly discernible line which "continues" the line $F(H)$ from $H > 0$. More significant: as N changes, the general pattern of levels does not remain fixed; two levels, however, the stable level and that which we identify as the metastable level do remain essentially unchanged. This is a

Fig. 2. (a). Branch points of $F^{(N)}(H)$ in complex H plane. For metastable continuation branch cuts connect the two branch points. To obtain the usual definition of free energy branch cuts are drawn which do not cross the real H axis.

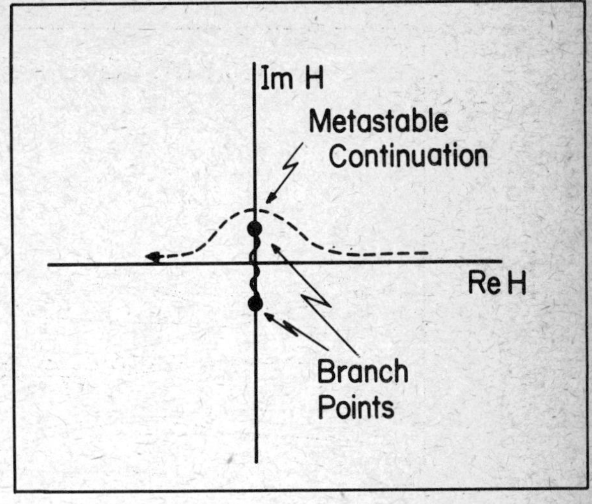

Fig. 2. (b). Detail of energy levels and passage from level to level by means of complex path (suggested by dotted lines).

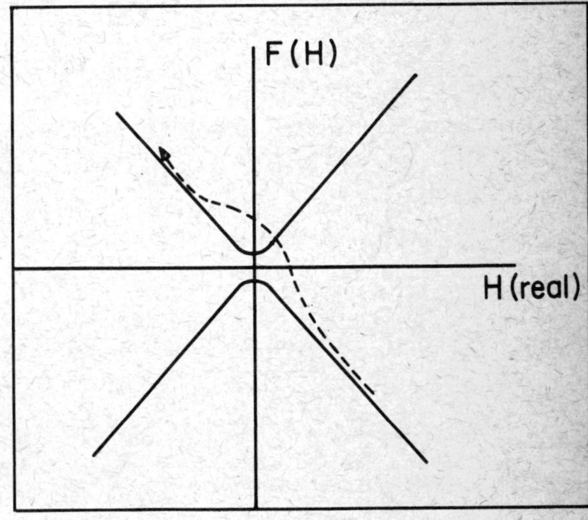

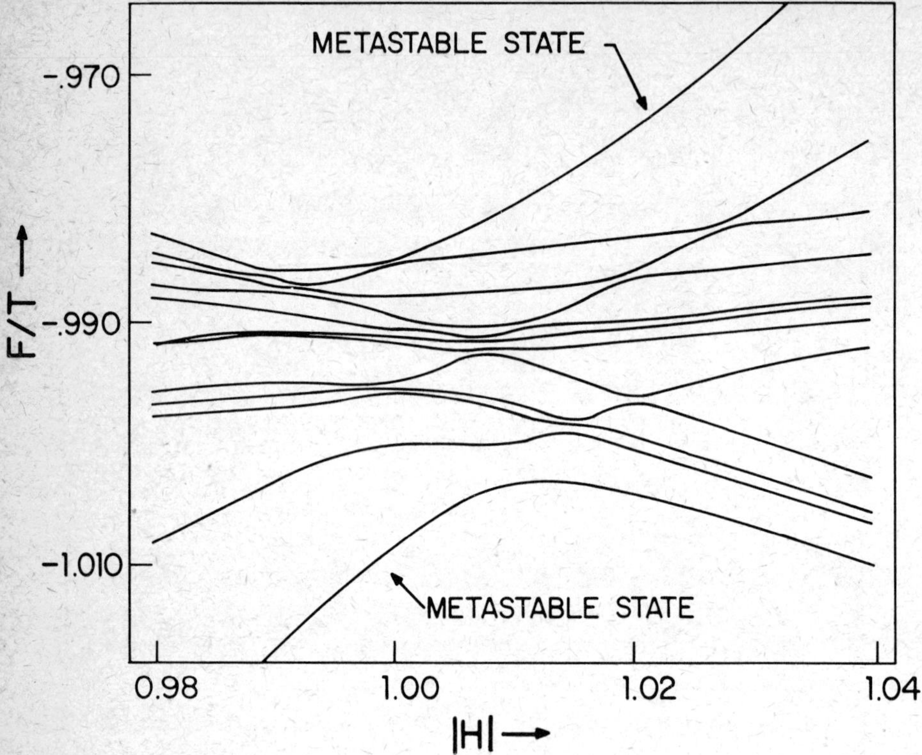

Fig. 3. Details of level crossing near ReH = — 1, for T = 1.0,N = 7 (i.e. the $2^7 \times 2^7$ transfer matrix). Energy levels for .98< - H < 1.04 with ImH = 0.

necessary condition for a thermodynamic limit (note the 1/N in eq. (1)).

This line - the "free energy of the metastable level" - provides, through its derivative, a prediction for the magnetization of the metastable system, away from the crossing regions. We shall now describe how this and other predictions have been tested by comparison with the low temperature expansion and with a dynamical Ising model.

Low Temperature Expansion

The low temperature expansion is probably nonconvergent but asymptotic in the metastable domain. In fact, taking only the first few terms of this expansion, excellent agreement is found[2] with transfer matrix metastable magnetization for small and moderate H at low T.

Dynamical Ising Model

The test of the transfer matrix that I find most satisfying physically is comparison with the dynamical Ising model[2]. This is as close as one can come to a laboratory experiment. Briefly, the dynamical Ising model is a spin system with stochastic dynamics, the probabilities for a spin flip to raise or lower energy being related to the Boltzman factor in such a way that this system evolves towards the Ising model equilibrium state. Under appropriate circumstances such a system can remain out of equilibrium for long times. The nonequilibrium state is found to have reproducible properties (e.g., magnetization) and moreover decays with an exponential decay law (if an ensemble of such systems is considered) with a definite lifetime. In Fig. 4 is shown a comparison of transfer matrix and dynamical model magnetization. For low H and away from the crossing regions agreement is quite good.

A notable feature of our dynamical model lifetimes is how well they conform to the droplet model of decay. Energy versus droplet size can be plotted and at large

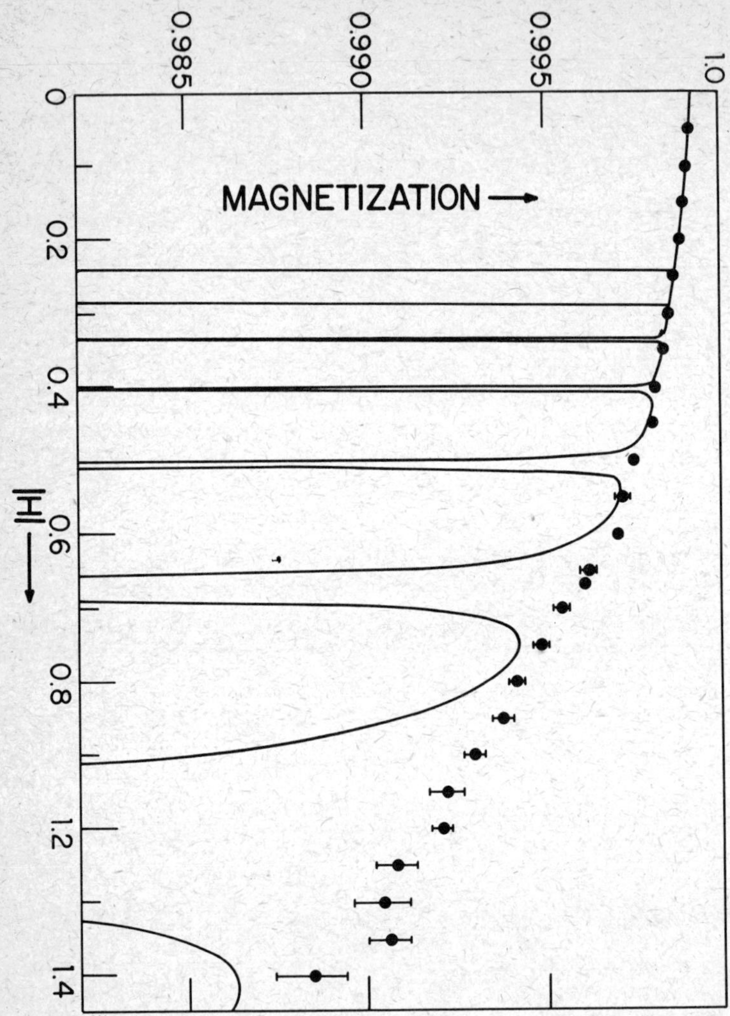

Fig. 4. Magnetization as a function of (-H) for T = 1. The points dynamic Monte Carlo numerical experiments. The curves are from the transfer matrix eigenvalue which yields the largest magnetization at the given H.

magnetic field the critical droplet can be identified. The picture of the system climbing through a mountain pass to escape from the metastable to the stable state is confirmed at low temperatures by quantitative measurements of lifetime.

A surprising result was that the measured lifetimes appear to be related to the location of the free energy singularities obtained from the transfer matrix. The severity of a crossing's effect on the metastable state can be measured either through the distance of the associated branch point from the real H axis or in the gap between the highest and lowest energies at the real H value of the crossing. Our data suggest that with increasing N the gaps do <u>not</u> close and the branch points do <u>not</u> coalesce at the real H axis (except at $H = 0$). Rather these quantities have a quantitative relation to the dynamical measured lifetimes. In particular we have found that the relation (gap) = $\exp(-A/T)$ where A depends strongly on H but weakly on T is well obeyed for low T. This temperature dependence of the gaps bears a close resemblance to the formula (rate) = (volume) $\exp(-B/T)$ with B relatively independent of T which we have found to hold for decays in the dynamical Ising model. Actually in ref. 2 we go somewhat beyond this confirmation of agreement on overall T dependence to find - by means of crude calculations - a relation for given H between the coefficients A and B arising from the transfer matrix and the dynamical Ising model respectively.

Anyone brought up on S matrix theory will certainly consider it reasonable that the distance of a singularity (in this case a branch point rather than a pole) off the real axis has something to do with lifetime. But it still seems to me remarkable that the transfer matrix, which ostensibly has no relation to dynamics, should have a bearing on lifetimes.

Remark on analytic continuation and metastability

This brings us back to the question of motivation: why should the eigenstates and eigenvalues of the transfer matrix have anything to do with the time dependent phenomenon of metastability? The transfer matrix eigenvector with largest eigenvalue (lowest free energy) is known to carry information about the physical stable state. The analytic continuation described above leaves us - as the sign of H is changed - in the metastable domain with an eigenvector resembling the ground state on the other side of $H = 0$. In terms of quantum mechanical time dependent perturbation theory the "adiabatic" approximation leaves the state label unchanged while the "sudden" approximation leaves the eigenstate unchanged. It is therefore appropriate to take a slight excursion into the complex H plane (cf. Fig. 2) to retain the form of the eigenstate. The distance off the real H axis one must go to keep the wavefunction unchanged presumably is related to just how sudden the physical process has to be to retain the wavefunction's form.

It should be mentioned that the association of lifetime with distance of a branch point from the real axis seems unknown in the literature. Stone, Wallace and I have looked at a simple quantum mechanical model[4] and have found that indeed such an association can be found in ordinary quantum mechanics.

Comments

Many questions remain, both in our own work and in the general problem of metastability. If as out estimates suggest, singularities in the free energy (F) gather just off the negative real H axis, the properties of physical metastable states may have to be deduced from the values of F a finite distance into the complex plane. The ambiguity inherent in that possibility may exactly match the choices of evolution rules for the dynamical Ising model.

With the definition of lifetime that we use (one critical droplet implies decay) the infinite volume limit still gives instant decay, so that even if decay rates (per unit volume) are built into the singularity structure of the thermodynamic free energy, still we have no way for an infinite volume system to be held in a metastable state.

This dilemma, discussed earlier, suggests that laboratory metastability is intrinsically a finite volume phenomenon. Because of the slowness of homogeneous nucleation in many physical systems, 10^{24} must be considered a finite number.[5] I do not find this view shocking and one must expect that for non equilibrium statistical mechanics some ideas - such as the idealization of an infinite volume dynamical system - must be compromised. We have tried to make our deviation from equilibrium formalism as small as possible, for example defining our metastable state in such a way that until it decays it has no memory of its earlier fluctuations nor does it have hysteresis. Our reward may be (if some of our foregoing conjectures prove to be correct) that the analytic continuation of the infinite volume free energy when used appropriately for finite volume systems give information on magnetization and lifetime.

Finally it should be observed that despite the speculative nature of our remarks and despite the great conceptual difficulties that surround our definition (and I believe anybody else's definition) of metastability in systems with short range forces, we can still return to the small H region of Fig. 4 for evidence that some grain of truth lies behind our approach.

1. C.M. Newman and L.S. Schulman, J. Math. Phys. 18, 23 (1977).
2. R.J. McCraw and L.S. Schulman, J. Stat. Phys., to appear.
3. S. Katsura, "On the Theory of Cooperative Phenomena", Prof. of the International Conf. of Theoretical Physics, Kyoto and Tokyo, Sept. 1953.
4. L.S. Schulman, M. Stone and D.J. Wallace, manuscript in preparation.
5. I was especially impressed by the example of Nickel mentioned in David Turnbull, Sci. Am. 212, No. 1, 38 (Jan. 1965).

CHAPTER 22

NEW RIGOROUS EXISTENCE THEOREMS FOR PHASE TRANSITIONS IN MODEL SYSTEMS

Barry Simon

Departments of Mathematics and Physics, Princeton University,

Princeton, N.J., U. S. A.

*Supported in part by U.S. National Science Foundation Grant MPS-75-11864

ABSTRACT

We describe a variety of new existence theorems for phase transitions in model systems including the classical Heisenberg and quantum spin 1/2 xy models in three or more dimensions. Two new tools are emphasized: the use of "chessboard estimates" to estimate contour probabilities in Peierls' type arguments and the use of "infrared bounds".

1. INTRODUCTION

In this note, I want to describe progress in the past two years in the rigorous theory of phase transitions involving some novel methods which have led to a variety of results including the first rigorous demonstration of a phase transition accompanied by a spontaneously broken continuous symmetry group. Two main themes are involved which, as we shall see, are intimately related. The first, involving a method for reducing estimates of contour probabilities in a Peierls' argument to bounding thermodynamic quantities, was developed by Glimm, Jaffe, and Spencer[1] to prove phase transitions in a model quantum field theory (equivalent to a classical statistical mechanics model) and extended to various lattice models including some quantum models by Fröhlich and Lieb.[2] The second, involving a new strategy of proving phase transitions was developed in the context of certain classical lattice gases and certain model quantum field theories by Fröhlich, Simon, and Spencer[3] and extended to certain quantum models by Dyson, Lieb, and Simon.[4] Extensions of both sets of ideas appear in a series of papers by Fröhlich, Israel, Lieb, and Simon.[5] For further discussion, the reader is referred to two lucid reviews of Fröhlich[6,7] and one of Fröhlich and Spencer.[8]

Both methods illustrate the rather close connection between constructive quantum field theory and rigorous statistical mechanics. The early years of this decade saw progress in the first subject due to the infusion of ideas from the second especially correlation inequalities[9,10] and high temperature[11] and related[12] expansions. More recently, there has been flux in the other direction. The particular high temperature expansion of Ref. 11 has been useful in some statistical mechanical models[13] and the methods discussed in this note owe a great deal to field theoretical intuition.

Models for which phase transitions were demonstrated for the first time with these methods include ($\nu \equiv$ number of space dimensions of lattice).

(a) Classical N-vector models (including the classical Heisenberg model: N=3) on nearest neighbor cubic lattices[3] and certain other lattices[5] if $\nu \geqslant 3$. The upper bound on the transition temperature in this case can be described by saying that with the normalization in which these models converge to the spherical model[14] as $N \to \infty$[15], the upper bound is just the transition temperature of the spherical model. The bound is surprisingly accurate, e.g. if $N = \nu = 3$, the bound is off by only 9%.

(b) Quantum spin 1/2 nearest neighbor, simple cubic xy model in $\nu \geqslant 3$ dimensions.[4]

(c) Quantum nearest neighbor, simple cubic Heisenberg antiferromagnet for spin $s \geqslant 1$ and $\nu \geqslant 3$ and for $s = 1/2$ and ν sufficiently large.[4]

(d) Classical Heisenberg antiferromagnet with Fisher Coupling (i.e. nearest neighbor antiferromagnetic and next nearest neighbor ferromagnetic) in $\nu \geqslant 3$ dimensions and in non-zero external field.[5]

(e) $\nu = 2$ dimensional quantum anisotropic Heisenberg antiferromagnet in simple cubic nearest models[2] with values of the anisotropy parameter, ε, (defined by $zz + \varepsilon(xx + yy)$ interaction) much larger than those treated by Ginbre[16] and Robinson[16] (and, in particular with $\varepsilon \to 1$ as $s \to \infty$).

(f) $\nu = 2$ dimensional classical Heisenberg models[5] with long-range interactions $J(n) \sim n^{-\alpha}$ with $2 < \alpha < 4$.

(g) $\nu = 1$ or 2 dimensional quanum lattice models[5] with long-range interactions[17] $J(n) \sim n^{-\alpha}$ $1 < \alpha < 2$ $(\nu=1)$; $2 < \alpha < 4$ $(\nu=2)$.

(h) Certain quantum field theories including ϕ_2^4 (Ref. 1) ϕ_3^4 (Ref. 3) and Y_2 (Ref. 6).

In addition to these new results, insight has been gained on various older results:

(i) Dyson's result[18] that one-dimensional Ising models with long-range interactions have a phase transition; see Ref. 5 and 33.

(j) Malyshev's result[19] that the two-dimensional anisotropic classical Heisenberg model has a phase transistion for any anisotropy $\varepsilon < 1$; see Ref. 2.

(k) Dobrushin's results that the Ising antiferromagnet has a phase transition in an external field region $|\mu| \lesssim 1 - \tilde{\beta}T$, see Ref. 5.

We should also mention two results which were announced (in Refs. 4 and 2, respectively) whose proofs have developed gaps; it is our belief that the necessary infrared and chessboard estimates hold in these cases but some of the intermediate steps used in the existing proofs of these bounds for models (a) - (k) unfortunately fail (see Section 4).

($\sim$a) $\nu \geqslant 3$ dimensional quantum nearest neighbor simple cubic Heisenberg ferromagnet REMAINS OPEN

(~b) $\nu = 2$ dimensional quantum nearest neighbor simple cubic anisotropic Hesenberg antiferromagnet REMAINS OPEN, except for very small ε[16].

Having summarized the results, we deal with methods in the remainder of this note: in Section 2, we illustrate (following Ref. 2) the improved contour estimates by sketching (j) above; in Section 3, we illustrate (following Ref. 3,5) the use of infrared bounds in model (a); in Section 4, a variety of additional remarks can be found.

2. OS POSITIVITY, CHESSBOARD ESTIMATES AND THE PEIERLS' ARGUMENT

The first two notions in the title of this section are ideas developed in constructive quantum field theory. In their fundamental paper[21] on Euclidean region axiomatics, Osterwalder and Schrader emphasized a positivity condition which was the translation to the imaginary time region of the positivity of the inner product in the physical Hilbert space. This positivity condition, called OS positivity, physical positivity,[22] or reflection positivity turns out to be enough to reconstruct the Hamiltonian semigroup, e^{-tH}, an object which is known[9] to be the analog of the transfer matrix in lattice models of statistical mechanics. In some sense, there will be an analog of OS positivity in all lattice theories having a suitable kind of self-adjoint transfer matrix.[23]

In the development of $P(\phi)_2$ theories[24] various important bounds[25] were realized[26] to be proven quite easily using the Markov property of Nelson[27] and Symanzik,[27] a property which implies OS positivity. In the middle of 1975, many people[28] realized approximately simultaneously that OS positivity was sufficient for these bounds, that they were capable of generalization, and that many new results could be thereby obtained; in particular, Glimm, Jaffe, and Spencer,[1] realized the importance of these bounds to proofs of phase transitions. The bounds were further applied and abstracted by Fröhlich and Simon[29] who dubbed them "chessboard estimates". The relevance of these two notions to lattice gases was realized in Refs. 2 - 5, especially Refs. 2 and 5.

To describe the notions consider an Ising chain of $2n$ spins at sites $\pm 1/2, \pm 3/2, \ldots, \pm (n-1/2)$. Let $\langle\cdot\rangle_0$ denote the usual free expectation $2^{-2n} \sum_{\sigma_i = \pm 1} \cdot$ and for a function F of the spins σ_i, $i > 0$ let θF be the same function of the spins σ_{-i}, i.e. if $F = f(\sigma_{1/2}, \sigma_{3/2} \ldots)$, then $\theta F = f(\sigma_{-1/2}, \sigma_{-3/2}, \ldots)$. OS positivity for the state $\langle\cdot\rangle_0$ is the inequality for real F's that

$$\langle(\theta F)F\rangle_0 \geqslant 0$$

which follows by noting that

$$\langle(\theta F)F\rangle_0 = \langle\theta F\rangle_0 \langle F\rangle_0 = \langle F\rangle_0^2 \geqslant 0 . \qquad (1)$$

Given a Hamiltonian $H(\sigma)$, we define as usual

$$\langle\cdot\rangle = \langle\cdot e^{-H}\rangle_0 / \langle e^{-H}\rangle_0 .$$

Notice that the nearest neighbor ferromagnet H, viz:

$$-H = \beta \sum_{i=-n+1/2}^{n-3/2} \sigma_i \sigma_{i+1} + \beta \sigma_{-n+1/2} \sigma_{n-1/2} \qquad \text{can be written}$$

$$-H = A + \theta A + C_1 \theta C_1 + C_2 \theta C_2 \quad \text{with} \quad A = \beta \sum_{i=1/2}^{n-3/2} \ldots, \; C_1 = \beta^{1/2} \sigma_{1/2} \; ; \; C_2 = \beta^{1/2} \sigma_{n-1/2} .$$

Therefore

$$\langle(\theta F)F\rangle \geqslant 0 \qquad (2)$$

since $\langle e^{-H}\rangle_0 \langle(\theta F)F\rangle = \langle\theta(Fe^A)[Fe^A](e^{C_1 \theta C_1})(e^{C_2 \theta C_2})\rangle_0$ is seen to be positive upon expanding $e^{C_i \theta C_i}$ and using (1). Equation (2) is OS positivity for the interacting model.

Notice that since we chose periodic boundary conditions, one could obtain OS positivity with a variety of θ's involving reflecting about various ways of cutting the circle in half. Moreover, (2) implies a Schwarz inequality which we can iterate by using these different θ's. For example, consider a chain of 4 spins: $\sigma_1, \sigma_2, \sigma_3, \sigma_4$ coupled by a Hamiltonian $-H = \beta(\sigma_1\sigma_2 + \sigma_2\sigma_3 + \sigma_3\sigma_4 + \sigma_4\sigma_1)$. Let $f_1, \ldots, f_4$ by 4 real valued functions on ± 1. Then

$$<f_1(\sigma_1)f_2(\sigma_2)f_3(\sigma_3)f_4(\sigma_4)> \leqslant <f_1(\sigma_1)f_1(\sigma_4)f_2(\sigma_2)f_2(\sigma_3)>^{\frac{1}{2}} <f_4(\sigma_1)f_4(\sigma_4)f_3(\sigma_2)f_3(\sigma_3)>^{\frac{1}{2}}$$

$$\leqslant \prod_{i=1}^{4} <f_i(\sigma_1)\ldots f_i(\sigma_4)>^{\frac{1}{4}}$$

where in the first step we use the Schwarz inequality for the breakup (12)(34) (i.e. $F(\sigma_1,\sigma_2) = f_1(\sigma_1)f_2(\sigma_2)$; $G(\sigma_1,\sigma_2) = f_4(\sigma_1)f_3(\sigma_2)$; $\theta(\sigma_1) = \sigma_4$, $\theta(\sigma_2) = \sigma_3$ and use $<F\theta G> \leqslant <F\theta F>^{\frac{1}{2}} <G\theta G>^{\frac{1}{2}}$) and in the second step the breakup (41)(23) . Clearly by repeating this argument and using it in different directions, we will obtain:

<u>Theorem 1</u> (Chessboard Estimates) Let Λ be a hypercube with sides $2^{n_1}\times\ldots\times 2^{n_\nu}$ in the lattice $\mathbb{Z}^\nu$ and for each α in Λ , let σ_α be an N-component spin. Let $d\mu$ be some measure on $\mathbb{R}^N$ (e.g. $\delta(|\sigma|-1)d^N\sigma$ for the N-vector model), let $H = \frac{1}{2}\beta \sum_{|\alpha-\gamma|=1, \mathrm{mod}\ \Lambda} (\sigma_\alpha-\sigma_\gamma)^2$ (mod Λ means periodic B.C.) , and let $<\cdot> = Z^{-1}\int \cdot\ e^{-H} \prod_{\alpha\in\Lambda} d\mu(\sigma_\alpha)$. Then, for any real valued functions $\{f_\alpha\}_{\alpha\in\Lambda}$ on $\mathbb{R}^N$:

$$< \prod_{\alpha\in\Lambda} f_\alpha(\sigma_\alpha)> \leqslant \prod_{\alpha\in\Lambda} < \prod_{\beta\in\Lambda} f_\alpha(\sigma_\beta)>^{1/|\Lambda|}$$

<u>Remark</u> By more effort,[2] one can allow any hypercube with $2\ell_1\times\ldots\times 2\ell_\nu$ sides.

<u>Example 1</u>[2] Take $\nu = 2$, $N = 3$; $d\mu(\sigma_\alpha) = \delta(|\sigma|-1)d^3\sigma$ and $-H = \beta[\varepsilon \sum_{|\alpha-\gamma|=1} \sigma_\alpha\cdot\sigma_\gamma + (1-\varepsilon)\sum_{|\alpha-\gamma|=1} \sigma_\alpha^{(3)}\sigma_\gamma^{(3)}]$, i.e. the anisotropic classical Heisenberg model. By scaling $\sigma_\alpha^{(3)}$, this can be brought into the form of Theorem 1. Let P_α^+ (resp P_α^-) be the projection onto those states with $\sigma_\alpha^{(3)} > 0$ (resp $\sigma_\alpha^{(3)} < 0$) and let Q_α^δ be the projection onto those states with $|\sigma_\alpha^{(3)}| < (1-\delta)$. By the simple checkerboard estimate at inverse temperature, β :

$$<Q_0^\delta>_\beta \leq <\prod_{\alpha\in\Lambda} Q_\alpha^\delta>_\beta^{1/|\Lambda|}$$

$$\leq \tau_\delta e^{2\beta a(\delta)} \Big/ 1/2\ (1-\tau_\lambda) e^{2\beta b(\lambda)}$$

where τ_δ is the volume of the spherical strip $\{\sigma | |\sigma^{(3)}| < 1-\delta\}$, $a(\delta)$ is the maximum value of $\varepsilon\sigma_1 \cdot \sigma_2 + (1-\varepsilon)\sigma_1^{(3)} \cdot \sigma_2^{(3)}$, when $|\sigma_i^{(3)}| < 1 - \delta$ and $b(\lambda)$ is the minimum of the same function when $\sigma_i^{(3)} > 1 - \lambda$. (The 1/2 comes from the fact that only one "cap" of the sphere is taken and the 2 from the fact that each pair is counted only once so there are $2|\Lambda|$ pairs.) Clearly, by choosing λ very small, we can arrange that $b(\lambda) > a(\delta)$, so $\lim_{\beta\to\infty} <Q_0^\delta> = 0$, i.e. at low temperature each spin likes to point mainly up or mainly down.

Next let $\alpha_1,\ldots,\alpha_n$ be any n distinct spins and let $\beta_1,\ldots\beta_n$ be their nearest neighbor directly to the right. Then

$$< \prod_{i=1}^{n} P_{\alpha_i}^{+} P_{\beta_i}^{-} > \leq <\text{conf}>^{n/|\Lambda|} \equiv C_\Lambda(\beta)^{n/|\Lambda|} \qquad (3)$$

where (conf) is the projection onto all states with $\sigma_\alpha^{(3)} > 0$ (resp. $\sigma_\alpha^{(3)} < 0$) for $\alpha_1 = 0,1,4,5,8,9$, etc.; (resp. $\alpha_1 = 2,3,6,7\ldots$). Equation (3) is obtained by using the reflection idea but with a two-site block rather than a single-site block. Thus, + - reflects to - + + - and the next time to - + + - - + + -, leading to the condition given. By considerations similar to the above $C_\Lambda(\beta)^{1/|\Lambda|} \to 0$ as $\beta \to \infty$ (uniformly in $|\Lambda|$) by crude estimates of free energies and entropies.

(3) is ideal for plugging into a Peierls argument. Explicitly, we will show that $<P_\alpha^+ P_\gamma^->\to 0$ as $\beta\to\infty$ uniformly in $|\alpha-\gamma|$ and Λ. Given that $<Q_0^\delta>\to 0$, this will imply the existence of long-range order for β large. In the usual way, one surrounds α with contours where all spins are up inside the contour and all spins immediately outside are down. If the contour has length ℓ, it must have at least $\ell/4$ pairs of the kind considered in (3) (or of

one of the other three $-+$, $\pm$, $\mp$ possibilities) so $\langle P_\alpha^+ P_\gamma^- \rangle \leq \sum_{\ell=4}^{\infty}$ (# of contours of length ℓ) $C(\beta)^{\ell/4} \to 0$ as $\beta \to \infty$. This proves that long-range order occurs.

The moral of the above is that estimates of contour can be reduced to estimates of "thermodynamic quantities" via chessboard estimates.

3. INFRARED BOUNDS

We want to illustrate the new method of Ref. 3 by proving a phase transition in the classical isotropic Heisenberg model in $\nu = 3$ dimensions; we use ideas from Refs. 4 and 5 as well as Ref 3. For a fixed model of this type (see Theorem 1) let $\langle \cdot \rangle_\Lambda$ denote a finite volume expectation. Griffiths[30] has proven that (for models even under $\sigma_\alpha \to -\sigma_\alpha$)

$$m^2 \geq \lim_{\Lambda\to\infty} \langle (\frac{1}{|\Lambda|} \sum_{\alpha \in \Lambda} \vec{\sigma}_\alpha)^2 \rangle_\Lambda \tag{4}$$

where m denotes the infinite volume spontaneous magnetization so that $m \neq 0$ implies long-range order and multiple phases: (4) is quite easy to prove (see also Ref. 4) and has obvious interpretations in terms of bulk fluctuations.

Let Λ^* be the dual volume to Λ, i.e. if Λ is a cube with $\ell_1 \times \ell_2 \times \ell_3$ sites then Λ^* is the set of p's with $p_i = 2\pi n_i/\ell_i$; $n_i = 0,1,\ldots,\ell_i - 1$. Introduce the Fourier spins: $\hat{\sigma}_p = \sum_{\alpha \in \Lambda} |\Lambda|^{-\frac{1}{2}} e^{-ip\cdot\alpha} \sigma_\alpha$. Then H becomes $H_\Lambda = \beta \sum_{p \in \Lambda^*} E_p \hat{\sigma}_p \cdot \hat{\sigma}_{-p}$ where E_p is the spin wave energy $E_p = 3 - \cos p_1 - \cos p_2 - \cos p_3$. Since $\{|\Lambda|^{-\frac{1}{2}} e^{-ip\cdot\alpha} | p \in \Lambda^*\}$ is an orthonormal basis for $\mathbb{C}^{|\Lambda|}$ we have the Plancherel relation:

$$\frac{1}{|\Lambda|} \sum_{p \in \Lambda^*} \langle \hat{\sigma}_p \hat{\sigma}_{-p} \rangle_\Lambda = 1 \tag{5}$$

since $\sigma_\alpha^2 = 1$. Equations (4) and (5) present a simple "spin wave" picture of phase transitions: Since $|\Lambda^*| = |\Lambda|$, (5) says that under "normal" circumstances, each $\langle \hat{\sigma}_p \hat{\sigma}_{-p} \rangle_\Lambda$ is to be expected to be $O(1)$ as $\Lambda \to \infty$. Equation (4) says that for a phase transition, we need $\langle \hat{\sigma}_{p=0}^2 \rangle_\Lambda$ to be $O(|\Lambda|)$ as $\Lambda \to \infty$.

The basic bound we require is

$$\langle \hat{\sigma}_p \hat{\sigma}_{-p} \rangle_\Lambda \leqslant 3/(2\beta E_p) \tag{6}$$

From this bound we see that

$$\lim_{|\Lambda|\to\infty} \left[\frac{1}{|\Lambda|} \sum_{p\neq 0} \langle \hat{\sigma}_p \hat{\sigma}_{-p} \rangle_\Lambda\right] \leqslant 3/2\beta \int_{|p_i|\leqslant\pi} E_p^{-1}\, d^3p/(2\pi)^3 \equiv 3/2\beta\, G_3(0)$$

since the Fourier sum approaches a Fourier integral. This says that if $\beta > 3/2\ G_3(0)$, then $\lim |\Lambda|^{-1} \langle \hat{\sigma}^2_{p=0} \rangle \neq 0$ so there is long-range order. To summarize the bound (6) and the sum rule (5) force macroscopic occupation of the $p = 0$ mode and thereby long-range order. $\nu \geqslant 3$ is essential for this argument (as it must be since[31] there is no spontaneous magnetization if $\nu = 2$) because the analog of $\int E_p^{-1}\, d^3p$ diverges if $\nu = 2$ since $E_p \sim 1/2\ p^2$ for p small.

Before proving (6) we make two remarks about it: looking at the formula for H and rewriting (6) as $E_p \langle \hat{\sigma}_p \hat{\sigma}_{-p} \rangle_\Lambda \leqslant 3/2\ kT$, we see that (6) is a kind of "equipartition inequality"(the 3 comes from the 3 degrees of freedom, σ^i_α, $i=1,2,3$); since the modes are far from uncoupled there is no reason for an equipartition theorem but an inequality holds nonetheless. Secondly, the field theoretic analog of (6) is trivial which is what suggested it in the first place (in the field theory case,[3] the analog of the sum rule (5) is non-trivial): it turns out to say that $\hat{F}(k)$, the Fourier transform of the two-point function, obeys $\hat{F}(k) \leqslant k^{-2}$; but $\hat{F}(k)$ has a Källén-Lehmann representation $\hat{F}(k) = \int d\rho(m^2)\,(k^2+m^2)^{-1}$ with $\int d\rho(m^2) = 1$ (for canonical theories) so that $(k^2+m^2)^{-1} \leqslant k^{-2}$ yields $\hat{F}(k) \leqslant k^{-2}$.

<u>Theorem 2</u> (Infrared Bounds) Under the hypotheses of Theorem 1:

$$\langle \hat{\sigma}_p \hat{\sigma}_{-p} \rangle \leqslant N/2\beta E_p$$

<u>Proof</u> Define for $(h_\alpha) \in (\mathbb{R}^N)^{|\Lambda|}$:

$$Z(h_\alpha) = \int e^{-H} \prod_{\alpha\in\Lambda} d\mu(\sigma_\alpha - h_\alpha)$$

We first claim that

$$Z(h_\alpha) \leqslant Z(0) \tag{7}$$

It suffices to prove (7) in case $d\mu(\sigma) = F(\sigma)d^N\sigma$ with F everywhere non-zero and then use a limiting argument. In that case, let $G_\alpha(\sigma) = F(\sigma - h_\alpha)/F(\sigma)$ so that

$$Z(0)^{-1}Z(h_\alpha) \equiv \langle \prod_{\alpha\in\Lambda} G_\alpha(\sigma_\alpha)\rangle \leq \prod_{\alpha\in\Lambda} \langle \prod_{\beta\in\Lambda} G_\alpha(\sigma_\beta)\rangle^{1/|\Lambda|} = 1$$

where the inequality follows from the chessboard estimates and the last step follows by noting that

$$\langle \prod_{\beta\in\Lambda} G_\alpha(\sigma_\beta)\rangle = \int e^{-H} \pi d\mu_\beta(\sigma_\beta - h_\alpha) / \int e^{-H} d\mu_\beta(\sigma_\beta) = 1$$

since H is only a function of $\sigma_\beta - \sigma_\gamma = (\sigma_\beta - h_\alpha) - (\sigma_\gamma - h_\alpha)$. This proves (7).

Now we rewrite (7) by making the change of variables, $\sigma_\alpha - h_\alpha \to \sigma_\alpha$ so $H(\sigma_\alpha) \to H(\sigma_\alpha + h_\alpha) = H(\sigma_\alpha) + \beta \sum_{|\alpha-\gamma|=1} (h_\alpha - h_\gamma)(\sigma_\alpha - \sigma_\gamma) + \beta/4 \sum_{|\alpha-\gamma|=1} (h_\alpha - h_\gamma)^2$. Thus (7) becomes:

$$\langle \exp[-\beta \sum_{|\alpha-\gamma|=1} (h_\alpha - h_\gamma)\cdot(\sigma_\alpha - \sigma_\gamma)]\rangle \leq \exp(\beta/4 \sum_{|\alpha-\gamma|=1} (h_\alpha - h_\gamma)^2) \tag{8}$$

from which we conclude that

$$\beta^2 \langle (\Sigma(h_\alpha - h_\gamma)\cdot(\sigma_\alpha - \sigma_\gamma))^2\rangle \leq \beta/2\, \Sigma(h_\alpha - h_\gamma)^2 \tag{9}$$

by taking $h_\alpha \to \lambda h_\alpha$ and expanding to second order about $\lambda = 0$. (9) has only been proven for h_α real but if $(\ )^2$ is interpreted as $|\cdot|^2$ it extends immediately to complex h_α. Take $h_\alpha = e^{ip\cdot\alpha}(1,0,\ldots)$ in (9) and one obtains

$$\langle \hat{\sigma}_p^{(1)} \hat{\sigma}_{-p}^{(1)}\rangle \leq 1/2\beta E_p$$

so the theorem follows by summing over components of σ. Q.E.D.

4. ADDENDA

Finally we make a series of remarks about extensions and problems:

1. The natural setting for proving theorems 1 and 2 is the following:[5] Let $\mathfrak{A}$ be an algebra (of observables), and $\mathfrak{A}_+$, $\mathfrak{A}_-$ two subalgebras with an automorphism $\theta : \mathfrak{A}_+ \to \mathfrak{A}_-$. An expectation $\langle\cdot\rangle_0$ on $\mathfrak{A}$ is said to obey

Generalized OS Positivity (GOS) (resp. Generalized Schwarz Inequality (GSI)) if and only if for all $A_1,\ldots,A_n \in \mathfrak{A}_+$ and $B_1,\ldots,B_n \in \mathfrak{A}_+$, we have $\langle A_1(\theta A_1)\ldots A_n(\theta A_n)\rangle \geqslant 0$ (resp. $\langle A_1(\theta B_1)..A_n(\theta B_n)\rangle \leqslant \langle A_1(\theta A_1)..A_n(\theta A_n)\rangle^{\frac{1}{2}} \langle B_1(\theta B_1)\ldots B_n(\theta B_n)\rangle^{\frac{1}{2}}$) . If $\langle\cdot\rangle$ obeys GOS, and $-H = A + \theta A + \Sigma\, B_i\theta B_i$, then by the argument in Theorem 1 and the Trotter product formula $\langle\cdot\rangle = Z^{-1}\langle \cdot e^{-H}\rangle_0$ is OS positive in the sense that $\langle A\theta A\rangle \geqslant 0$ so that with translation invariance, one can obtain Chessboard type estimates. If $\langle\cdot\rangle_0$ obeys GSI, then

$$\langle\exp(A+\theta B+\Sigma\, C_i\theta D_i)\rangle^2 \leqslant \langle\exp(A+\theta A+\Sigma\, C_i\theta C_i)\rangle \langle\exp(B+\theta B+\Sigma\, D_i\theta D_i)$$

by using the Trotter product formula, expanding the exponential in $\exp((\Sigma\, C_i\theta D_i)_n)$ and using GSI. This bound and translation invariance imply analogs of Equation (8).

2. When can one prove GOS and GSI? Clearly, if $\mathfrak{A}_+$ and $\mathfrak{A}_-$ commute with each other and if $\langle\cdot\rangle_0$ is OS positive, we have $\langle A_1\theta A_1\ldots A_n\theta A_n\rangle = \langle A_1\ldots A_n\theta(A_1\ldots A_n)\rangle \geqslant 0$ and similarly GSI. Three cases of interest occur: (i) (Classical systems; reflections about planes containing no bounds[3]) This is the case discussed in Sections 2 and 3; $\mathfrak{A} = L^\infty(M\times M)$; $\mathfrak{A}_+ = L^\infty(M)$ as $F(m,m') = F(m)$; $\mathfrak{A}_- = L^\infty(M)$ as $F(m,m') = F(m')$ and $(\theta F)(m,m') = f(m')$ if $F(m,m') = f(m)$. $\langle\cdot\rangle_0 = \int \cdot\, d\mu(m)d\mu(m')$; $\mathfrak{A}_+$ commutes with $\mathfrak{A}_-$ and $\langle f(\theta f)\rangle = \int F(m)F(m')d\mu(m)\, d\mu(m') = (\int F(m)d\mu(m))^2 \geqslant 0$. (ii) (Classical systems; reflections about planes containing bounds[5]) This allows one to deal with lattices like the face centered cubic where the perpendicular bisectors of bounds contain lattice sites $\mathfrak{A} = L^\infty(M\times N\times M)$; $\mathfrak{A}_+$ = functions of m and n , $\mathfrak{A}_-$ = functions of n and m' ; $\theta: F(m,n) \mapsto F(m',n)$; $\langle\cdot\rangle_0 = \int \cdot\, d\mu(m)d\nu(n)d\mu(m')$ and $\langle f(\theta f)\rangle = \int d\nu(n) (\int f(m,n)d\mu(m))^2 \geqslant 0$. (iii) ("Real" Quantum Systems; reflections about planes containing no sites.) Here $\mathfrak{A}$ is an algebra of matrices on $\mathcal{H}\otimes\mathcal{H}$ which are all simultaneous <u>real</u> , $\mathfrak{A}_+$ (resp. $\mathfrak{A}_-$) is matrices of the form $A\otimes 1$ (resp. $1\otimes A$) and $\theta(A\otimes 1) = 1\otimes A$, $\langle\cdot\rangle_0 = \mathrm{Tr}(\cdot)$. Then $\mathfrak{A}_+$ commutes with $\mathfrak{A}_-$ and $\langle B\theta B\rangle = \mathrm{Tr}(B)^2 > 0$ <u>since Tr(B) is real</u> . This allows one to treat models like

the x,y model and also, in effect, the antiferromagnet (where the Hamiltonian becomes, after suitable rotations, $-H = \sigma_x\sigma_x' + \sigma_z\sigma_z' + (i\sigma_y)(i\sigma_y')$ and $\sigma_x, \sigma_z, i\sigma_y$ are simultaneously real) but <u>not</u> the ferromagnet.

3. There are two cases of interest in which GOS fails: (i) ("Real" quantum systems, reflection about planes with sites). Here $\mathfrak{A}$ is an algebra of real matrices on $\mathcal{H}\otimes\mathcal{K}\otimes\mathcal{H}$ with $\mathfrak{A}_+$ = matrices of $A\otimes B\otimes 1$ and $\theta(A\otimes B\otimes 1) = 1\otimes B\otimes A$. Here $\mathfrak{A}_+$ and $\mathfrak{A}_-$ do not commute. If $\langle\cdot\rangle_0 = \mathrm{Tr}$ it always happens that $\langle\cdot\rangle_0$ is OS positive but[5] GOS may fail. (ii) (General Quantum Systems) Note that if $\mathfrak{A}_+$ commutes with $\mathfrak{A}_-$ and $\langle\cdot\rangle_0$ is OS positive, then $\langle(A_1\theta A_1+\ldots+A_n\theta A_n)^k\rangle \geq 0$. For two quantum Pauli spins $\mathrm{Tr}((\vec{\sigma}_1\cdot\vec{\sigma}_2)^3) < 0$ so OS positivity fails; in fact $\langle(\sigma_x\sigma_y\sigma_z)(\sigma_x'\sigma_y'\sigma_z')\rangle_0 = -1$. We should emphasize that we do not know that OS positivity restricted to functions of σ_z fails for the interacting expectation in either case (this is all that is missing to extend Example 1 to quantum ferromagnets); all we know is that the proofs thus far used, all of which expand an exponential, do fail.

4. One can ask when $\sum_{n\neq m} J(|n-m|)s_{n+\frac{1}{2}}s_{m+\frac{1}{2}}$ is of the form $A+\theta A+\Sigma\, C_i\theta C_i$ in terms of the θ of Section 2. For finite range J's, only for $J(n) = \delta_{n1}$ but infinite range σ's of the form $|n|^{-\alpha}$ are also OK.[5] In particular, one obtains infrared bounds in that case. If one notes that $\int dp\, E_p^{-1} < \infty$ only for $\alpha < 2$, one has a simple proof[5] of Dyson's result [18] (see Ref. 33).

5. There is one major step in pushing the argument of Section 3 through in the quantum case (like the xy model) that we haven't yet mentioned. The sum rule (5) (with suitable replacements for "1") is true for the thermal expectations. But the infrared bound, (6), holds for a different kind of "two point function", namely $(\hat{\sigma}_p, \hat{\sigma}_{-p})$ where (A,B) is the "DuHamel two-point function":

$$(A,B) = \mathrm{Tr}(e^{-H})^{-1}\int_0^1 \mathrm{Tr}(e^{-xH}A\,e^{-(1-x)H}B)\,dx$$

This enters since going from (7) to (9) involves an expansion of $Z(\lambda h)$ to second order and

$$\mathrm{Tr}(e^{-H+\lambda A}) = \mathrm{Tr}(e^{-H})\ [1+\lambda \langle A\rangle + 1/2\ \lambda^2 (A,A) + 0(\lambda^3)]$$

To complete the proof of phase transitions one needs an upper bound on $g(A) = 1/2 \langle A^*A+AA^*\rangle$ in terms of $b(A) = (A^*,A)$. The following bound of Bruch and Falk[32] (rediscovered in Ref. 4) does the trick: $b(A) \geq g(A)f(c(A)/4g(A))$ where $c(A) = \langle [A^*,[H,A]]\rangle$ and f is the function $f(x \tanh x) = x^{-1} \tanh x$.

6. Among the important open questions are the following: (i) Do infrared bounds hold for the nearest neighbor isotropic quantum Heisenberg ferromagnet? (ii) Is the nearest neighbor anisotropic Heisenberg ferromagnet OS positive for functions of the σ_z's alone? (iii) For which interactions $J(n)$ do the infrared bounds hold in the sense $\langle \hat{\sigma}_p \hat{\sigma}_{-p}\rangle \leq N/2\beta E_p$ with $E_p = 1/2 \sum_n J(n)(1-\cos np)$ with $(n \in Z^\nu)$?

FOOTNOTES

1. J. Glimm, A. Jaffe, and T. Spencer, Commun. Math. Phys. 45 (1975) 203.
2. J. Fröhlich and E.H. Lieb, Phys. Rev. Lett. 38 (1977) 440, and in prep.
3. J. Fröhlich, B. Simon, and T. Spencer, Phys. Rev. Lett 36 (1976) 804 and Commun. Math. Phys. 50 (1976) 79.
4. F. Dyson, E.H. Lieb, and B. Simon, Phys. Rev. Lett. 37 (1976) 120 and J. Stat. Phys., to appear.
5. J. Fröhlich, R. Israel, E.H. Lieb, and B. Simon, in prep.
6. J. Fröhlich, 1976 Schadlming Winter School Lectures, "Current Problems in Elementary Particle and Mathematical Physics", ed. P. Urban, Springer, 1976.
7. J. Fröhlich, Bull. Am. Math. Soc., to appear.
8. J. Fröhlich and T. Spencer, 1976 Cargese Summer School Lectures.
9. F. Guerra, L. Rosen, and B. Simon, Ann. Math. 101 (1975) 111.
10. B. Simon and R. Griffiths, Commun. Math. Phys. 33 (1973) 145.
11. J. Glimm, A. Jaffe, and T. Spencer, 1973 Erice Summer School Lectures; Springer Physics Lectures No. 25 and Ann. Math. 100 (1974) 585.
12. T. Spencer, Commun. Math. Phys. 39 (1974) 63; J. Glimm, A. Jaffe, and T. Spencer, Ann. Phys. 101 (1976) 610, 631.
13. Various lattice models with unbounded spins are discussed in G. Sylvester, 1977 Rockefeller preprint; lattice gauge theories are discussed in K. Osterwalder and E. Seiler, 1977 Harvard preprint; quantum lattice systems are discussed by J. Frohlich, Ref. 7 and in prep.

14. T. Berlin and M. Kac, Phys. Rev. 86 (1952) 821.
15. E. Stanley, Phys. Rev. 176 (1968) 718; M. Kac, Quart. Appl. Math. 30 (1972) 17-29.
16. J. Ginibre, Commun. Math. Phys. 14 (1969) 205; D.W. Robinson, Commun. Math. Phys. 14 (1969) 195.
17. R. Israel in Commun. Math. Phys. 43 (1975) 59-68 proved the existence of some pair interacting quantum Heisenberg ferromagnet with spontaneous magnetization. His method is so abstract that no details on the behavior of J(n) as $n \to \infty$ are available.
18. F. Dyson, Commun. Math. Phys. 12 (1969) 91.
19. V.A. Malyshev, Commun. Math. Phys. 40 (1975) 75.
20. R.L. Dobrushin, Funct. Anal. Appl. 2 (1968) 292; 302.
21. K. Osterwalder and R. Schrader, Commun. Math. Phys. 31 (1974) 83
22. In the field theory case, this positivity condition is just positively of the physical inner product, but the term "physical positivity" is inappropriate in the statistical mechanical models.
23. It must be emphasized that "a suitable kind" of transfer matrix is much more general than the usual kind of transfer matrix; in particular, even for the Ising case, one can have infinite dimensional "transfer matrices" when, e.g. suitable long-range forces are present.
24. This theory is reviewed in B. Simon, The $P(\phi)_2$ Euclidean Quantum Field Theory, Princeton University Press, 1974.
25. J. Glimm and A. Jaffe, J. Math. Phys. 13 (1972) 1558; J. Fröhlich, Helv. Phys. Acta 47 (1974) 265.
26. F. Guerra, L. Rosen, and B. Simon, Commun. Math. Phys. 27 (1972) 107 and B. Simon, 1973 Erice Summer School Lectures
27. K. Symanzik, unpubl. 1964 report; E. Nelson, J. Func. Anal. 12 (1973) 97-112.
28. Ref. 1; F. Guerra, 1975 Marseille Conf. Proc.; O. McBryan, 1975 Marseille Conf. Proc.; Y. Park, J. Math. Phys. 12 (1976) 1073; 1143; E. Seiler and B. Simon, Ann. Phys. 97 (1976) 470.
29. J. Fröhlich and B. Simon, Ann. Math., to appear.
30. R. Griffiths, Phys. Rev. 152 (1966) 240.
31. N.D. Mermin, J. Math. Phys. 8 (1967) 1061; R.L. Dobrushin and S.B. Shlosman, Commun. Math. Phys. 42 (1975) 31.
32. H. Falk and L. Bruch, Phys. Rev. 180 (1969) 442-444.
33. R. Israel, Proceedings of this Conference

CHAPTER 23

PERCOLATION CLUSTERS

E. Stoll, C. Domb* and T. Schneider

IBM Zürich Research Laboratory, 8803 Rüschlikon, Switzerland

*Permanent Address: Physics Dept., King's College,

London WC2R 2LS, U. K.

ABSTRACT

The film illustrates percolation in a two-dimensional s.q. lattice in the neighborhood of the critical concentration. A preliminary version was shown at the Banff Summer School on Critical Phenomena last year, and a number of suggestions for improvement were put forward as a result. The present version incorporates these suggestions, and adds the teaching slides as a result of requests by many in the audience at Banff for an introduction to the basic concepts.

1. INTRODUCTION

The problem of percolation involving randomness in a medium has application to many disciplines including solid-state physics, statistics, polymer chemistry, and electrical engineering. Since the exact mathematical information available is meagre, the subject is particularly amenable to illustration by computer simulation. The present film is derived from a one-spin-flip Ising model. A finite square net of size 85×85 with a cyclic boundary condition is used to simulate an infinite medium. At sufficiently high temperatures the coupling between spins can be ignored and the system corresponds to a random mixture, the ratio of components being controlled by the magnetic field.

The film is designed to illustrate the rapid growth of large clusters as the concentration p approaches the critical percolation probability p_c, and the emergence of an infinite (spanning) cluster as p passes through p_c. The percolating particles are colored green (•), the largest cluster in any snapshot is colored red (+), and a spanning cluster which extends two-dimensionally to the edges of the lattice is colored yellow (*).

The introductory slides are meant to illustrate the following concepts:

1. Randomness in a medium as opposed to the more usual randomness in particles.
2. Percolation of a fluid in a medium. Bond percolation processes.
3. Finite size effects. Difference between finite and infinite networks.
4. Critical bond concentration. Formation of an infinite cluster.
5. Amorphous semiconductors and switches.
6. Randomness in sites. Dilute magnets.
7. Site percolation processes. Formation of an infinite cluster.
8. Relation between bond and site percolation processes. Bond → site transformation. Covering lattices.
9. Statistics of critical behavior.

2. ILLUSTRATIVE FILM SNAPSHOTS

We show in Figs. 1-3 typical snapshots illustrating different stages in the growth of clusters. The critical concentration p_c for a site process on an SQ lattice is $\simeq 0.59$. Figure 1 corresponds to $p/p_c = 0.60$. The largest cluster is

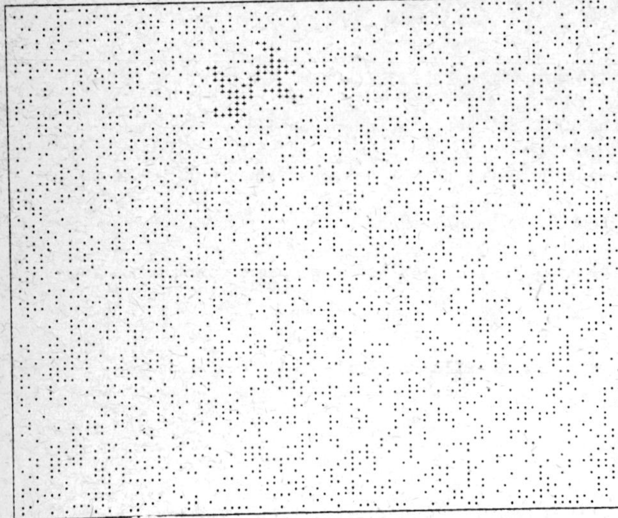

Fig. 1. A small maximum cluster for $p/p_c = 0.60$ (marked by + signs, red in film). The dots represent percolating particles (green in film).

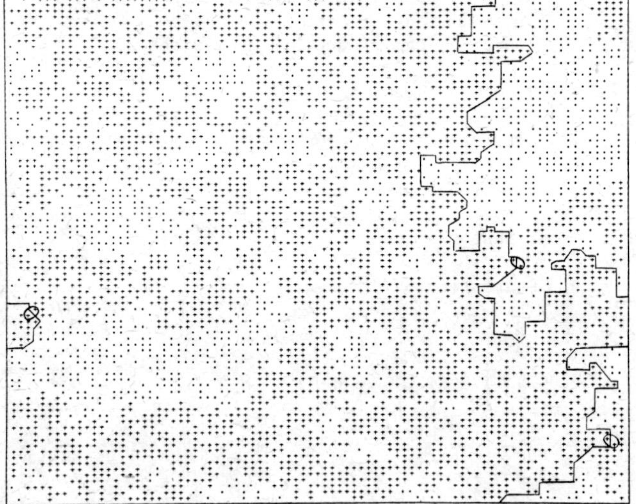

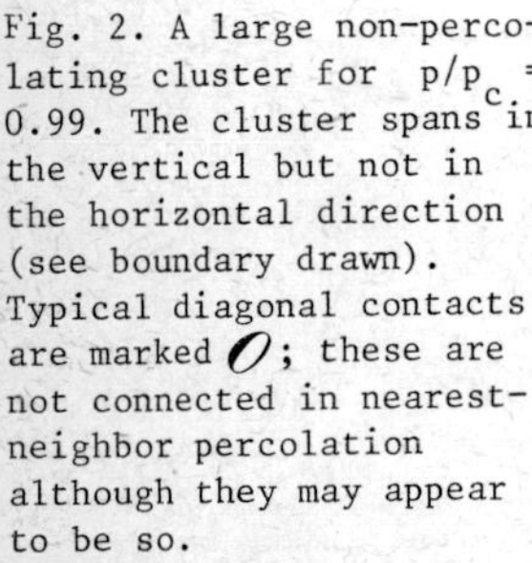

Fig. 2. A large non-percolating cluster for $p/p_c = 0.99$. The cluster spans in the vertical but not in the horizontal direction (see boundary drawn). Typical diagonal contacts are marked $\varnothing$; these are not connected in nearest-neighbor percolation although they may appear to be so.

still quite small and is comparable in size to the other clusters; there is little evidence of aggregation among the clusters. Figure 2 corresponds to $p/p_c = 0.99$. A dominant cluster has clearly emerged much larger than the others, but it is not yet a spanning cluster. In Fig. 3, corresponding to $p/p_c = 1.04$, a spanning cluster has formed and it contains nearly all of the percolating particles.

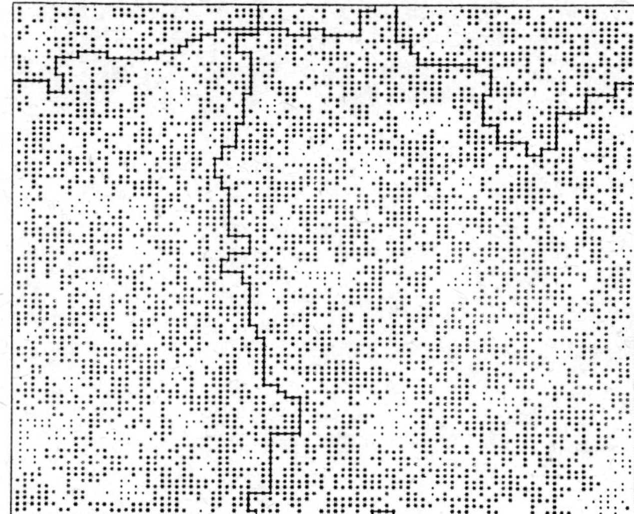

Fig. 3. A percolating cluster for $p/p_c = 1.04$ (marked by *, yellow in film). Typical horizontal and vertical spanning paths are shown.

3. FINAL SLIDES

The last group of slides is designed to illustrate some more complex problems arising in percolation theory. If second, third, and higher neighbor "contacts" are taken into account on a lattice, the problem of "continuum" percolation in a random medium is approached. If we assume that site occupation is not random but that short-range correlations exist, we are led to the problem of percolation in Ising systems.

The kinetic Ising-model simulations can be used equally well to explore percolation with short-range correlations; this is a useful asset since other techniques are not readily available for this case.

CHAPTER 24

OVERVIEW AND APPLICATIONS OF PERCOLATION THEORY

Richard Zallen

Xerox Webster Research Center, Webster, N. Y., U. S. A.

Percolation theory provides an intuitively-satisfying and well-defined mathematical model for geometrically-statistical situations. Spatially-random phenomena which can be viewed as percolation processes include a variety of composition-dependent metal↔insulator transitions: macroscopic mosaics of conducting and insulating particles, microscopic dispersions of alkali atoms in rare-gas solids, and Anderson localization in amorphous semiconductors. A survey of applications to heterogeneous systems will be discussed, as will recent developments such as polychromatic percolation.

1. INTRODUCTION

Percolation theory,[1] as a glance at recent issues of Phys. Rev. reveals, is a burgeoning topic. There are good reasons for this. For one thing,percolation is a pretty subject, with much intellectual appeal and some almost game-like mathematical aspects. For another, it provides a well-defined, transparent, and intuitively-satisfying model for spatially-random phenomena. Situations in which geometric disorder or (worse yet) topological disorder play a dominant role, as in amorphous solids or in composite materials, are noted for their stubborn intractability. Any added theoretical tool with which to attack such problems is obviously welcome. However, the single most seductive aspect from the perspective of solid state physicists (who have largely taken over the subject) is the inherent presence in percolation processes of a sharp phase transition, at which long-range connectivity suddenly appears. It is this transition, which occurs with increasing density or occupation or concentration, which makes percolation a natural model for a diversity of phenomena.

For some degree of self-containment, we mention the typical percolation setting. We start with a regular geometric object, a periodic lattice made up of sites (vertices) and bonds (edges, pairwise connections between sites). Now a nongeometric two-state property, which carries the statistical character of the problem, is randomly assigned to each site or bond. (Later in this paper we will discuss a generalization, called polychromatic percolation, in which the nongeometric property may take on more than two values.) Thus, in site-percolation processes, each site is either filled or empty (with probabilities p and 1-p, respectively), while in bond percolation each bond is either unblocked or blocked (with corresponding probabilities). Adjacent filled sites (or unblocked bonds) are regarded as linked, and when p exceeds a critical value p_c (the percolation threshold) the clusters of linked sites (or bonds) are no longer all isolated and an unbounded cluster appears. This onset of the infinite cluster or percolation path is sketched in Fig. 1 in the form of the function P(p), which specifies the probability of a site or bond belonging to that extended cluster.

Note that the spatial disorder in the described situation arises via the superposition of a random two-state variable on an essentially crystalline substrate—a regular lattice. Such a model is appropriate for cases like dilute magnets, with magnetic atoms distributed substitutionally on the sites of a nonmagnetic host lattice. Most available percolation-theory results deal with situations like this. However, there are many situations of interest in which the underlying structure bears no resemblance to a periodic lattice, but is itself intrinsically disordered. The distinction here is between a mild form of disorder—compositional disorder on a regular lattice, and a severe form—topological disorder. The former is exemplified by a mixed crystal such as Ge_xSi_{1-x}, in which we 'know' where each atom is (on a diamond-structure lattice in this case) but we don't know what it is (i.e. Ge or Si). The

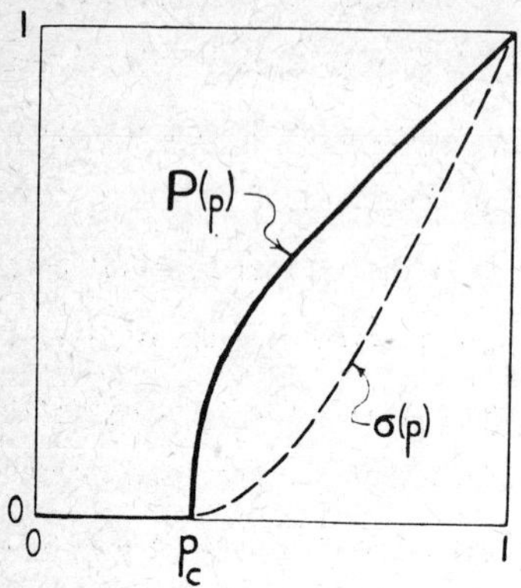

Fig. 1. Behavior of the percolation probability P and the conductivity σ as a function of the concentration p .

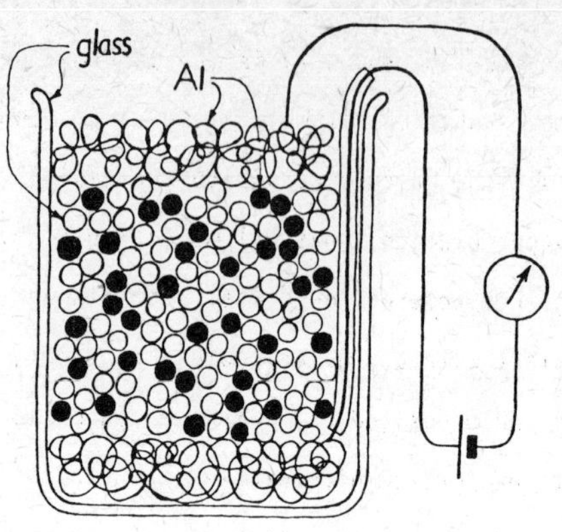

Fig. 2. Schematic of the table-top experiment of Fitzpatrick, Malt and Spaepen (Ref. 2)

latter is exemplified by an amorphous solid, such as a-Ge. In a mixed glass such as a-Ge_xSi_{1-x}, compositional disorder is superimposed on topological disorder. This is the analog of a site-percolation structure ($x \leftrightarrow p$) on a disordered lattice. In recent years, percolation theory has been extended to deal with processes occurring on amorphous substrates (i.e. topologically-disordered lattices), and also to processes occurring on continua.

In this paper, representative recent applications of percolation theory will be briefly reviewed, in a case-history approach. Also, a few other recent issues involving percolation will be mentioned, albeit sketchily because of space limitations. Table I lists a dozen typical applications of percolation theory, half of them referring to macroscopic phenomena and the other half to microscopic phenomena.

2. 'pressed down and shaken together'

The homely experiment schematically represented in Fig. 2, which was reported (as an undergraduate student project!) by Fitzpatrick, Malt, and Spaepen,[2] illustrates

Table I. Applications of percolation theory

phenomenon or system	transition
flow of liquid in a porous medium	local/extended wetting
spread of blight in an orchard	containment/epidemic
communication or resistor networks	disconnected/connected
conductor-insulator composite materials	insulator/metal
composite superconductor-metal materials	normal/superconducting
thin vapor-grown metal films	insulator/metal
thin He films on surfaces	normal/superfluid
polymer gelation	liquid/gel
dilute magnets	para/ferromagnetic
metal-atom dispersions in insulators	insulator/metal
mobility edge in amorphous semiconductors	localized/extended states
variable-range hopping in amorph. semicond.	(resistor-network analog)

many of the possible complicating and interesting features simultaneously. The experimental procedure is essentially as follows. Crumpled Al foil is pressed into the bottom of a glass beaker, and is connected to a battery via an insulated wire. A mixture of small plastic and metallic balls of equal size is now poured into the beaker. A second Al-foil electrode is crumpled and pressed down onto the top of the mixture, and the current circuit is completed through an ammeter to the battery. Current is measured as a function of the composition (fraction of balls which conduct) of the conductor/insulator mixture. No current flows until a critical composition, a minimum fraction of metal balls, is reached.

The mixture in the beaker is a prototype of a macroscopically heterogeneous composite material, and the conductivity threshold is a percolation process since it is controlled by the geometric connectivity of the material. The substrate on which percolation is superimposed is a topologically-disordered structure of special significance, the random close-packed (rcp) structure. This drastically-disordered (but not 'random', of course) structure, which results when hard spheres rapidly aggregate under compression (provided the container avoids plane surfaces, since these encourage crystalline domains) is very reproducible, has been extensively characterized,[3] and is currently believed to provide the best model for the microscopic structure of amorphous metals.[4] The rcp structure also has an ancient history, since for a very long time volume has served as a measure of grain (for example) in barter and commerce, volume occupied by disordered arrays of macroscopic particles in contact with one another. The biblical quotation which provides the heading for this section (the key terms in the definition of 'good measure', St. Luke, VI, 38) reveals the existence of an early and widespread appreciation of random close packing.

Not only are percolation and topological disorder simultaneously present in this simple example, but it also exemplifies another important class of problem in which we are interested, namely the conductivity on a random resistor network. Although the onset of long-range conductivity occurs at the percolation threshold, the subsequent increase in σ is much softer than that for P(p). This has been indicated by the dashed curve in Fig. 1. The reason for this is that near p_c the percolation paths are extremely tortuous and contain many dead ends (which contribute to P but not to σ). The critical exponents, to be discussed later, are different for σ and P.

Well then, what is the critical composition for the mixture of Fig. 2, as provided by percolation theory? Clearly this is a site-percolation situation. If all of the balls were neatly stacked in a cubic close-packed structure, then the answer could be provided by the known p_c for site percolation on the fcc lattice. But the actual structure is quite different, and no explicit percolation calculations are available for the rcp structure. However the answer is provided by the construct of a critical volume fraction, discussed next.

3. Critical volume fraction and continuum percolation

The critical volume fraction $\phi_c(d)$ may be defined as $\langle p_c^j(d) f_j(d) \rangle$, where $p_c{}^j(d)$ is the site-percolation threshold for the simple (nearest-neighbor bonds only) lattice j in d dimensions.[5,6] Here, $f_j(d)$ is the filling factor of the lattice corresponding to the packing of equal touching nonoverlapping d-dimensional spheres centered on the lattice sites. The utility of $\phi_c(d)$ in two and three dimensions is based on the empirical rule that, while p_c^j varies strongly with j, the product $p_c^j f_j$ (equal to the fraction of space which, at the site-percolation threshold, is occupied by the spheres containing filled sites) is nearly <u>independent</u> of j for given d.[5] Thus, at least for d=2 and d=3, $p_c^j f_j$ is an approximate 'dimensional invariant' (a form of universality) which depends on dimensionality but does <u>not</u> depend on the details of lattice geometry: $p_c^j(d) f_j(d) \approx \phi_c(d)$. While the values of p_c for site percolation on the d=3 fcc, bcc, sc, and diamond-structure lattices vary between 0.198 and 0.43, the values of $p_c^j f_j$ for all four fall within the range 0.156 ± 0.010. Recent work by the author indicates that the idea of a critical volume fraction retains validity for d=4 and d=5 (with $\phi_c(d{=}4) \approx 0.061$ and $\phi_c(d{=}5) \approx 0.023$), but not for d>9.[7]

For the problem posed in the preceding section, all that we need to know is the filling factor f_{rcp} for the random close-packed structure. The percolation threshold is then $\phi_c(d{=}3)/f_{rcp} = 0.156/0.637 = 0.24$, a value which agrees well with the experimental observation for the fraction of metal balls (about 25%) at the onset of conductivity. This example demonstrates that the construct of a critical volume fraction applies to systems in which the underlying structure is topologically disordered.

Davidson and Tinkham[7½] have studied a related situation in which superconducting Nb_3Sn filaments are embedded in a copper matrix. Here the transition involved is normal-metal$\leftrightarrow$superconductor rather than insulator$\leftrightarrow$metal, but again the critical volume fraction for percolation is found to apply.

From these examples we can see that percolation theory has an important application to a broad class of composite materials(many of technological significance) which are macroscopic mosaics of different phases, materials which traditionally present great difficulties for mathematical analysis. It should be noted however that the observed values of the critical volume fraction is not always near 15% but is sometimes substantially larger, as in the work reported by Abeles et al on $W\text{-}A\ell_2O_3$ granular metal films.[8] A percolation explanation of those data has been proposed by

Cohen et al on the basis of a bias toward the formation of isolated metallic regions.[9] Even in such high-threshold cases, it is significant that the above-threshold composition dependence of the conductivity is ubiquitously observed to follow the expected percolation-on-a-resistor-network behavior.

It is evident that the introduction of a critical volume fraction leaves behind the necessity of a lattice and is a big step toward dealing with percolation on a continuum. Zallen and Scher[6] discussed this idea in the context of the localization-delocalization transition of a classical particle in a random potential (i.e. a classical analog of the Anderson transition).[9½] Webman, Jortner, and Cohen[10] have carried out a series of three-dimensional resistor-network calculations in which they simulate continuum percolation by introducing correlations between bonds. Their result for ϕ_c is 0.145$\pm$0.005, satisfyingly close to the previously-cited (and very differently obtained)[5,6] estimate of 0.156$\pm$0.010. A 'round-number' value of 15% for the critical volume fraction satisfactorily fits both estimates.

4. Atomic dispersions

Thus far we have emphasized macroscopically-heterogeneous systems, because these constitute an important new direction for percolation. But more traditional microscopic applications continue to arise, and two will be mentioned here in connection with recent experimental work.

Birgeneau and co-workers have carried out extensive experiments on the dilute antiferromagnets $Rb_2Mn_pMg_{1-p}F_4$.[11] Because of the crystal-structure anisotropy and the dominance of nearest-neighbor exchange, this system corresponds to two-dimensional site percolation on a simple square lattice. The critical concentration of Mn atoms is p_c=0.59. These authors discuss their work in the context of the interpretation of the point $p=p_c$, T=0 as a multicritical point in the p,T plane.[12]

Phelps and Flynn[13] have made measurements on atomically-mixed dispersions, rare-gas-alkali-metal solid films, which they interpret in a manner completely analogous to the discussion of the macroscopic-mixture experiment of Fig. 2. These authors present arguments for these alloys as 'paradigms of percolative processes on an atomic scale'.

Their results for the volume fraction of alkali atoms at the conductivity onset are 0.14 (for Rb in Kr) and 0.17 (for Cs in Xe), in good agreement with ϕ_c=0.15.

5. Critical exponents, marginal dimensionality, and RGT

This section collects some results which bear on the critical behavior of the percolation transition. Table II lists the present 'best' estimates of exponents which characterize several functional dependences, near p_c, as culled by me from various sources describing numerical simulations.[14] The first three exponents are analogous to those for magnetic systems, e.g. the percolation probability $P\sim(p-p_c)^\beta$ is the analog of the magnetization $M\sim(T_c-T)^\beta$. The quantities S and L are the mean cluster size and pair-connectedness correlation length, respectively.

In addition to the values in two and three dimensions, the exact values known for the Bethe lattice (no loops) are shown in the column labelled $d\geqslant6$. A Bethe lattice is a $d\rightarrow\infty$ object and should correspond to the mean-field limit. From the Bethe-lattice values and the scaling law $\nu d=2\beta+\gamma$, Toulouse[15] 'nonimplausibly conjectured' that the marginal dimensionality (above which, the mean-field exponents apply) is d^*=6 for percolation processes. This has since been confirmed numerically by computer calculations and series expansions for hypercubic lattices,[16] and also by renormalization group theory.[17] (The latter was made possible by the recently-established mapping of percolation onto a generalized Ising model.[18]) Percolation processes, incidentally, provide a very transparent setting for renormalization-group ideas. (This is a pedagogic application of percolation.) p_c appears as a fixed-point solution in an RGT analysis. Note that the large d^* accounts for the pronounced critical behavior (i.e. small β, very different from the mean-field value of unity) of the percolation order parameter P in d=2 and d=3.

Table II. Critical exponents.

function	exp.	d=2	d=3	$d\geqslant6$
$P\sim(p-p_c)^\beta$	β=	0.14	0.39	1
$S\sim(p_c-p)^{-\gamma}$	γ=	2.3	1.8	1
$L\sim(p_c-p)^{-\nu}$	ν=	1.3	0.9	1/2
$\sigma\sim(p-p_c)^t$	t=	1.2	1.5	2

The exponent most relevant to our preceding discussion is t(d=3) which governs the percolation conductivity and has a theoretically-estimated value near 1.5. Cohen et al[9] show that this value applies to the observations of Abeles et al[8] on granular metal films. Also, Coutts[19] has assembled a set of reported results for this exponent in other mixed insulator-conductor composites, and the observed values do indeed cluster around 1.5.

6. Polychromatic percolation

The thrust of this short review has been primarily practical up to this point. We now change direction to briefly discuss a generalization of percolation in which the statistically-assigned state function (e.g. filled or empty), which is superimposed on a regular geometry to yield a percolation process, is permitted to take on three or more discrete values, rather than just two.[7] Ordinary percolation is a two-species problem in black and white (although we usually focus on one and ignore the other), and it is attractive (and alliterative) to call the multi-species generalization polychromatic percolation.[7] In the simplest type of polychromatic site percolation, each site is occupied by a 'particle' of some color, and adjacent sites occupied by particles of the same color are linked. Similarly, for bond percolation, each bond is permeable to a 'fluid' of some color, and adjacent bonds of (permeable to the fluid of) the same color are linked.

Composition diagrams characterize polychromatic percolation processes. Examples are shown in Fig. 3 for a p_c of 0.2, which is the value for site percolation on the fcc lattice. The upper diagram shows the linear composition field of ordinary (bichromatic) percolation, while the lower diagram represents the triangular composition field with three species present. With C colors present the composition figure is a simplex in C-1 dimensions. The hyperplane $x_i=p_c$ is the locus of compositions for which the concentration x_i of species i equals the percolation threshold, and there are C such hyperplanes which partition the composition field into regions in which different numbers of colors, C_p, percolate. Figure 3 illustrates this for C=2, p_c=0.2 and C=3, p_c=0.2.

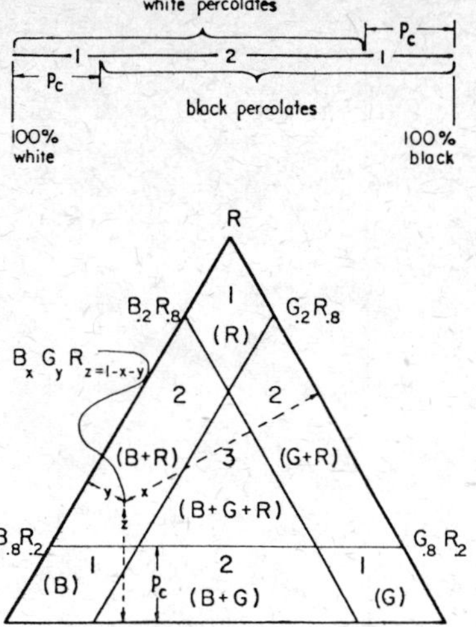

Fig. 3. Composition fields for site-percolation processes on the fcc lattice, shown for ordinary (black-and-white) percolation and for polychromatic percolation with three colors present (Ref. 7)

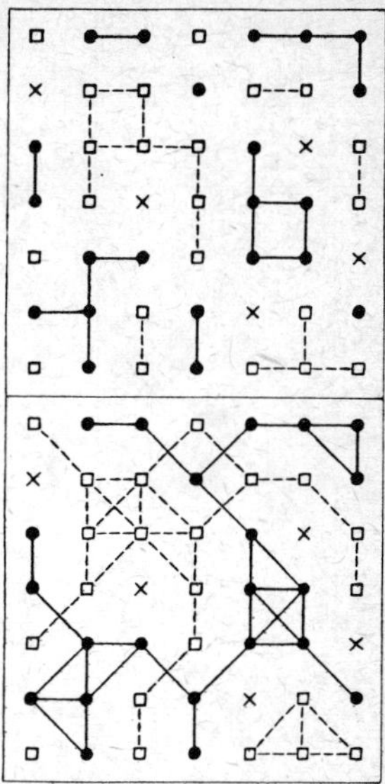

Fig. 4. Multiple percolation via increased range of interaction

Each of these two examples exhibits a central simplex-shaped region in which all species percolate. Such a panchromatic regime occurs whenever $C<p_c^{-1}$, occupying a fraction $(1-Cp_c)^{C-1}$ of the composition field.[7]

The possibility of having many colors or species present, rather than just two, is most interesting in situations in which p_c is low, since in such cases it is possible for several species to simultaneously percolate. Increasing the range of interaction by introducing bonds to more distant neighbors is one way of obtaining a low threshold. A graphic illustration is displayed in Fig. 4, in which three species occupy the sites of a square lattice in concentrations of roughly 45%, 45%, and 10%. Adding bonds to second neighbors permits two colors to percolate, while none can with only nearest-neighbor bonds present.

Close-packed (cp) lattices in higher dimensions provide examples of lattices of very high connectivity. Empirical estimates of site-percolation thresholds for cp

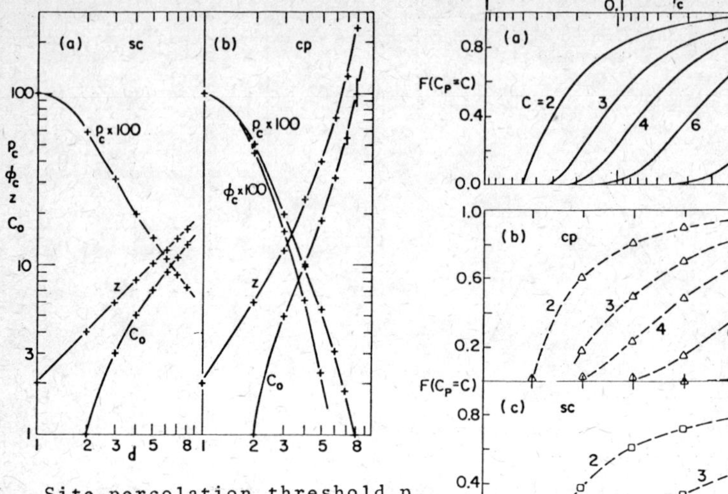

Fig. 5. Site-percolation threshold p_c, coordination number z, and maximum number of percolating species C_o for d-dimensional simple cubic (a) and close-packed (b) lattices. Included in (b) is the lattice-independent critical volume fraction ϕ_c

Fig. 6. Behavior of the panchromatic percolation fraction

lattices up to d=8 have been discussed by us elsewhere.[7] These estimates are plotted versus dimensionality in Fig. 5, along with $p_c(d)$ values for simple-cubic (sc) lattices. Also shown is the d dependence of the coordination number z (the simplest measure of connectivity) and the maximum number of percolating colors C_o. The latter is an over-simplified indicator of the scope for multiple percolation; a much better measure is shown in Fig. 6 in the form of $F(C_p=C)$. This quantity is the panchromatic fraction of the composition field over which all colors percolate. Figure 6 presents $F(C_p=C)$ versus p_c for a general percolation process, as well as for site percolation in particular on d-dimensional cp and sc lattices.

7. Concluding comment

Percolation theory is (to coin a phrase) an idea whose time has come. It is hoped that some of the reasons for this have been communicated in this brief review.

REFERENCES

1. S.R. Broadbent and J.M. Hammersley, Proc. Cambridge Philos. Soc. 53, 629 (1957); H.L. Frisch and J.M. Hammersley, J. Soc. Ind. Appl. Math 11, 894 (1963); V.K.S. Shante and S. Kirkpatrick, Adv. Phys. 20, 325 (1971); J.W. Essam, in Phase transitions and Critical Phenomena, edited by C. Domb and M.S. Green (Academic, London, 1972), Vol. 2, p. 197.

2. J.P. Fitzpatrick, R.B. Malt, and F. Spaepen, Phys. Lett. A 47, 207 (1974).

3. J.L. Finney, J. de Physique 36, C2-1 (1975).

4. G.S. Cargill, Solid State Physics 30, 227 (1975).

5. H. Scher and R. Zallen, J. Chem. Phys. 53, 3759 (1970).

6. R. Zallen and H. Scher, Phys. Rev. B 4, 4471 (1971).

7. R. Zallen, Phys. Rev. B, Aug. 15, 1977.

7½. A. Davidson and M. Tinkham, Phys. Rev. B 13, 3261 (1976).

8. B. Abeles, H.L. Pinch, and J.I. Gittleman, Phys. Rev. Lett. 35, 247 (1975).

9. M.H. Cohen, J. Jortner, and I. Webman, Phys. Rev. B, in press.

9½. See also: M. H. Cohen, in Proc. 10th Intl. Conf. on the Physics of Semiconductors (U.S. At. En. Comm., Springfield, 1970), p. 645.

10. I. Webman, J. Jortner, and M.H. Cohen, Phys. Rev. B 11, 2885 (1975); Phys. Rev. B 14, 4737 (1976).

11. R.J. Birgeneau, R.A. Cowley, G. Shirane, and H.J. Guggenheim, Phys. Rev. Lett. 37, 940 (1976); H.E. Stanley, R.J. Birgeneau, P.J. Reynolds, and J.F. Nicoll, J. Phys. C 9, L553 (1976).

12. T.C. Lubensky, Phys. Rev. B 15, 311 (1977); D. Stauffer, Z. Phys. B 22, 161 (1975).

13. D.J. Phelps and C.P. Flynn, Phys. Rev. B 14, 5279 (1976).

14. Representative reviews of information on the exponents are given for β, γ, and ν by A.B. Harris and S. Kirkpatrick, Phys. Rev. B 16, 542 (1977); and for t by I. Webman, J. Jortner, and M.H. Cohen, Phys. Rev. B, in press.

15. G. Toulouse, Nuovo Cim. B 23, 234 (1974).

16. S. Kirkpatrick, Phys. Rev. Lett. 36, 69 (1976); D.S. Gaunt, M.F. Sykes, and H. Ruskin, J. Phys. A 9, 1899 (1976).

17. A.B. Harris, T.C. Lubensky, and J-H. Chen, Phys. Rev. Lett. 36, 415 (1976).

18. C.M. Fortuin and P.W. Kasteleyn, Physica (Utrecht) 57, 536 (1972).

19. T.J. Coutts, Thin Solid Films 38, 313 (1976).

CHAPTER 25

PANEL DISCUSSION: SYSTEMS OF LOW DIMENSIONALITY

Moderator: Mark Kac, Rockefeller University, New York, N. Y., U. S. A.

Panellists: H. Gutfreund, Hebrew University, Jerusalem, Israel

Leo P. Kadanoff, Brown University, Providence, R. I., U. S. A.

J. Solyom, Central Research Institute for Physics, Budapest, Hungary

F. Y. Wu, Northeastern University, Boston, Mass., U. S. A.

J. Zittartz Universitat Köln, Germany

THE ONE-DIMENSIONAL FERMI GAS

H. Gutfreund
Rakah Institute of Physics
The Hebrew University, Jerusalem, Israel

ABSTRACT*

The one-dimensional Fermi-gas model is based on a Hamiltonian with a linearized kinetic energy and forward as well as backward scattering interactions, characterized by corresponding coupling parameters. The aim of the model is to establish which types of order occur in various regions of the coupling-parameter space. It is done by calculating the low-frequency behavior of various two-particle correlation functions, or generalized susceptibilities, corresponding to the different types of order.

The main purpose of the discussion is to point out that this model is extremely rich, and deserves more attention at statistical physics conferences. It is intimately related to many celebrated models of statistical mechanics and field theory, such as the sine - Gordon model, the massive Thirring model, the planar harmonic rotator model, the classical two-dimensional Coulumb gas, and the xyz Baxter models. Moreover, the model may also have something to do with reality. This connection with a physical system poses new interesting problems provoked by experimental results. For example, the recent discovery of the $4k_F$ instability in the linear chain conductor TTF - TCNQ stimulated the investigation of four-particle correlation functions - an investigation which yielded a richness of new and interesting results.

*The Editors regret that pressing printers' deadlines prevented us from waiting for a full manuscript.

LATTICE COULOMB-GAS REPRESENTATION OF TWO-DIMENSIONAL PROBLEMS

Leo P. Kadanoff
Department of Physics and Materials Research Laboratory,
Brown University, Providence, Rhode Island 02912, U.S.A.

ABSTRACT

Many of the standard two-dimensional problems of statistical physics can be transformed into "coulomb gas" problems in which there are two kinds of "charges" represented by integers n and m. Such a transformation works for the Ising model, the three and four state Potts models, the Ashkin-Teller model, and many others. In general the n-n and m-m interactions have the coulombic character in which the interaction is, for large separations, proportional to the logarithm of the distance. On the other hand, the $n(\vec{r})$-$m(\vec{R})$ interaction is, for large distances, proportional to i times the angle $\theta(\vec{r} - \vec{R})$ which measures the angular position of $\vec{R}$ relative to $\vec{r}$. This latter interaction is akin to that between a magnetic monopole and an electric charge.

I. Introduction

A. Goals

In this paper, we should reformulate many of the standard problems of statistical physics in terms of a pair of integer variables $n(\vec{r})$ and $m(\vec{R})$ which describe respectively quantum numbers for "charges" and "magnetic monopoles" confined to lattice sites $\vec{r}$ and $\vec{R}$. In the recent literature there have appeared several formulations in which a single set of charges $n(\vec{r})$ interact via coulomb-like potentials. (See, for example, Anderson and Yuval [1969], Kosterlitz and Thouless [1973], Kosterlitz [1974], and also the review of Kosterlitz and Thouless [1977]). However, in a recent publication Jose, Kadanoff, Kirkpatrick, and Nelson (1977) have stressed the usefulness of having two kinds of quantum numbers. (This approach was also apparently derived independently by Villain, whose work was described to me by Kosterlitz. For parallel discussions of the two dimensional case and extension to higher dimensionality see Banks, Myerson, and Kogut [1977] and Savit [1977]). These appear in an action (minus the Hamiltonian divided by kT) which takes the form

$$A[n,m] = \sum_{\vec{r}} Y(n(\vec{r})) + \sum_{R} Y(m(\vec{R})) + A_o$$
$$+ \frac{1}{2} \sum_{\vec{r}\vec{r}'} n(\vec{r})V_{nn}(\vec{r},\vec{r}')n(\vec{r}')$$
$$+ \frac{1}{2} \sum_{\vec{R}\vec{R}'} m(\vec{R})V_{mm}(\vec{R},\vec{R}')m(\vec{R}')$$
$$+ \sum_{\vec{r}\vec{R}} m(\vec{R})V_{mn}(\vec{R},\vec{r})n(\vec{r}) \qquad (1.1)$$

when

$$\sum_{\vec{r}} n(\vec{r}) = 0$$
$$\sum_{\vec{R}} m(\vec{R}) = 0 \qquad (1.2)$$

Thus, the action (1.1) applies to the case in which the total electric and magnetic charges vanish. When they do not, the action is minus infinity. The statistical mechanics problem of calculating the partition function is then equivalent to doing the sum

$$Z = \sum_{\{n(\vec{r})\}} \sum_{\{m(\vec{R})\}} e^{A[n,m]} \tag{1.3}$$

with the sums restricted by the conditions (1.2).

The special character of the formulation (1.1) lies in the long-ranged nature of the interactions V. For large separations $V_{nn}(\vec{r},\vec{r}')$ and $V_{m,m}(\vec{r},\vec{r}')$ are each proportional to $\ell n|\vec{r}-\vec{r}'|$. This is, of course, the usual form of coulomb interactions in two dimensions. The "electric"-"magnetic" interaction $V_{m,n}(\vec{r},\vec{r}')$ also has a long ranged character. For large separations, it is proportional to $\Theta(\vec{r}-\vec{r}')$, where Θ is the usual angular variable defined by

$$\tan\Theta(\vec{r}-\vec{r}') = \frac{y-y'}{x-x'} \tag{1.4}$$

As we should see in Chapter III a large collection of the standard problems in two dimensional statistical physics can be phrased in this manner, in which the difference between the problems are mostly reflected in charges in the parameters in the interactions, V. This unification is one of the advantages of the coulombic formulation. Another advantage lies in the simplicity of the duality transformation (Kramers and Wannier, [1941] in this language. The Kramers-Wannier duality is simply the replacement of n by m. If $V_{mm} = V_{nn}$, the problem is self-dual. As we shall see, many of the standard duality statements are made self-evident in this formulation.

There is an additional possible advantage of this formulation. In the usual handling of critical phenomena, long-ranged correlations are built up from short-ranged interactions. The interactions in (1.1) are - from the first - long ranged. Hence, one might be able to gain an insight into the nature of the long-ranged correlations via the action (1.1).

B. Methodology

Eqs. (1.1)-(1.3) define a description of our statistical mechanics problem which is, in some sense, analogous to a two-dimensional electrodynamics with charges and magnetic monopoles. We shall henceforth call this the "electrodynamic" description of the problem. Our job is to connect this electrodynamic description to a "standard" or Ising description of the very same problem. In this latter representation, the basic statistical variables are called $\sigma(\vec{r})$, which take on some preassigned set of values - for example $\sigma(\vec{r}) = \pm 1$. Then the action includes from all the sites a contribution

$$A_{site}[\sigma] = \sum_{\vec{r}} h(\sigma(\vec{r})) \tag{1.5a}$$

and a set of interactions fron nearest neighbor bonds

$$A_{bond}[\sigma] = \sum_{<\vec{r}\vec{r}'>} K(\sigma(\vec{r})- (\vec{r}')) \tag{1.5b}$$

The partition function is then, of course, just the sum over all possible values of σ of the form

$$Z = [\prod_{\vec{r}} \sum_{\sigma(\vec{r})}]\, e^{A_{bond}[\sigma]+A_{site}[\sigma]} \tag{1.6}$$

We wish to show how the very same partition function can be equivalently represented in the forms (1.3) and (1.4). To do this we go through an intermediate "gauge-theoretical" representation of this problem. (References on "gauge-theory" models in statistical mechanics include Wegner [1971], Balian, Drouffe and Itzykson [1974; 1975a and 1975b], Migdal [1975a and 1975b], and Kadanoff [1975a and 1975b].) This third representation includes a set of variables and interactions defined on sites, on bonds, and on plaquettes. Each site contains a variable $\theta(\vec{r})$ which can range from $-\pi$ to π and a variable $n(\vec{r})$ which takes on integral values. The interaction on each site is of the form:

$$A_{site}[n,\Theta] = \sum_{r} Y(n(\vec{r})) + i\Theta(\vec{r})\cdot n(\vec{r}) \tag{1.7a}$$

Each nearest neighbor bond contains a variable $\ell(\vec{r},\vec{r}\,') = -\ell(\vec{r}\,',\vec{r})$ which takes on integral values. The interaction on the bond takes the form

$$A_{bond}[\Theta,\ell] = -\sum_{\langle rr'\rangle} \frac{2\pi}{2}\left[\frac{\Theta(\vec{r})-\Theta(\vec{r}\,')}{2\pi} - \ell(\vec{r},\vec{r}\,')\right]\cdot X\cdot\left[\frac{\Theta(\vec{r})-\Theta(\vec{r}\,')}{2\pi} - \ell(\vec{r},\vec{r}\,')\right] \tag{1.7b}$$

Finally, each square on the lattice forms a plaquette bounded by four bonds. (See Fig. 1). On each plaquette we define a variable $m(\vec{R})$, $\vec{R}$ being the center of the plaquette, which is essentially the circulation of the ℓ's around the plaquette. That definition makes $m(\vec{R})$ an integer variable. The plaquette interaction is given by

$$e^{A_{plaquette}[m,\ell]} = \prod_{R} \delta_{m(\vec{R}),\,\ell_{12}+\ell_{23}+\ell_{34}+\ell_{41}}\; e^{\tilde{Y}[m(\vec{R})]} \tag{1.7c}$$

In some cases we shall wish m, n, ℓ and Θ to have M components. Hence the dots on Eq. (1.7a) and (1.7b) which indicate a scalar product. In this situation the interaction constant will be an M by M matrix.

Our third description of the partition function will then be one in which Z is a sum over n, m, ℓ and an integral over Θ, i.e.

$$Z = \left(\prod_{\vec{r}}\int_{-\pi}^{\pi}\frac{d\Theta(\vec{r})}{2\pi}\sum_{n(\vec{r})}\right)\left(\prod_{\langle\vec{r}\vec{r}\,'\rangle}\sum_{\ell(\vec{r},\vec{r}\,')}\right)\left(\prod_{\vec{R}}\sum_{m(\vec{R})}\right)e^{A[n,\Theta,\ell,m]} \tag{1.8}$$

Here the total action is, of course,

$$A[n,\Theta,m,\ell] = A_{site}[n,\Theta]+A_{bond}[m,\ell]+A_{plaquette}[m,\ell] \tag{1.9}$$

we call this third formulation of the action a gauge-theoretical one because $A[n,\Theta,m,\ell]$ has an invariance under the replacement

$$\begin{aligned}\Theta(\vec{r}) &\to \Theta(\vec{r})+2\pi q(\vec{r})\\ \ell(\vec{r},\vec{r}\,') &\to \ell(\vec{r},\vec{r}\,')+q(\vec{r})-q(\vec{r}\,')\end{aligned} \tag{1.10}$$

for any set of integers $q(\vec{r})$.

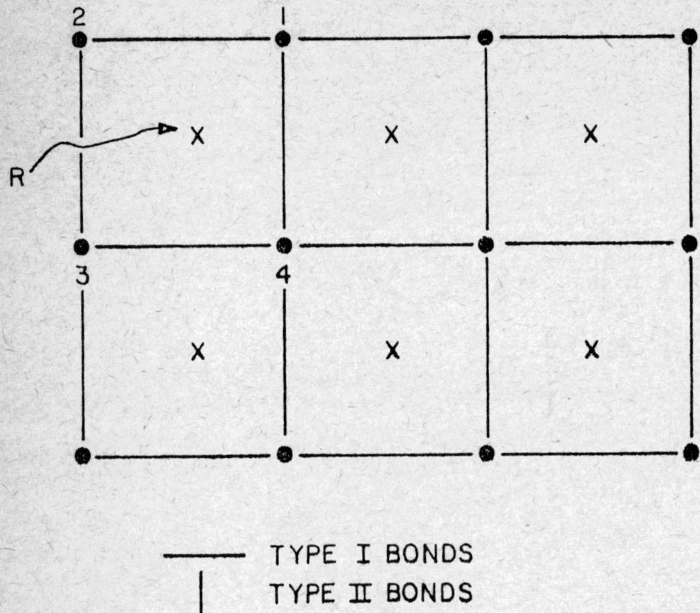

Fig. 1. The square lattice. The dots represent lattice sites, $\vec{r}$. The x's indicate sites of the dual lattice $\vec{R}$. The definition of the plaquette variable shown is $m(\vec{R}) = \ell_{12} + \ell_{23} + \ell_{34} + \ell_{41}$.

We shall generate both the "standard" description of the problem and the electrodynamic description by doing partial summations on the gauge theoretical formulation. In particular, we shall show that the sum over the ℓ's and the Θ's yeilds an action of the form (1.1), i.e.

$$e^{A[n,m]} = \left(\prod_{\langle \vec{r}\vec{r}'\rangle} \sum_{\ell(\vec{r},\vec{r}')}\right)\left(\prod_{\vec{r}} \int_{-\pi}^{\pi} \frac{d\Theta(\vec{r})}{2\pi}\right) e^{A[n,\Theta,\ell,m]} \tag{1.11}$$

This correspondence between the electrodynamic and the gauge theoretical formulations is developed in Chapter III.

The connection with the standard formulation of the problem is developed in Chapter II. Finally, in Chapter IV all the results are brought together and summarized.

II. Connections to the Standard Problems

In this chapter, we reduce the gauge theoretic formulation of our statistical mechanics problem to the standard formulation, which involves variables $\sigma(\vec{r})$ and nearest neighbor interactions.

A. Elimination of $m(\vec{R})$ and $\ell(\vec{r},\vec{r}')$

The first step in this reduction is to sum expression (1.8) over the variables $m(\vec{R})$ and $\ell(\vec{r},\vec{r}')$. If one makes the specific choice

$$\tilde{Y}(m) = 0 \tag{2.1}$$

this sum is very simple indeed. We then find that

$$(\prod_{\vec{R}} \sum_{m(\vec{R})}) e^{A_{\text{plaquette}}[m,\ell]} = 1 \tag{2.2}$$

The sum over ℓ's on different bonds are then independent of one another. We find that each bond has a coupling of the Villian [1975] form

$$e^{K(\theta(\vec{r})-\theta(\vec{r}'))} = \sum_{\ell} \exp\left\{-\pi \left[\frac{\theta(\vec{r})-\theta(\vec{r}')}{2\pi} - \ell\right]\cdot X\cdot \left[\frac{\theta(\vec{r})-\theta(\vec{r}')}{2\pi} - \ell\right]\right\} \tag{2.3}$$

We wish to use Eq. (2.3) in three cases. In the first case the bonds are the usual nearest neighbor bonds of the square lattice and θ, ℓ, and X are just simple real numbers. In the second case, we still have the same bond structure but $\tilde{\theta}$ and $\tilde{\ell}$ are each vectors with M components while X is a symmetrical M by M matrix. Then, Eq. (2.3) will become

$$e^{K(\tilde{\theta}-\tilde{\theta}')} = \sum_{\ell_1 \ell_2 \cdots \ell_M} \exp\{-\pi \sum_{j,k} (\frac{\theta_j-\theta_j'}{2\pi} - \ell_j) X_{jk} (\frac{\theta_k-\theta_k'}{2\pi} - \ell_k)\} \tag{2.4}$$

The Ashkin-Teller [1943] model can be represented by a situation in which M=2. Then X_{jk} is a 2x2 matrix which we write as

$$\tilde{X} = \begin{pmatrix} X_o & X_1 \\ X_1 & X_o \end{pmatrix} = X_o + X_1 \tau_1 \tag{2.5}$$

Yet another two component case arises when $\theta(\vec{r})$ has two components which appear on two different sublattices, each of which is a simple square lattice. See Figure (2). We denote one sublattice by giving it the lattice sites $\vec{r}$ and the other by giving it the site names $\vec{R}$. Then the basic coupling terms indicated in Fig. (2) is $K(\theta_1(\vec{r})-\theta_1(\vec{r}'),\theta_2(\vec{R})-\theta_2(\vec{R}'))$ where

$$e^{K(\theta_1,\theta_2)} = \sum_{\ell_1 \ell_2} \exp -\pi \sum_{jk} \left[\left(\frac{\theta_j}{2\pi} - \ell_j\right) X_{jk} \left(\frac{\theta_k}{2\pi} - \ell_k\right) \right]$$

Eq. (2.6) represents a coupling structure appropriate for the 8-vertex model. (For references on the Ashkin-Teller and the 8-vertex models see Baxter [1971 and 1972], Barber and Baxter [1973], Wegner [1972], Fan [1972], Fan and Wu [1970], Knops [1975], Wu [1977], Wu and Lin [1974], and Kadanoff and Wegner [1971]).

Now we have bond structures entirely similar to that of the standard problems.

B. Going from $n(\vec{r})$ and $\theta(\vec{r})$ to $\sigma(\vec{r})$

In the gauge theoretical formulation, we have at each site a summation over $n(\vec{r})$ and an integration over $\theta(\vec{r})$ of the form:

$$I = \sum_{n(\vec{r})} \int_{-\pi}^{\pi} \frac{d\theta(\vec{r})}{2\pi} \left[e^{in(\vec{r})\cdot\theta(\vec{r}) + Y(n(\vec{r}))} \right] f(\theta(\vec{r})) \tag{2.7}$$

Here $f(\theta)$ is some very complicated function determined by the integrations over all other variables.

We wish to reduce this integration to a discrete summation over some set of values of $\theta(\vec{r})$, i.e.,

$$\theta(\vec{r}) = \sigma(\vec{r}) \frac{2\pi}{p} - \pi \tag{2.8a}$$

with

$$\sigma(\vec{r}) = 1,2,\ldots,p \tag{2.8b}$$

This goal is easily accomplished with the aid of the Poisson sum formula which states

$$\sum_{\ell} e^{i\ell\theta} = 2\pi \sum \delta(\theta - 2\pi k) \tag{2.9}$$

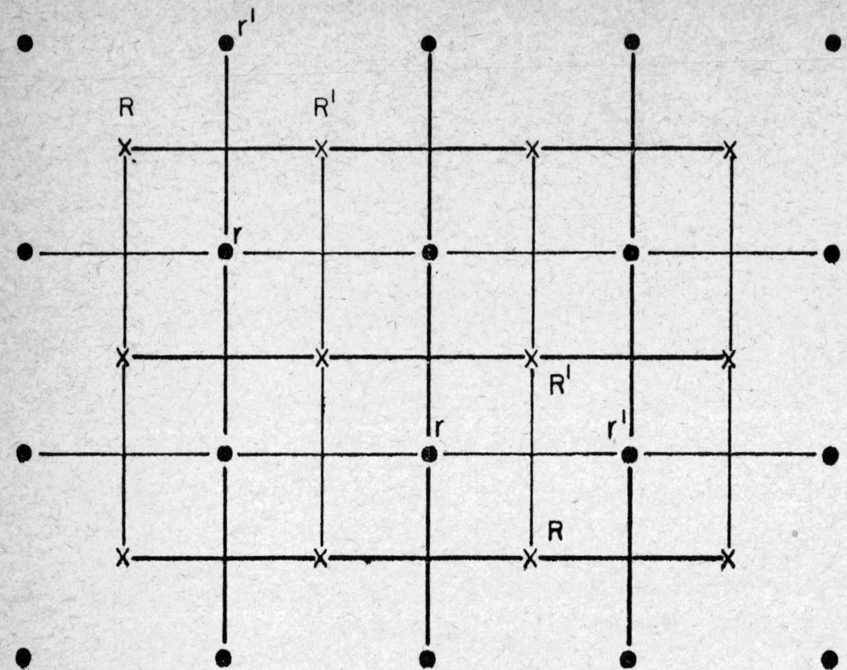

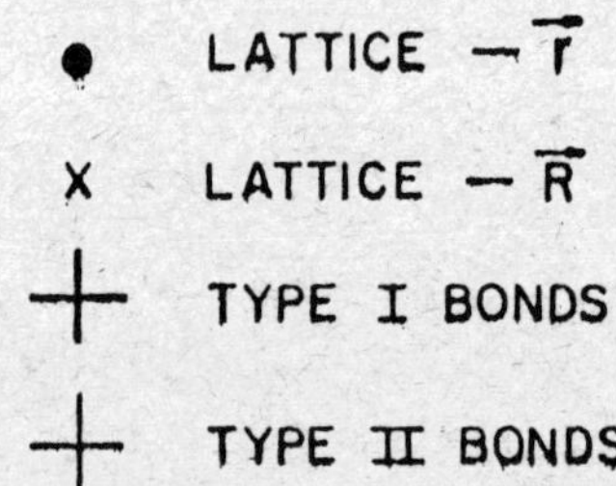

Fig. 2. The lattice used for the 8-vertex model

Then simply choose

$$e^{Y(n)} = \sum_{\sigma=1}^{p} e^{-in(\frac{\sigma 2\pi}{p} - \pi)+h(\sigma)} \tag{2.10}$$

Then expression (2.7) reduces to

$$I = \sum_{\sigma(\vec{r})=1}^{p} e^{h(\sigma(\vec{r}))} f(\frac{2\pi}{p}\sigma(\vec{r})-\pi) \tag{2.11}$$

Here $h(\sigma(\vec{r}))$ is the site weight function of Eq. (1.5a).

In short, we have succeeded in converting the θ-integrals to discrete sums and the bond interactions to simple nearest neighbor couplings. In gross, the gauge problem has been reduced to the standard problem.

C. Examples

1. The Ising Model

For the Ising case $\sigma=1,2$. Let $h(\sigma) = +h$ if $\sigma=2$ and $-h$ if $\sigma=1$. Then

$$e^{Y(n)} = \begin{cases} 2\cosh h & \text{for n even} \\ -2\sinh h & \text{for n odd} \end{cases} \tag{2.12a}$$

If the magnetic field, h, is zero, then

$$e^{Y(n)} = \begin{cases} 2 & \text{for n even} \\ o & \text{for n odd} \end{cases} \tag{2.12b}$$

The bond interaction is

$$e^{K((\sigma-\sigma'))} = \sum_{\ell} e^{-\pi(\frac{\sigma-\sigma'}{2} \bullet \ell)^2 X}$$

The standard way of writing this interaction is to take it to be $\exp(K_o \pm K_1)$ depending upon whether σ is equal or unequal to σ' then

$$e^{K_o+K_1} = \sum_{\ell} e^{-\pi X \ell^2}$$

$$e^{K_o-K_1} = \sum_{\ell} e^{-\pi X(\ell-1/2)^2} \tag{2.13}$$

2. The p-state model

The p-state (planar-Potts) model is obtained with unit vertex weight if we choose

$$e^{Y(n)} = \begin{cases} p \text{ if } n/p = \text{an integer} \\ 0 \text{ otherwise} \end{cases} \qquad (2.14)$$

Then θ takes on the values (2.8) and the vertex function $h(\sigma)$ vanishes. In this case, the coupling on each bond takes the form

$$e^{K(\sigma(r)-\sigma(r'))} = \sum_{\ell} e^{-\pi X\left[\frac{\sigma(r)-\sigma(r')}{p} -\ell\right]^2} \qquad (2.15)$$

For p=3 the coupling (2.15) is a representation of the three-state Potts model. For p=4 and above, we have special cases of the planar Potts models.

3. Ashkin-Teller model

For the Ashkin-Teller model there are two variables σ_1=1,2 and σ_2=1,2 at each site. Accordingly we require a two component θ-variable and a two component n-variable. In order to achieve a vanishing on site interaction, we choose:

$$e^{Y(n_1,n_2)} = \begin{cases} 4 \text{ if } n_1 \text{ and } n_2 \text{ are even} \\ 0 \text{ otherwise} \end{cases} \qquad (2.16)$$

The inclusion of odd-n terms with a Y chosen so that $Y(n_1,n_2)$ is periodic under $n_1 \to n_1+2$ and/or $n_2 \to n_2+2$ will generate magnetic field terms coupled to σ_1, σ_2, and $\sigma_1\sigma_2$.

The standard Ashkin-Teller interaction can be generated via equations (2.4) and (2.5), which yield a coupling structure

$$e^{K(\sigma_1-\sigma_1',\sigma_2-\sigma_2')} = \sum_{\ell_1,\ell_2} \exp\{-\pi[(\frac{\sigma_1-\sigma_1'}{2}-\ell_1)^2+(\sigma_2-\sigma_2'-\ell_2)^2]X_0$$
$$-2\,[(\frac{\sigma_1-\sigma_1'}{2}-\ell_1)(\frac{\sigma_2-\sigma_2'}{2}-\ell_2)]X_1\} \qquad (2.17)$$

The usual way of representing this coupling structure is to make use of the variables

$$S = 1-(\sigma_1-\sigma_1')^2/2$$
$$\tau = 1-(\sigma_2-\sigma_2')^2/2 \qquad (2.18)$$

which take on the values ±1. Then the coupling is written

$$e^{K_o+K_1(S+\tau)+K_3S\tau} = \sum_{\ell_1\ell_2} \exp\{-\pi X_o[(\ell_1-\frac{1-S}{4})^2+(\ell_2-\frac{1-\tau}{4})^2]$$
$$-2\pi X_1(\ell_1-\frac{1-S}{4})(\ell_2-\frac{1-\tau}{4})\} \qquad (2.19)$$

Clearly X_o represents a coupling within the set of variables σ_1 and within the set σ_2 while X_1 represent a σ_1-σ_2 coupling.

Notice that the coupling (2.19) is unchanged under a change in sign of X_1. (To see this replace ℓ_2 by $-\ell_2+\frac{1-\tau}{2}$.)

4. The 8-vertex Model

The spin formulation of the 8-vertex model (Kadanoff and Wegner [1971], Fan and Woo [1970]) is exactly the same as that of the Ashkin-Teller model except that the variables σ_1 θ_1 n_1 and σ_2 θ_2 n_2 appear on different sublattices. On each site, therefore, there is one n-variable, one θ and one σ. If the magnetic field terms are zero, Y(n) takes the form (2.12b). Eqs. (2.17) and (2.18) applies equally well to the 8-vertex model and the Ashkin-Teller model. However the standard notation for Eq. (2.19) is slightly different in that the left hand side of this equation is often written

$$e^{K_o+K(S+\tau)+\lambda S\tau} \qquad (2.20)$$

(See Kadanoff and Wegner (1971).)

D. Duality Statements

All the models mentioned here obey duality relations which take a simple form when the on-site fields, h, vanish. The duality statements are essentially that the partition functions of the models in question are left unchanged when the couplings are replaced by the coefficients in their fourier series expansions. To be specific, Eq. (2.4) can be written in terms of the variables (2.8) as

$$e^{K(\sigma-\sigma')} = \sum_{\ell_1 \cdots \ell_M} \exp -\pi \sum_{j,k} \left(\frac{\sigma_j-\sigma_j'}{p} - \ell_j\right) X_{jk} \left(\frac{\sigma_k-\sigma_k'}{p} - \ell_k\right) \tag{2.21}$$

This equation has a Fourier series representation

$$e^{K(\sigma)} = \sum_{\mu_1=1}^{p} \sum_{\mu_2=1}^{p} \cdots \sum_{\mu_M=1}^{p} \frac{1}{p^{M/2}} e^{\tilde{K}(\mu)+2\pi i \sum_k \sigma_k \mu_k / p} \tag{2.22}$$

The duality statement is that the partition function for the problem with couplings $K(\sigma)$ and that with couplings $\tilde{K}(\sigma)$ are identical.

Now calculate $\tilde{K}(\sigma)$. By inverting the Fourier transform we find

$$e^{\tilde{K}(\mu)} = \frac{1}{p^{M/2}} \sum_{\sigma_1 \cdots \sigma_2} e^{K(\sigma)-2\pi i\, \sigma\cdot\mu/p}$$

$$= \frac{1}{p^{M/2}} \sum_{\sigma=1}^{p} \sum_{\ell=-\infty}^{\infty} e^{-2\pi i\, \sigma\cdot\mu/p}\, e^{-\frac{2\pi}{2}\left(\frac{\sigma}{p}-\ell\right)\cdot X\cdot\left(\frac{\sigma}{p}-\ell\right)}$$

Now the sum over ℓ and the sum over σ can be combined into a single sum over a new variable $k = \sigma - p\ell$ to give

$$e^{\tilde{K}(\mu)} = \frac{\cdot 1}{p^{M/2}} \sum_{k=1-\infty}^{\infty} e^{-2\pi i\, k\cdot\mu/p - \frac{2\pi}{2} k\cdot \frac{X}{p^2}\cdot k}$$

But the Poisson sum formula implies that

$$\sum_{k=-\infty}^{\infty} f(k) = \int_{-\infty}^{\infty} dk \sum_{\ell=-\infty}^{\infty} e^{2\pi i \ell k} f(k)$$

Thence, after a bit more calculation we find that

$$e^{\tilde{K}(\mu)} = \frac{(2\pi)^{M/2} p^{M/2}}{(\det X)^{\frac{1}{2}}} \sum_{\ell=-\infty}^{\infty} e^{-\pi(\mu/p-\ell)\cdot p^2 X^{-1}\cdot(\mu/p-\ell)} \tag{2.23}$$

Expression (2.23) for $\tilde{K}$ is very similar indeed to expression (2.21) for K. Except for a change in prefactors, the coupling is exactly of the same form with X replaced according to

$$X \to \tilde{X} = p^2 X^{-1} \tag{2.24}$$

For the Ising model, p=2 and the duality transform is

$$X \to \tilde{X} = 4/X \tag{2.25}$$

The model is self-dual at $X=2$. This self-duality point is, in fact, the critical point of the Ising model!

For the Ashkin-Teller and Baxter models the duality statement is, once again, Eq. (2.25) but now X and $\tilde{X}$ are two by two matrices. If we represent X as before, in the form (2.5), then

$$\tilde{X} = \frac{4(X_o - X_1\tau_1)}{X_o^2 - X_1^2} \tag{2.26}$$

The model will be self-dual if

$$X_o^2 - X_1^2 = 4 \tag{2.27}$$

This is indeed a simple statement of the duality conditions for the Ashkin-Teller and the 8-vertex models.

III. To the Electrodynamic Representation

A. From gauge theory to a gaussian integral

This chapter is directed toward the derivation of the electrodynamic representation of our problem starting from the gauge theoretical representation. In particular, we seek to calculate

$$Z[n,m] = \left(\prod_{\vec{r}} \int_{-\pi}^{\pi} \frac{d\theta(\vec{r})}{2\pi}\right)\left(\prod_{\langle \vec{r}\vec{r}'\rangle} \sum_{\ell(\vec{r},\vec{r}')}\right) e^{A_{site}[\theta,n]+A_{bond}[\theta,\ell]} \; e^{A_{plaquette}[\ell,m]} \tag{3.1}$$

We would like to prove that Z[n,m] is the exponential of the action described in Eqs. (1.1). Since the terms $Y(n(\vec{r}))$ and $\tilde{Y}(m(\vec{R}))$ move through the sums and integrals in Eq. (3.1) quite without change, we can simplify our notation by simply ignoring these terms.

To evaluate the sums and integrals in Eq. (3.1), we shall make use of the gauge invariance of the theory, which is the statement that the integrand is invariant under the transformations (1.10).

$$\begin{aligned} \theta(\vec{r}) &\rightarrow \theta(\vec{r})+2\pi q(\vec{r}) = 2\pi\phi(\vec{r}) \\ \ell(\vec{r},\vec{r}') &\rightarrow \ell(\vec{r},\vec{r}')+q(\vec{r})-q(\vec{r}') = \ell'(\vec{r},\vec{r}') \end{aligned} \tag{3.2}$$

where the q's are integers. Notice that our lattice shown in Fig. 1 - can each be decomposed into two types of bonds. On the I-type bonds choose $q(\vec{r})$ so that $\ell'(\vec{r},\vec{r}')$ vanishes. Then rewrite the sum (3.1) in terms of the ϕ's and the ℓ''s, which still exist on the II-type bonds. The result is

$$Z[n,m] = \prod_{r}\left[\int_{-\pi}^{\pi} \frac{d\theta(\vec{r})}{2\pi} \sum_{q(n)} e^{in(\vec{r})\phi(\vec{r})}\right] \cdot \left[\prod_{\text{I-type bonds}} e^{-\pi[\phi(\vec{r})-\phi(\vec{r}')]\cdot X\cdot[\phi(\vec{r})-\phi(\vec{r}')]}\right]$$

$$\cdot\{\prod_{\text{II-type bonds}} \sum_{\ell'(\vec{r},\vec{r}')} \exp[-\pi[\phi(\vec{r})-\phi(\vec{r}')-\ell'(\vec{r},\vec{r}')]\cdot x\cdot[\phi(\vec{r})-\phi(\vec{r}')-\ell'(\vec{r},\vec{r}')]]\}$$

$$\cdot \prod_{\vec{R}} \delta_{\ell'_{2,3}-\ell'_{1,4},m(\vec{R})} \tag{3.3}$$

Notice how the integrand in Eq. (3.3) does not depend upon $\theta(\vec{r})$ or $q(\vec{r})$ separately but only upon the combination $\phi = \theta/2\pi+q$. Thence we can convert the integral into gaussian form by making the replacement

$$\int_{-\pi}^{\pi} \frac{d\theta(\vec{r})}{2\pi} \sum_{q(\vec{r})} \to \int_{-\infty}^{\infty} d\phi(\vec{r}) \tag{3.4}$$

If there are N sites on the $\vec{r}$-lattice Eq. (3.3) contains N summations over ℓ' and N delta symbols which define $\ell'(\vec{r},\vec{r}')$ in terms of the $m(\vec{R})$. Thus one can solve for the ℓ's in terms of the m's in the form

$$\ell'(\vec{r},\vec{r}') = \ell(\vec{r},\vec{r}',[m]) \tag{3.5}$$

$$= -\sum_{k=o}^{\infty} m((\frac{\vec{r}+\vec{r}'}{2})+(k+\frac{1}{2})\hat{e}_1$$

Here $\hat{e}_1$ is a unit vector in the x direction and $\vec{r}$ lies just above $\vec{r}'$.

At this point, Eq. (3.3) reduces to just a gaussian integral:

$$Z[n,m] = [\prod_{\vec{r}} \int d\phi(\vec{r})]e^Q$$

$$Q = \sum_r in(\vec{r})\phi(\vec{r})2\pi - \sum_{\text{I-bonds}} \pi[\phi(\vec{r})-\phi(\vec{r}')]\cdot X\cdot[\phi(\vec{r})-\phi(\vec{r}')]$$

$$- \sum_{\text{II-bonds}} \pi[\phi(\vec{r})-\phi(\vec{r}')-\ell(r,r';[m])]\cdot X\cdot[\phi(\vec{r})-\phi(\vec{r}')-\ell(\vec{r},\vec{r}';[m])] \tag{3.6}$$

Such a gaussian integral can of course be evaluated immediately. In general the evaluation of a gaussian integral is given by

$$I = (\pi_i \int_{-\infty}^{\infty} d\phi_i)\exp[-\frac{1}{2}\sum_{ij}\phi_i G_{ij}^{-1}\phi_j+\sum_i \alpha_i\phi_i]$$

$$= [\det(2\pi G)]^{\frac{1}{2}}\exp\left[\frac{1}{2}\sum_{ij}\alpha_i G_{ij}\alpha_j\right] \tag{3.7}$$

where G is the matrix inverse of G^{-1}. Expression (3.6) has a form very similar to that of (3.7), except that there are extra quadratic terms in $\ell[m]$. Hence we conclude that the result of doing the integral is that $\ell n\ Z[n,m]$ takes the form of a constant plus a quadratic form in n and m. In this way, we see that expression (1.1) is indeed justified. Next we must turn to evaluating the parameters A_o, V_{nn}, V_{mm}, V_{nm} and V_{mn} in that expression.

B. The green's function

Expression (3.6) has the ϕ's coupled together by an inverse green's function of the form

$$G_{jk}^{-1}(r,r') = 2\pi\ X_{jk}(4\delta_{r,r'}-\sum_{\hat{e}}\delta_{r,r'+\hat{e}}) \tag{3.8}$$

where the $\hat{e}$ are the nearest neighbor vectors on the lattice. Fourier transformation enables us to calculate the green's function immediately. We write the result as

$$G(\vec{r},\vec{r}') = G(0,0) - \frac{1}{(2\pi)^2}X^{-1}V_R(\vec{r},\vec{r}') \tag{3.9}$$

Here G(0,0) is the infinite constant

$$G(0,0) = \frac{X^{-1}}{2\pi}\int_{-\pi}^{\pi}\frac{dq_x}{2\pi}\int_{-\pi}^{\pi}\frac{dq_y}{2\pi}\frac{1}{4-2\cos q_x-2\cos q_y} \tag{3.10}$$

while $V(\vec{r},\vec{r}')$ is the finite quantity

$$V_R(\vec{r}-\vec{r}') = \int_{-\pi}^{\pi} \frac{dq_x}{2\pi} \int_{-\pi}^{\pi} dq_y \, \frac{1-e^{iq\cdot(\vec{r}-\vec{r}')}}{4-2\cos q_x-2\cos q_y} \tag{3.11}$$

Notice that for large separations, $|\vec{r}-\vec{r}'|$,

$$V_R(\vec{r},\vec{r}') \to \ell n|\vec{r}-\vec{r}'| \tag{3.12}$$

In our later analysis, we shall need another function $V_I(\vec{R},\vec{r})$ to describe the n-m couplings. Here $\vec{R}$ is a site on the dual lattice and V_I is defined in terms of $V_R(\vec{r},\vec{r}') = V_R(x-x', y-y')$ by

$$V_I(\vec{R}-\vec{r}) = \lim_{\varepsilon\to o^+} \sum_{j=o}^{\infty} \varepsilon^{-j}[V_R(X-x-\tfrac{1}{2}-j, Y-\tfrac{1}{2}-y)-V_R(X-x-\tfrac{1}{2}-j, Y+\tfrac{1}{2}-y)]$$

$$= \lim_{\varepsilon\to o^+} \int_{-\pi}^{\pi} \frac{dq_x}{2\pi} \int_{-\pi}^{\pi} dq_y \, \frac{e^{i\vec{q}\cdot(\vec{R}-\vec{r})}}{\left(4-2\cos q_x-2\cos q_y\right)} \frac{(-2i)\sin q_y/2}{\left(e^{-\varepsilon}e^{iq_x/2}-e^{-iqx/2}\right)} \tag{3.13}$$

The form of $V_I(\vec{R}-\vec{r})$ is relatively simple. For large values of the separation $\vec{R}-\vec{r}$ we have

$$\frac{\partial V_I}{\partial x}(\vec{r}) = -\frac{\partial V_R}{\partial y}(\vec{r}) = -\frac{y}{x^2+y^2} \tag{3.14}$$

This expression is true except when $\vec{r}$ lies on the negative x axis. On that axis V_I has a discontinuity of strength 2π. Furthermore, $V_I(\vec{r})$ changes sign when $y \to -y$. All these properties together imply that $-V_I(\vec{r})$ is, for large $\vec{r}$, just the angle $\Theta(\vec{r})$ which obeys

$$\tan\Theta(\vec{r}) = y/x \tag{3.15}$$

with the conditions that $\Theta(\vec{r}) \to 0$ as $y \to o$ when x is positive and that $\Theta(\vec{r})$ jumps from π to $-\pi$ as y passes through zero from above with x negative. So in the asymptotic limit

$$V_I(\vec{R}-\vec{r}) \to -\Theta(\vec{R}-\vec{r}) \tag{3.16}$$

Actually this asymptotic form even holds quite well for small values of $\vec{R}-\vec{r}$. When $\vec{r} = (\frac{1}{2},\frac{1}{2}), (-\frac{1}{2},\frac{1}{2}), (-\frac{1}{2},-\frac{1}{2}), (\frac{1}{2},-\frac{1}{2})$, $V_I(\vec{r})$ equals respectively $-\pi/4$, $-\pi/4, -3\pi/4, 3\pi/4$, and $\pi/4$ - just as one would get from the asymptotic expression.

C. Coupling Terms

With this Green's function evaluated, one can start doing the integral (3.6). When all the m's are zero, one finds by applying (3.9) that

$$\ell nZ[n,o] = -\frac{1}{2}\sum_{\vec{r}}[\ell n \det X + \int \frac{dq}{(2\pi)^2} \ell n(4-2\cos q_x - 2\cos q_y)]$$

$$-\frac{1}{2}\sum_{jk}(\sum_{\vec{r}} n_j(\vec{r}))(2\pi)^2 G_{jk}(\vec{0},\vec{0})(\sum_{\vec{r}'} n_k(\vec{r}'))$$

$$+\frac{1}{2}\sum_{jk}\sum_{rr'} n_j(\vec{r})(X^{-1})_{jk} V_R(\vec{r}-\vec{r}')n_k(\vec{r}') \qquad (3.17)$$

The first term in Eq. (3.17) is the constant term A_o in Eq. (1.1). Since $G_{jk}(\vec{0},\vec{0}) = \infty$, the second term reduces $Z[n,m]$ to zero unless the first of conditions (1.2), i.e.

$$\sum_{\vec{r}} n_j(\vec{r}) = 0 ,$$

is satisfied. The third term in Eq. (3.17) establishes the value of $V_{nn}(\vec{r},\vec{r}')$ to be

$$V_{nn}(\vec{r},\vec{r}') = (X^{-1})_{jk} V_R(\vec{r}-\vec{r}') \qquad (3.18)$$

The next step is to consider the terms bilinear in n and ℓ which emerge from Eq. (3.8). These terms in $\ell n\, Z[n,m]$ are

$$2\sum_{\vec{R}\vec{r}} m(\vec{R})\cdot V_{mn}(\vec{R},\vec{r})\cdot n(\vec{r}) = \sum_{<r_1r_2>}\sum_{r}\sum_{j,k,p} i\, \ell_j(\vec{r}_1,\vec{r}_2;[m])X_{jk}$$

$$[G_{kp}(\vec{r}_1,\vec{r})-G_{kp}(\vec{r}_2,\vec{r})]n_p(\vec{r})$$

A substitution of the expression for ℓ in terms of m [Eq. (3.6)] immediately gives an evaluation of the next coupling term as

$$V_{mn}(\vec{R},\vec{r}) = i\,\delta_{j,k}\,V_I(\vec{R}-\vec{r}) \tag{3.19}$$

The final term in $\ell nZ[n,m]$ is bilinear in $\ell(\vec{r},\vec{r}')$ and hence bilinear in $\ell(\vec{r},\vec{r}')$. This term has the form

$$-\pi \sum_{<rr'>} \ell(\vec{r},\vec{r}')\cdot X\cdot \ell(\vec{r},\vec{r}')$$

$$+\frac{1}{2}\sum_{<\vec{r}_1\vec{r}_1'>}\sum_{<\vec{r}_2\vec{r}_2'>} \ell(\vec{r}_1,\vec{r}_1')\cdot X\cdot(2\pi)^2$$

$$[G(\vec{r}_1,\vec{r}_2)+G(\vec{r}_1',\vec{r}_2')-G(\vec{r}_1,\vec{r}_2')-G(\vec{r}_2,\vec{r}_1')]\cdot \ell(\vec{r}_2,\vec{r}_2') \tag{3.20}$$

Here the sums are over II-type bonds. This term may be rewritten as a sum over m's in the form

$$-\frac{1}{2}\sum_{\vec{R},\vec{R}'} m(\vec{R})\cdot F(\vec{R},\vec{R}')m(\vec{R}') \tag{3.21}$$

with

$$F(\vec{R},\vec{R}') = 2\pi\,X\,\delta_{\hat{e}_2\cdot\vec{R}-\vec{R}',0}\,|(R-R')\cdot\hat{e}_1|-(2\pi)^2\sum_{\ell=o}^{\infty}\sum_{\ell'=o}^{\infty} X\cdot\{\}\cdot X \tag{3.22a}$$

and

$$\{\} = \sum_{\sigma=\pm1}\sum_{\sigma'=\pm1}\sigma\sigma'\,G(\vec{R}-\ell\hat{e}_1-\frac{1}{2}(\hat{e}_1+\sigma\hat{e}_2),\vec{R}'-\ell'\hat{e}_1-\frac{1}{2}(\hat{e}_1+\sigma'\hat{e}_2)) \tag{3.22b}$$

But notice that F(R,R') obeys

$$4F(\vec{R},\vec{r}')-\sum_{\hat{e}} F(\vec{R}+\hat{e},\vec{R}') = \delta_{\vec{R},\vec{R}'}\,2\pi X$$

Therefore $F(\vec{R},\vec{R}')$ is $X\cdot G(\vec{R},\vec{R}')\cdot X$, where G is the Green's function which we defined in Eq. (3.10).

In summary, this final term in $\ell nZ[n,m]$ is

$$\ell nZ[0,m] = \ldots - \frac{1}{2} \sum_{RR'} m(\vec{R})\cdot X\cdot G(\vec{R},\vec{R}')\cdot X\cdot m(\vec{R}') \tag{3.23}$$

With any further work we can conclude that $Z[n,m]$ vanishes unless

$$\sum_{\vec{R}} m_k(\vec{R}) = 0 \tag{3.24}$$

and that

$$V_{mm}(\vec{R},\vec{R}') = X_{jk} V_R(\vec{R}-\vec{R}') \tag{3.25}$$

Thus all terms in Eq. (1.1) have been evaluated.

D. Another lattice

To show how the 8-vertex model fits into this picture, we make use of the two interpretating square lattices shown in Figure 2. Each lattice site of the first lattice, labeled by $\vec{r}$, contains variables $\theta_1(\vec{r})$, $n_1(\vec{r})$ and $m_2(\vec{r})$. The other lattice has sites labeled by $\vec{R}$ and has on each site variables $\theta_2(\vec{R})$, $n_2(\vec{R})$ and $m_1(\vec{R})$. Notice that there are two types of bonds indicated in Figure 2. Both bonds have a coupling structure

$$-\pi\psi\cdot X\cdot\psi \tag{3.26}$$

For the type II bond, where X is a 2 by 2 matrix given by Eq. (2.5).

$$\psi_1 = \frac{\theta_1(\vec{r}+\hat{e}_1)-\theta_1(\vec{r})}{2\pi} - \ell_1(\vec{r}+\hat{e}_1,\vec{r})$$
$$\psi_2 = \frac{\theta_2(\vec{R}+\hat{e}_2)-\theta_2(\vec{R})}{2\pi} - \ell_2(\vec{R}+\hat{e}_2,\vec{R}) \tag{3.27a}$$

On the other hand, for the type I_ bond, we choose

$$\psi_1 = \frac{\theta_1(\vec{r}+\hat{e}_2)-\theta_1(\vec{r})}{2\pi} - \ell_1(\vec{r}+\hat{e}_2,\vec{r})$$
$$\psi_2 = \frac{\theta_2(\vec{R})-\theta_2(\vec{R}+\hat{e}_1)}{2\pi} - \ell_2(\vec{R},\vec{R}+\hat{e}_1) \tag{3.27b}$$

Here, $\hat{e}_1$ and $\hat{e}_2$ unit vectors in the x and y direction.

Now we are ready to repeat the calculation which led from Eq. (3.1) to Eq. (3.25). Start by choosing a gauge in which $\ell_1(\vec{r},\vec{r}')$ vanishes on type II and $\ell_2(\vec{r},\vec{r}')$ vanishes on type I bonds. Then, for both types of bonds, in this gauge,

$$\ell_k(\vec{r}+\hat{e}_2,\vec{r}) = -\sum_{\ell=o}^{\infty} m_k(\vec{r}+\hat{e}_2/2+(\ell+\tfrac{1}{2})\hat{e}_1) \tag{3.28}$$

Rewrite Eq. (3.6) for this new case. Denote by s a position variable which can take on either $\vec{r}$ or $\vec{R}$-values. Let, for example, $n(\vec{s})$ be $n_1(\vec{r})$ when $\vec{s}$ is on the first sublattice and be $n_2(\vec{R})$ when $\vec{s}$ is on the second. With this convention

$$Z[n,m] = \pi_s \int de(s) e^{Q_o+Q_1}$$

$$\begin{aligned} Q_o = & \sum_{\vec{s}} 2\pi i\, n(\vec{s})\, \phi(\vec{s}) \\ & - \pi X_o \sum_{\vec{s}} \Big\{ [\phi(\vec{s}) - \phi(\vec{s}+\hat{e}_1)]^2 \\ & + [\phi(\vec{s}) - \phi(\vec{s}+\hat{e}_2)]^2 \\ & + [\ell(\vec{s}+\hat{e}_2,\vec{s})]^2 \Big\} \end{aligned}$$

$$\begin{aligned} Q_1 = & 2\pi \sum_{\vec{r}} \ell_1(\vec{r}+\hat{e}_2,\vec{r}) \{X_o[\phi_1(\vec{r}+\hat{e}_2)-\phi_1(\vec{r})] - X_1[\phi_2(r+\frac{\hat{e}_2}{2}+\frac{\hat{e}_1}{2}) - \phi_2\left(r+\frac{\hat{e}_2}{2}-\frac{\hat{e}_1}{2}\right)]\} \\ & + 2\pi \sum_{R} \ell_2(\vec{R}+\hat{e}_2,\vec{R}) \{X_o[\phi_2(\vec{R}+\hat{e}_2)-\phi_2(\vec{R})] + X[\phi_1(R+\frac{\hat{e}_1}{2}+\frac{\hat{e}_2}{2}) - \phi_1(R+\frac{\hat{e}_2}{2}-\frac{\hat{e}_1}{2})]\} \end{aligned} \tag{3.29}$$

Repeat the previous analysis. Then one finds, once more, a coupling with a structure given by Eq. (1.1) with terms

$$A_o = -\sum_{\vec{r}} \int \frac{dq_x}{2\pi} \frac{dq_y}{2\pi} \ln\{X_o[4-2\cos q_x - 2\cos q_y]\} \tag{3.30}$$

an n-n coupling

$$V_{nn}(\vec{s},\vec{s}') = \delta_{j,k} X_o^{-1} V_R(\vec{s}-\vec{s}') \tag{3.31}$$

an m-m coupling

$$V_{mm}(\vec{s},\vec{s}') = \delta_{jk} \frac{X_o^2 - X_1^2}{X_o} V_R(\vec{s}-\vec{s}')$$

and an n-m coupling

$$V_{nm}(\vec{s},\vec{s}') = \begin{pmatrix} V_I(\vec{s}-\vec{s}')\ i & \frac{X_1}{X_o} V_R(\vec{s}-\vec{s}') \\ \frac{X_1}{X_o} V_R(\vec{s}-\vec{s}') & i\ V_I(\vec{s}-\vec{s}') \end{pmatrix} \tag{3.32}$$

IV. Representations for Specific Systems

In this chapter, we summarize the results of the previous chapters by describing how the electrodyamic representations work out for the standard problems described in Chapter 2.

A. The p-state case

Let $\sigma(\vec{r})$ take on integer values between 1 and p as described in Eq. (2.8a). Let the coupling between nearest neighbor σ's be that described by Eq. (2.15). Take the vertex function $h(\sigma)$ to be zero so that there is no breaking of the symmetry under the displacement of $\sigma(\vec{r})$ by an integer, q, i.e.

$$\sigma(\vec{r}) \rightarrow \sigma(\vec{r})+q$$

According to Eq. (2.14) this latter condition forces $n(\vec{r})$ to be p times an integer. For this reason we define

$$N(\vec{r}) = pn(\vec{r})$$
$$M(\vec{R}) = m(\vec{R}) \tag{4.1}$$

and say that the p-state model has a partition function which is given by Eqs. (1.1)-(1.3), i.e.

$$Z = \sum_{\{M,N\}} e^{A[N,M]} \tag{4.2a}$$

with

$$A[N,M] = -\infty, \text{ if } \sum_{\vec{r}} N(\vec{r}) \neq 0 \text{ or } \sum_{\vec{R}} M(\vec{R}) \neq 0 \tag{4.2b}$$

and, otherwise,

$$A[N,M] = V[N,M]+A_o \tag{4.2c}$$

In this case

$$V[N,M] = \frac{1}{2} \sum_{\vec{r}\vec{r}'} N(\vec{r}) V_R(r-r') N(\vec{r}')\; p^2/X$$

$$+ i \sum_{r,R} M(\vec{R}) V_I(\vec{R}-\vec{r}) n(\vec{r})\; p$$

$$+ \frac{1}{2} \sum_{\vec{R},\vec{R}'} M(\vec{R}) V_R(\vec{R}-\vec{R}') M(\vec{R}')\; \mathbf{X} \qquad (4.3)$$

In Eq. (2.24), we mentioned that the p-state model had a dual symmetry in which the partition function had a very simple transformation under the replacement of the coupling X by $\tilde{X} = p^2/X$. This symmetry is manifestly demonstrated in Eq. (4.3) as a symmetry of the entire problem under

$$N \leftrightarrow M$$
$$X \leftrightarrow p^2/X \qquad (4.4)$$

Therefore if the partition function has a singularity at some value of X, say X_c, then it also has a singularity at

$$\tilde{X}_c = p^2/X_c$$

For p=2 and 3, there is but one singularity at the self-dual point for which $X_c = \tilde{X}_c$ or

$$X = X^* = p \qquad (4.5)$$

B. Ashkin-Teller and 8-Vertex Models

For the Ashkin-Teller model, n and m have two components and p=2. Thus, we define $N(\vec{r})$ and $M(\vec{R})$ by

$$N_k(\vec{r}) = 2n_k(\vec{r})$$
$$M_k(\vec{R}) = m_k(\vec{R}) \qquad (4.6)$$

Here $\mathbf{X}$ is a two by two matrix of the form

$$X = X_o + X_1\tau_1$$

so that, in this case, instead of Eq. (4.3), we find

$$V[N,M] = \frac{1}{2}\sum_{\vec{r}\vec{r}'} N(\vec{r})\cdot V_R(\vec{r}-\vec{r}')(X_o - X_1\tau_1)N(\vec{r}')\,\frac{4}{X_o^2 - X_1^2}$$

$$+ 2i \sum_{\vec{r}\vec{R},k} N_k(\vec{r})V_I(\vec{r}-\vec{R})M_k(R)$$

$$+ \frac{1}{2}\sum_{\vec{R}\vec{R}'} M(\vec{R})\cdot V_R(\vec{R}-\vec{R}')(X_o + X_1\tau_1)\cdot M(\vec{R}') \qquad (4.7)$$

The duality transformation for this problem is

$$N_1(\vec{r}) \leftrightarrow M_1(\vec{R})$$

$$N_2(\vec{r}) \leftrightarrow -M_2(\vec{R})$$

$$X_o \leftrightarrow 4X_o/(X_o^2 - X_1^2)$$

$$X_1 \leftrightarrow 4X_1/(X_o^2 - X_1^2) \qquad (4.8)$$

For real values of X_1 in the region

$$X_1^2 < X_o^2$$

the problem has a critical line in which the behavior is a function of X_1. This line occurs at critical values defined by

$$(X_o^*)^2 - (X_1^*)^2 = 4 \qquad (4.9)$$

We can then parametrize the critical line by writing

$$X_o = 2 \cosh u$$

$$X_1 = 2 \sinh u \qquad (4.10)$$

so that on the critical line, V[N,M] takes the form

$$V[N,M] = \sum_{\vec{r}\vec{r}'} N(\vec{r})\cdot e^{-u\tau_1} V_R(\vec{r}-\vec{r}')\cdot N(r')$$

$$+ 2i \sum_{\vec{r}\vec{R}} N(\vec{r})\cdot V_I(r-R)M(\vec{R})$$

$$+ \sum_{\vec{R}\vec{R}'} M(\vec{R})e^{u\tau_1} V_R(\vec{R}-\vec{R}')M(\vec{R}') \tag{4.11}$$

A rather similar scheme holds for the 8-vertex model. In this case, $n_1(\vec{r})$ and $n_2(\vec{R})$ take on even integral values. Then define

$$N_1(\vec{r}) = 2n_1(\vec{r})$$

$$N_2(\vec{r}) = m_2(\vec{r})$$

$$M_1(\vec{R}) = m_1(\vec{R})$$

$$M_2(\vec{R}) = 2n_2(\vec{R}) \tag{4.12}$$

From this definition and Eqs. (3.31) - (3.33), we find

$$V[N,M] = \frac{1}{2} \sum_{rr'} N(\vec{r})\cdot V_R(\vec{r}-\vec{r}') \begin{pmatrix} \frac{4}{X_o} & i\,\frac{2X_1}{X_o} \\ i\,\frac{2X_1}{X_o} & \frac{X_o^2-X_1^2}{X_o} \end{pmatrix} \cdot N(\vec{r}')$$

$$+ 2i \sum_{\vec{r}\vec{R}} N(\vec{r})\cdot V_I(\vec{r}-\vec{R})M(\vec{R})$$

$$+ \frac{1}{2} \sum_{\vec{R}\vec{R}'} M(\vec{R})\cdot V_R(\vec{R}-\vec{R}') \begin{pmatrix} \frac{X_o^2-X_1^2}{X_o} & -i\,\frac{2X_1}{X_o} \\ -i\,\frac{2X_1}{X_o} & \frac{4}{X_o} \end{pmatrix} \cdot M(\vec{R}') \tag{4.13}$$

The self-duality and criticality condition for this model is once again Eq. (4.9). Once more the duality statement is simple representable by an interchange of N and M. In this critical domain, one can rewrite Eq. (4.13) in a form exactly parallel to that of Eq. (4.11) by using instead of (4.10) the parametrization:

$$X_o = \frac{2}{\cos \tilde{u}}$$

$$X_1 = 2 \tan \tilde{u} \tag{4.14}$$

which will once again ensure the satisfaction of Eq. (4.10). In this critical domain,

$$\begin{aligned} V[N,M] &= \frac{1}{2} \sum_{rr'} N(\vec{r}) V_R(\vec{r}-\vec{r}') e^{i\tilde{u}\tau_1} N(\vec{r}') \\ &+ 2i \sum_{rR} N(\vec{r}) V_I(\vec{r}-\vec{R}) M(\vec{R}) \\ &+ \frac{1}{2} \sum_{RR} M(\vec{R}) V_R(\vec{R}-\vec{R}') e^{-i\tilde{u}\tau_1} M(\vec{R}') \end{aligned} \tag{4.15}$$

C. Correlation Functions for the Ising Model

Eq. (2.12a) shows that in the Ising model a non-zero magnetic field generates terms with odd n. If there were no magnetic field, then only terms with even n would appear. Thus observation translates into a simple rule for calculating a multiple spin correlation function. Let us represent the Ising model in the standard way with $\sigma(\vec{r}) = \pm 1$, being the possible values of the on-site variable. Then according to Eq. (2.12) a coupling

$$\sum_r \sigma(\vec{r}) h(\vec{r})$$

translates into a vertex function

$$\sum_{\vec{r}} Y(n(\vec{r}),\vec{r})$$

with

$$e^{Y(h(\vec{r}),n(\vec{r})} = \begin{cases} 2 \cosh h(\vec{r}) & \text{for } n(\vec{r}) \text{ even} \\ -2 \sinh h(\vec{r}) & \text{for } n(\vec{r}) \text{ odd} \end{cases} \tag{4.16}$$

On one hand, we can calculate the multiple spin correlation function containing L spins as

$$C(\vec{r}_1,\vec{r}_2, \ldots \vec{r}_L) = \left< \prod_{i=1}^{L} \sigma(\vec{r}_i) \right>$$

$$= \left. \frac{\left\{ \prod_i \frac{\partial}{\partial h(\vec{r}_i)} \right\} Z[h]}{Z[h]} \right|_{h=o} \tag{4.17}$$

where $Z[h]$ is the partition function in the presence of a magnetic field. But, notice that Eq. (4.16) states

$$\left. \frac{\partial}{\partial h} e^{Y(h,n)} \right|_{h=o} = \begin{cases} 0 & \text{for } n \text{ even} \\ -1 & \text{for } n \text{ odd} \end{cases}$$

Consequently, we find that

$$C(\vec{r}_1 \ldots \vec{r}_L) = (-1)^L \frac{Z'}{Z} \tag{4.18}$$

Here Z is the partition function defined by Eqs. (4.2) and (4.3) and Z' is the corresponding partition function defined with some of the N-values displaced by half-integral values. (Remember that n odd implies N half-integral for the Ising model.) In particular

$$Z' = \sum_{\{M,N\}} e^{A[N+\Delta, M]} \tag{4.19}$$

with

$$\Delta(\vec{r}) = \frac{1}{2} \sum_i \delta_{\vec{r},\vec{r}_i} \tag{4.20}$$

This result can be generalized one step further. Recall that in the two-dimensional Ising model there exists a variable dual to $\sigma(\vec{r})$ called $\mu(\vec{R})$ (Kadanoff and Ceva [1971]).

Define a correlation function involving L-$\sigma(\vec{r})$ variables and I dual $\mu(\vec{R})$ variables as

$$C(\vec{r}_1 \dots \vec{r}_L, \vec{R}_1 \dots \vec{R}_I) = \langle [\prod_{i=1}^{L} \sigma(\vec{r})][\prod_{j=1}^{I} \mu(\vec{R}_j)] \rangle = (-1)^L Z'/Z \tag{4.21}$$

Because of the dual relation between N and M a natural definition of such correlation function is to replace (4.18) by

$$Z' = \sum_{\{M,N\}} e^{A[N+\Delta, M+\Lambda]} \tag{4.22}$$

with

$$\Lambda(\vec{R}) = \frac{1}{2} \sum_j \delta_{\vec{R},\vec{R}_j} \tag{4.23}$$

This definition of a σ-μ correlation function is precisely equivalent to the standard definition.

References

Anderson, P. W. and Yuval, G. [1969] Phys. Rev. Lett. 23, 89.

Ashkin, J. and Teller, E. [1943] Phys. Rev. 64 178.

Balian, R., Drouffe, J. and Itzykson, C. [1974] Phys. Rev. D10, 3396.

Balian, R., Drouffe, J. and Itzykson, C. [1975b] Phys. Rev. D11, 2104.

Balian, R., Drouffe, J. and Itzykson, C. [1977a] Phys. Rev. D11, 2098.

Banks, T., Myerson, R. and Kogut, J. [1977] preprint, June 1977.

Barber, M. N. and Baxter, R. J. [1973] J. Phys. C6, 2913.

Baxter, R. J. [1971] Phys. Rev. Lett. 26, 832.

Baxter, R. J. [1972] Ann. Phys. (N.Y.) 70, 193.

Fan, C. [1972] Phys. Lett. 39A, 136.

Fan, C. and Woo, F. Y. [1970] Phys. Rev. B2 723.

Jose, J., Kadanoff, L. P., Kirkpatrick, S., and Nelson (1977), Phys. Rev. B, to be published.

Kadanoff, L. P. [1977a] Annals of Physics (N.Y.) 100, 359.

Kadanoff, L. P. [1977b] Rev. Mod. Phys. 49, 267.

Kadanoff, L. P. and Ceva, H. [1971] Phys. Rev. B3, 3918.

Kadanoff, L. P. and Wegner, F. [1971] Phys. Rev. B4, 3989.

Knops, H. J. F., [1975] J. Phys. A8, 1508.

Kosterlitz, J. M. [1974], J. Phys. C7 181 [1974].

Kosterlitz, J. M. and Thouless, D. J., [1973], J. Phys. C6 1181.

Kosterlitz, J. M. and Thouless, D. J. [1977] Proceedings of Low Temperature Physics, to be published.

Kramers, H. A. and Wannier, G. H. [1941] Phys. Rev. 60 252.

Migdal, A. A. [1975a] Zh. Eksp. Theor. Fiz. 69, 810.

Migdal, A. A. [1975b] Ibid, 1457.

Savit, R. [1977] Phys. Rev. Lett. 39, 55.

Villain, J. [1975] J. Physique 36,581.

Wegner, F. [1971] J. Math. Phys. 12, 2259.

Wegner, F. [1972] J. Phys. C5, L181.

Wu, F. Y. [1977] J. Math. Phys. 18 611.

Wu, F. Y. and Lin, K. Y. [1974] J. Phys. C7, 181.

THE ONE-DIMENSIONAL FERMI GAS MODEL

J. Sólyom

Central Research Institute for Physics, Budapest, Hungary

ABSTRACT

A review of our present understanding of the properties of the one-dimensional Fermi gas model is presented. Exact results are known for the Tomonaga model and for a particular case of the backward scattering model. Scaling arguments are used to extend these results for general values of the backward scattering amplitude. The equivalence of the one-dimensional Fermi gas model to other models is shown and its consequences are explored. The limitations of scaling in view of the recent results for the soliton solutions are discussed. Extensions of the model to a set of coupled chains are also reviewed.

1. INTRODUCTION

In recent years we have witnessed a revival of interest in one-dimensional and two-dimensional systems. One reason for that is that physicists realized what was well known to chemists for a long time already that there exist in nature organic crystals built up of chains or stacks of large, flat molecules. The motion of electrons is confined mostly to these chains with rare hopping between the chains. Therefore the electronic properties of these systems are well described by one-dimensional models. A brief review of the models relevant to these systems is given in the preceding paper by H. Gutfreund.[1]

One of the models which has been widely studied is the one-dimensional (1-d) Fermi gas model. The electron-phonon coupling and the effect of phonons is neglected in this model. The electrons moving along a particular chain can interact with other electrons on the same chain only. In this talk I want to review briefly our present knowledge on this model and to point out those problems which are still unsolved and where the relation of this model to other 1-d and 2-d models can be used to clarify the situation.

First I will define the model for the strictly one-dimensional case (one chain problem). This model can be solved exactly for particular values only of the coupling constants. The main features of these solutions are presented here. Then a renormalization group argument is used to extend the results for general values of the coupling constants. Quite recent results about quantum solitons in these systems cast some doubt on the applicability of scaling argument for strong attractive coupling. This problem and the problem of the corresponding regions in equivalent models like 2-d Coulomb plasma, 2-d XY model etc. are discussed. Finally the extensions of the model to a set of coupled chains are also mentioned.

2. THE MODEL

The 1-d Fermi gas model is a model of weakly interacting fermions (electrons) confined to one spatial dimension. The dispersion relation of the free particles is approximated by

$$\varepsilon_k = v_F (|k| - k_F) . \qquad (1)$$

It consists of two branches with velocities v_F and $-v_F$ for the electrons around the Fermi points k_F and $-k_F$, respectively.

As a consequence of the two branches of the dispersion relation, there may be four basically different interaction terms as shown in Fig. 1.

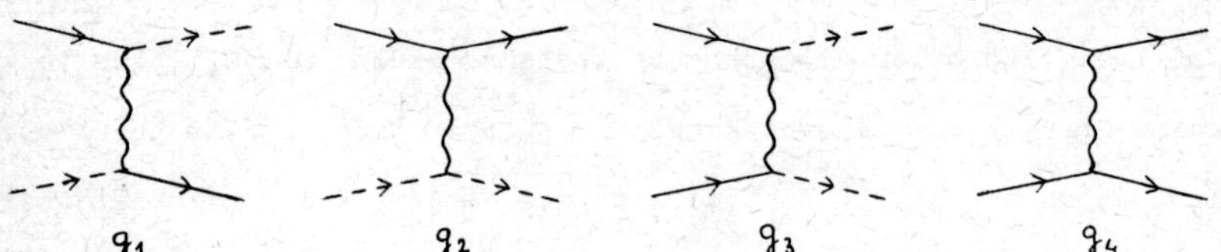

Fig. 1. Possible interaction processes.

The solid and dashed lines of the diagrams represent electrons with momenta around k_F and $-k_F$, respectively. The interaction terms with coupling constant g_1 and g_3 correspond to large momentum transfer, while the terms with g_2 and g_4 correspond to small momentum transfer. The term with g_3 is an Umklapp process and is important only if the band is half-filled. Though the physical situation usually corresponds to spin independent coupling, a spin dependence is often introduced formally. In this way the relation of this model to other models becomes more transparent. An index $\parallel$ is used if the spins of the electrons are parallel and an index $\perp$ if the spins are opposite.

For the precise definition of the model a cut-off has to be introduced. It is convenient to use a bandwidth cut-off. In this case all the momenta are restricted to a range around the Fermi points. Another usual cut-off procedure is to introduce a momentum transfer cut-off around 0 and $2k_F$ momentum transfers. This, however, is not sufficient (except for the Tomonaga model) and in general also a bandwidth cut-off has to be introduced. Different choices of the cut-off may lead to physically different behaviour of the systems.

3. EXACT SOLUTION OF THE TOMONAGA MODEL

The Tomonaga model is that particular case of the Fermi gas model when the large momentum transfer interactions, g_1 and g_3 are neglected. The interesting feature of the model is that its dynamics can be solved exactly. The Green's function and the response functions can be calculated exactly and they show non-trivial behaviour.

One way to perform the calculation is to sum up exactly all the diagrams using Ward identities which are valid only if there is no large momentum transfer. For the Green's function e.g. we get[2]

$$G(x, t>0) = \frac{1}{2\pi} \frac{1}{(x - v_F t + i/\Lambda)^{1/2} (x - ut + i/\Lambda)^{1/2}} \times \left[\Lambda^2 (x - ut + i/\Lambda)(x + ut - i/\Lambda)\right]^{\alpha}, \qquad (2)$$

where

$$u^2 = \left(v_F + \frac{1}{\pi} g_4\right)^2 - \left(\frac{1}{\pi} g_2\right)^2,$$

$$\alpha = \frac{1}{4u}\left(v_F + \frac{1}{\pi} g_4 - u\right),$$

and Λ is the momentum transfer cut-off.

Similar expressions can be obtained for the response functions mentioned in the preceding talk[1]. The charge density and spin density response functions behave in the same way, if the interactions are spin independent, the singlet and triplet pairing functions are again similar but their divergences appear in complementary regions of the coupling constants[3].

4. THE LUTHER-EMERY SOLUTION OF THE BACKWARD SCATTERING PROBLEM

It was a major progress in our understanding of the behaviour of the Fermi gas model when Luther and Emery[4] have shown that the model with backward scattering (large momentum transfer) can again be solved for a particular value of the other couplings. This solution uses extensively the representation of fermions in terms of boson operators, the charge density ρ_i and spin density σ_i operators:

$$\psi_{is}(x) \rightarrow \frac{1}{\sqrt{2\pi\alpha}} \exp\left\{\pm\left[ik_F x + \frac{\sqrt{2}\,\pi}{L}\sum_k \frac{e^{-\frac{\alpha}{2}|k|-ikx}}{k}\left(\rho_i(k) + s\sigma_i(k)\right)\right]\right\}, \tag{3}$$

where i=1,2 refer to the two branches of the free particle spectrum. This representation is exact if the limit $\alpha \rightarrow 0$ is performed at the end of the calculation. In the actual calculations of Luther and Emery α is identified with the cut-off and is kept finite. It is hoped that this does not lead to non-physical results.

Using this representation, the Hamiltonian splits into two commuting parts, one depending on σ_i only

$$\begin{aligned} H_\sigma &= \frac{2\pi}{L}\sum_k v_1\left[\sigma_1(k)\,\sigma_1(-k) + \sigma_2(-k)\,\sigma_2(k)\right] \\ &+ \frac{1}{L}\sum_k \left(g_{2\parallel} - g_{2\perp} - g_{1\parallel}\right)\sigma_1(k)\,\sigma_2(-k) \\ &+ \frac{g_{1\perp}}{(2\pi\alpha)^2}\int dx\left[\exp\frac{2\sqrt{2}\,\pi}{L}\sum_k \frac{e^{-\frac{\alpha}{2}|k|-ikx}}{k}\left(\sigma_1(k)+\sigma_2(k)\right) + h.c.\right], \end{aligned} \tag{4}$$

with $v_1 = v_F + \frac{1}{2\pi}(g_{4\parallel} - g_{4\perp})$, and another part H_ρ depending on the charge density ρ_i. H_ρ has a similar structure as H_σ with the replacements

$$\begin{aligned} &v_1 \rightarrow v_2 = v_F + \frac{1}{2\pi}\left(g_{4\parallel} + g_{4\perp}\right), \\ &g_{2\parallel} - g_{2\perp} - g_{1\parallel} \rightarrow g_{2\parallel} + g_{2\perp} - g_{1\parallel}, \\ &g_{1\perp} \rightarrow g_3 . \end{aligned} \tag{5}$$

The energy spectrum can be calculated exactly for the case when

$$g_{2\parallel} \mp g_{2\perp} - g_{1\parallel} = \frac{6}{5}\pi v_F ,$$

and a gap appears with the value

$$\Delta_\sigma = \frac{g_{1\perp}}{2\pi\alpha} \qquad \Delta_\rho = \frac{g_3}{2\pi\alpha} \tag{6}$$

Using this energy spectrum, the response functions can be calculated[4-7].

5. THE RENORMALIZATION GROUP APPROACH

We know the solution of the model for particular values of the couplings. The renormalization group or scaling arguments can be helpful to extend these results for other values[8].

A remarkable feature of this model is that under a renormalization transformation when the degrees of freedom near the cut-off are eliminated, no relevant new coupling appears (at least to third order in the renormalized couplings) and the scaling trajectories have a very simple form. There are two independent sets of scaling equations, one relating $g_{2\parallel}-g_{2\perp}-g_{1\parallel}$ and $g_{1\perp}$, the other relating $g_{2\parallel}+g_{2\perp}-g_{1\parallel}$ and g_3. This separation corresponds to the separation of H into H_σ and H_ρ. The corresponding scaling curves in the lowest and next approximation are shown in Fig. 2.

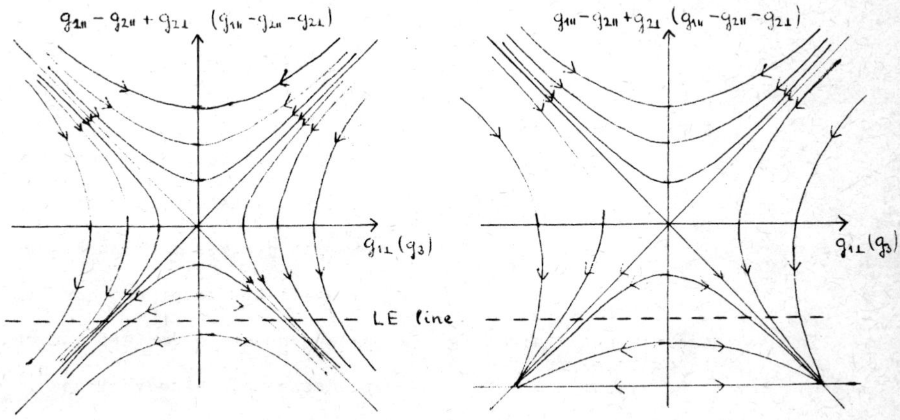

Fig. 2. Flow diagrams of the renormalized couplings in first and second order renormalization.

These scaling curves show that the region $g_{1\parallel}-g_{2\parallel}+g_{2\perp}\geq|g_{1\perp}|$ is scaled to $g_{1\perp}=0$, where the solution is known from the Tomonaga model. Outside of this region the behaviour is different, the problem is equivalent to the Luther-Emery model. The behaviour below the LE line is questionable because the approximate scaling equations are not valid there. I will return to this problem later on.

6. THE PHASE DIAGRAM

The aim of the study of the Fermi gas model is to understand what kind of instabilities may occur in the 1-d systems. These systems do not have a phase transition, instabilities may, however, appear at T=0 and they determine the character of the ground state.

I plot in Fig. 3 the phase diagram in the (g_1, g_2) plane[5,9]. In the phase diagram the possible ground state of the system is indicated for different values of the couplings. CDW and SDW refer to charge and spin density wave state, respectively, while SS and TS refer to singlet and triplet superconductivity, respectively. The phase diagram becomes much more complicated if the couplings are spin dependent or if Umklapp and g_4 processes are also included[7].

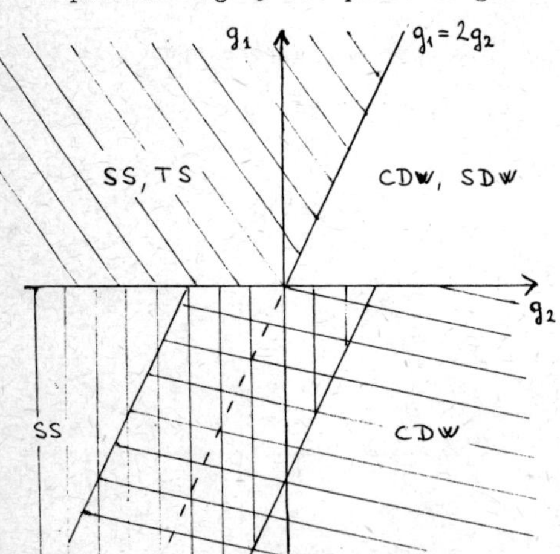

Fig. 3. Phase diagram indicating the possible ground state of the system.

7. BOUND SOLITONS IN THE FERMI GAS MODEL

We know already the behaviour of the system for couplings which are above the LE line. As I mentioned, the behaviour below the LE line may be different. In fact Luther[10] produced a remarkable solution for the energy spectrum in this region. This solution starts from a mapping of the 1-d X-Y-Z Hamiltonian onto an equivalent continuum Fermi gas model, which is the backward scattering model. Since the energy spectrum of the X-Y-Z Hamiltonian is known[11], the low lying energies of the backward scattering model can also be obtained. The result is that below the LE line bound states appear within the gap Δ with energies

$$\Delta_n = \Delta \sin \tfrac{1}{2} n\pi \left(\frac{\Theta}{1-\Theta}\right) , \quad n = 1, 2, \ldots < \Theta^{-1} - 1 , \tag{7}$$

where

$$\Theta = \left(\frac{1 + \frac{g_{1\parallel}}{2\pi v_F}}{1 - \frac{g_{1\parallel}}{2\pi v_F}} \right)^{1/2} .$$

This shows that if the equivalence is in fact exact, the solution below the LE line has new features which are not present above it and one cannot use scaling arguments in this region.

8. THE RELATION OF THE FERMI GAS MODEL TO OTHER MODELS

Starting from eq. (4) one can show after a few transformation steps[10] that this Hamiltonian is equivalent to the sine-Gordon Hamiltonian, which in turn is equivalent to the massive Thirring model[12].

A more useful analogy is obtained[13] if we write the partition function corresponding to H_σ in a series in powers of $g_{1\perp}$. Z will have the form

$$Z = \sum_n \frac{1}{(n!)^2} \left[\frac{g_{1\perp}}{(2\pi\alpha)^2} \right]^{2n} \int \left(\prod_i^{2n} d^2 r_i \right) \exp \left[\sum_{i>j} s_i s_j \, 2\Theta \ln \frac{|r_i - r_j|^2}{\alpha^2} \right] , \tag{8}$$

where s_i and s_j take values ± 1. This expression is the same as the partition function of a two-dimensional Coulomb gas with charges $s_i = \pm 1$, if the fugacity and the temperature are identified as

$$e^{-\beta\mu} = \frac{g_{12}}{(2\pi)^2} , \qquad \beta s^2 = 4\Theta . \tag{9}$$

The three different regimes of the Fermi gas model have their analogues in the Coulomb gas. These are the insulating region, the fully ionized plasma and a transition region inbetween. There is a lot of activity on the Coulomb gas and one hopes that the results obtained there will provide us with a better understanding of the Fermi gas model.

I want to mention in this respect what will be shown in the following talks that the 2-d XY model can also be related to the 2-d Coulomb gas[14] and thereby to the 1-d Fermi gas model.

9. THREE-DIMENSIONAL SYSTEM OF COUPLED CHAINS

Until now I have spoken of strictly 1-d systems only. The physically interesting systems are built up of chains and there is always a coupling between the chains. This coupling may stabilize an ordered phase at a finite temperature, but it may lead to a different type of order than what one would expect from the properties of the single chain.

In fact it turns out[15-18] that a Coulomb coupling between the chains leads always to a charge density wave state, the phase relations between the neighbouring chains are determined by the interchain couplings. A superconducting state can occur only if hopping between the chains is also taken into account[15,17]. In more realistic models the difference between the donor and acceptor chains should also be considered.

10. SUMMARY

In this talk I tried to review briefly the current status of the Fermi gas model. This model is interesting for solid state physicists because it allows to describe many features of the quasi-one-dimensional materials which are very intensively studied nowadays. On the other hand physicists working on statistical physical problems may find it interesting because it is related to several other problems, like the 2-d Coulomb gas, the 2-d XY model, the 1-d X-Y-Z model, the sine-Gordon equation and the Thirring model. A development in one of these theories may reveal something new for the other models. The next talks will mainly be concerned with the relation of these models and what we know about the XY model. It would be interesting to return to the problem how exact these equivalences are and what they really teach us.

1. H. Gutfreund, preceding paper.
2. I.E. Dzyaloshinsky and A.I. Larkin, Zh. Eksp. i Teor. Fiz. 65, 411 (1973).
3. H.C. Fogedby, J. Phys. C: Solid St. Phys. 9, 3757 (1976).
4. A. Luther and V.J. Emery, Phys. Rev. Lett. 33, 589 (1974).
5. P.A. Lee, Phys. Rev. Lett. 34, 1247 (1975).
6. H. Gutfreund and R.A. Klemm, Phys. Rev. B14, 1073 (1976).
7. V.N. Prigodin and Yu.A. Firsov, Zh. Eksp. i Teor. Fiz. 71, 2252 (1976).
8. N. Menyhárd and J. Sólyom, J. Low Temp. Phys. 12, 529 (1973).
9. J. Sólyom, J. Low Temp. Phys. 12, 547 (1973).
10. A. Luther, Phys. Rev. B15, 403 (1977).
11. J.D. Johnson, S. Krinsky and B. McCoy, Phys. Rev. A8, 2526 (1973).
12. S. Coleman, Phys. Rev. D11, 2088 (1975).
13. S.-T. Chui and P.A. Lee, Phys. Rev. Lett. 35, 315 (1975).
14. J. Zittartz, Z. Physik B23, 63 (1976).
15. L.P. Gorkov and I.E. Dzyaloshinsky, Zh. Eksp. i Teor. Fiz. 67, 397 (1974).
16. L. Mihály and J. Sólyom, J. Low Temp. Phys. 24, 579 (1976).
17. R.A. Klemm and H. Gutfreund, Phys. Rev. B14, 1086 (1976).
18. P.A. Lee, T.M. Rice and R.A. Klemm, Phys. Rev. B15, 2984 (1977).

PACS: 05.50.+q; 63.80.+n

SOME EXACT RESULTS ON LATTICE MODELS IN TWO DIMENSIONS*

F. Y. Wu

Department of Physics, Northeastern University

Boston, Massachusetts 02115, U.S.A.

ABSTRACT

Exact results on the Ashkin-Teller and Potts models in two dimensions and related problems are reviewed.

I would like to present an overview on some exact results on lattice models in two dimensions. Since Professor Simon and Professor Baxter have already given their reviews on exact and rigorous results for large classes of classical and quantum mechanical lattice systems, I shall speak of mainly two models that they did not specifically consider. These are the Potts and Ashkin-Teller models that have been of increasing recent interest. Some of the results to be mentioned are also valid in higher-than-two dimensions.

There have been many attempts in generalizing Onsager's Ising solution.[1] One direction of generalization is the Z-invariant model Professor Baxter described earlier in this Conference.[2] However, two of the first attempts of generalization of the Ising model, the Ashkin-Teller (AT)[3] and the Potts models,[4] have remained unsolved to this date. It has been recognized in recent years that these

*Work supported in part by NSF Grant No. DMR76-20643.

models are much more profound than they thought to be, and this had led to renewed interests in their considerations.

The AT model is a 4-state spin model of spin states, say, A, B, C and D. The nearest-neighbor interaction is ε_0, ε_1, ε_2 and ε_3 between, respectively, the same state, AB or CD, AC or BD, and AD or BC. In two dimensions the AT model possesses a duality relation [3,5] which, unfortunately, does not locate its critical point. It is now known that the AT model exhibits two phase transitions [6,7] and, consequently, the usual Kramers-Wannier argument does not hold for locating the transition point. The situation is best seen by construct its critical surface in the $(\omega_1, \omega_2, \omega_3)$ space where $\omega_i = \exp[(\varepsilon_0 - \varepsilon_i)/kT]$. A schematic plot of such a critical surface for the triangular lattice is shown in Fig. 1. As the temperature rises from 0 to ∞ the parameters $(\omega_1, \omega_2, \omega_3)$ trace a definite path from (0, 0, 0) to (1, 1, 1). A phase transition occurs in the AT model whenever the path intersects with the critical surface. It is seen that unless $\varepsilon_i = \varepsilon_j < \varepsilon_k$ for which the system is degenerate, the path always intersects the critical surface at two distinct points and, consequently, two phase transitions occur. Nothing is known about the nature of the transitions nor the exact shape of the critical surface. However, it has been established that the transition is continuous in two dimensions in the special case of $\varepsilon_1 = \varepsilon_2 = \varepsilon_3$.

The AT model has an Ising representation which, in two dimensions, can be used to transform the model into an eight-vertex model.[8] The exact relationship of this transformation for the square lattice is given in Table 1. In its Ising representation the AT model is composed of two <u>different</u> Ising lattices coupled via 4-spin interactions. This distinguishes it from the Baxter eight-vertex model whose Ising representation consists of the coupling of two <u>identical</u> Ising lattices. This has the consequence that the Baxter model possesses a unique transition.

The Potts model [4] is another generalization of the Ising model, to q states, with a nearest-neighbor interaction $-J\delta(\sigma,\sigma')$ between spin states σ and σ', where $\sigma,\sigma' = 1,2,\ldots,q$. In two dimensions the Potts model also possesses a duality relation which, in conjunction with the uniqueness assumption, locates the critical point. [4,9,10] This leads to the critical conditions

$$x = 1\ , \qquad \text{for the square lattice,}$$

and

$$\sqrt{q}\, x^3 + 3x^2 = 1\ , \qquad \text{for the triangular lattice.} \tag{1}$$

Here $x = v/\sqrt{q}$ and $v = e^K - 1$. The uniqueness of the transition for $q > 4$ has since been rigorously established [11] so Eq. (1) is now exact. Again, not much is known about the nature of transition except that, in two dimensions, the transition has a latent heat for $q > 4$ and is continuous for $q \leq 4$. [10,12]

The Potts model in two dimensions can be formulated as a six-vertex problem. [13] The exact transformation for a square lattice is given in Table 1.

It was pointed out by Kasteleyn and Fortuin [14] that the Potts model generates other well-known lattice statistical problems. They showed that the bond percolation is generated from the Potts model with nearest-neighbor interactions. To complete the description we introduce an external field to the interactions between, say, states 1 so that the Hamiltonian reads [15]

$$-\mathcal{H}/kT = K \sum_{\langle i,j\rangle} \delta(\sigma_i,\sigma_j) + L \sum_i \delta_{ij1}\ , \tag{2}$$

where $\delta_{ij1} = 1$ if $\sigma_i = \sigma_j = 1$, and $\sigma_{ij1} = 0$ otherwise. Define the generating function

$$h(v,L) = \left[\frac{\partial}{\partial q} \lim_{N_b\to\infty} N_b^{-1} \ln Z\right]_{q=1}\ , \tag{3}$$

where Z is the partition function defined by (2) and N_b is the total number of edges in the lattice. We then obtain for the <u>bond</u> percolation

$$S(p) = \text{the mean square of cluster size}$$

$$= \left[\frac{\partial^2}{\partial L^2} h(ve^{-L}, L)\right]_{L=0+} \tag{4}$$

with $p = 1 - e^{-K} = v/(1 + v)$.

To generate the site percolation on a lattice one considers a Potts model on its covering lattices with a Hamiltonian[16]

$$-\mathcal{H}/kT = K \sum_{\text{site}} \delta_{ij\ldots\ell} + L \sum_{\text{site}} \delta_{ij\ldots\ell 1} \quad . \tag{5}$$

The Hamiltonian (5) describes a Potts model with many-body interactions. One may then derive (4) and the expression[16]

$$P(p) = \text{the percolation probability}$$

$$= 1 - p^{-1}\left[\frac{\partial}{\partial L} h(ve^{-L}, L)\right]_{L=0+} \quad , \tag{6}$$

where h is defined by (3) for (5) with the total number of sites N in place of N_b.

The tree problems on a lattice are generated by the Potts model in the $q \to 0$ limit.[14] We mention here only one exact results valid in two dimensions: There exists a one-to-one correspondence between the spanning trees on <u>any</u> planar graph G and the dimer configurations on a related planar graph G^*.[17] Since the latter can be counted by standard means, this correspondence leads to an exact counting of the spanning trees on any (finite or infinite) planar graph. In particular, this leads to the number of spanning trees on lattices which was first evaluated from the $q \to 0$ limit of the Potts partition function.[18] The relation between G^* and G is as follows: The site set of G^* contains all sites of G, except any one site on the boundary, all sites of D, the dual of G, and the midpoints of the edges of G. The edge set of G^* consists of all half edges of G and the lines connecting the sites of D to the midpoints of the surrounding edges of G. An example is shown in Fig. 2.

To conclude this review, we summarize in Table 1 the vertex model formulations of the AT and Potts models as compared with the Baxter 8-vertex model for a square lattice. We see that both the AT and the Potts models will be solved if we can solve a more general staggered eight-vertex problem. As a matter of fact, as indicates in the Table, the staggered model also includes the long outstanding problem of Ising model in an external fields as a special case.[19] Evidently, a major breakthrough will be needed for making progress along this direction.

Baxter	a	b	c		d
AT	a	b	c d		d (Sublattice A) c (Sublattice B)
Potts	1 x	x 1	$1+xe^{\theta}$ $x+e^{\theta}$	$1+xe^{-\theta}$ $x+e^{-\theta}$	0 (Sublattice A) (Sublattice B)
Ising in a field	0 e^{-K}	e^{-K} 0	$e^{K-L/2}$ $e^{K+L/2}$	$e^{K+L/2}$ $e^{K-L/2}$	0 (Sublattice A) (Sublattice B)

Table 1. Vertex weights for the Baxter, Ashkin-Teller, Potts model and the Ising model in an external field.

AT: $a = (\omega_0 + \omega_1)/\sqrt{2}$, $b = (\omega_2 - \omega_3)/\sqrt{2}$,

$c = (\omega_2 + \omega_3)/\sqrt{2}$, $d = (\omega_0 - \omega_1)/\sqrt{2}$.

Potts: $x = (e^{K} - 1)/\sqrt{q}$, $\cosh\theta = \sqrt{q}/2$.

Ising: $K = J/kT$, $L = H/kT$.

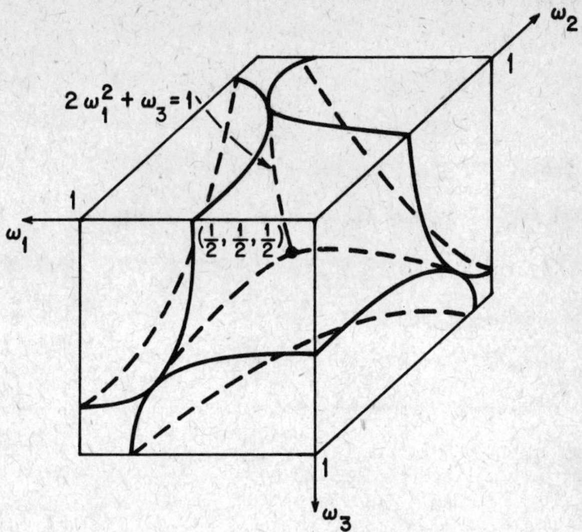

Fig. 1 Schematic plot of the critical surface for the triangular AT model. [There is a misprint in Ref. 7. Equation (9) in Ref. 7 should read $\omega_i^2 + \omega_k = 1$.]

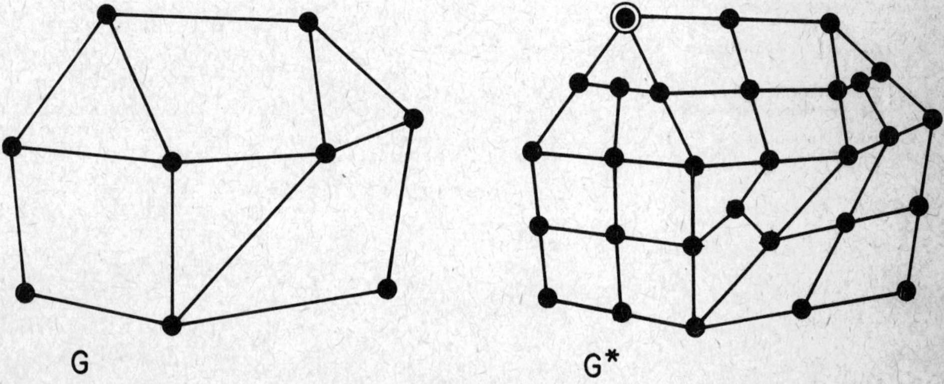

Fig. 2 The spanning tree lattice G and the dimer lattice G^*. The circled site on the boundary of G^* is deleted.

1. L. Onsager, Phys. Rev. 65, 117 (1944).

2. R. J. Baxter, this volume.

3. J. Ashkin and E. Teller, Phys. Rev. 64, 178 (1943).

4. R. B. Potts, Proc. Camb. Phil. Soc. 48, 106 (1952).

5. F. Y. Wu, J. Math. Phys. 18, 611 (1977).

6. F. Y. Wu and K. Y. Lin, J. Phys. C 7, L181 (1974).

7. F. Y. Wu, J. Phys. C 10, L23 (1977).

8. F. Y. Wu, J. Math. Phys. 18, 611 (1977).

9. D. Kim and R. I. Joseph, J. Phys. C 7, L167 (1974).

10. R. J. Baxter, H. N. V. Temperley and S. E. Ashley, preprint.

11. A. Hintermann, H. Kunz and F. Y. Wu, to be published.

12. R. J. Baxter, J. Phys. C 6, L445 (1973).

13. R. J. Baxter, S. B. Kelland and F. Y. Wu, J. Phys. A 9, 397 (1976).

14. P. W. Kasteleyn and C. M. Fortuin, J. Phys. Soc. Japan 26 (Suppl.) 11 (1969).

15. F. Y. Wu, to be published in J. Stat. Phys.

16. H. Kunz and F. Y. Wu, to be published.

17. I am indebted to Professor H. N. V. Temperley for first pointing out to me this correspondence for the square lattice.

18. F. Y. Wu, J. Phys. A 10, L113 (1977).

19. F. Y. Wu and K. Y. Lin, Phys. Rev. B12, 419 (1975).

PANEL DISCUSSION : INFRARED PROBLEMS IN LOW- DIMENSIONAL SYSTEMS

J. Zittartz, Universität Köln, Germany

In my contribution to this discussion I would like to emphasize the unifying aspects in the description of physical properties of low-dimensional systems. It has turned out that in many different situations, some of them already discussed in this panel, one observes the same mathematical structure in physical quantities, although the physics described by them is sometimes quite different due to the interpretation of the parameters involved.

Phase transitions are characterized by the occurence of singularities in quantities such as for instance the free energy. Now low-dimensional systems do not tend to interact cooperatively or to order as well as systems in higher dimensions, however, once they do so, they usually display a richer manifold of singular behaviour. The lack of ordering is for instance expressed by those theorems which state that a continuous symmetry is <u>not</u> broken spontaneously in low dimensions, i.e. at a continous symmetry we have no infinitly long range order and no order parameter. Nevertheless one finds critical behaviour as described by power law singularities of the form $\lambda^{\varkappa}$ where λ is some field (temperature, magnetic field, coupling constant) and the singular exponent $\varkappa$ usually varies continuously with other parameters of the system. This is in contrast to the situation in usual phase transitions in higher dimensions where the critical exponents mostly assume universal values.

The explanation for such $\lambda^{\varkappa}$-behaviour which very often can be interpreted as a phase transition of continuous order $\varkappa$ is what I might call a " persistent infrared problem ". Consider a large system with Hamiltonian

$$H = H_o + \lambda H'$$

where for $\lambda = 0$ H_o displays a relevant continuous symmetry which is lifted by H' (example: magnetic field λ in a Heisenberg ferromagnet). H_o then has low-lying excitations (no gap) with spectrum $\varepsilon(q) \sim q^{\sigma}$ $(q \to 0)$ and some dispersion coefficient σ . H' will introduce a gap of order λ . Now in the calculation of some physical quantity by expansions one encounters d-dimensional integrals with energy denominators

$$\int_0 \frac{d^d q}{\varepsilon(q,\lambda)} \, |\text{matr. el.}|$$

Such integrals may lead to singularities in λ ($\lambda \to 0$) which for $d \neq \sigma$ are usually algebraic and depend only on d and σ (universality). For $d = \sigma$, however, one finds logarithms of the form

$$\varkappa \cdot \ln \lambda \Rightarrow \ln \lambda^{\varkappa}$$

where the prefactor $\varkappa$ depends on system parameters. If one is lucky the expansions sum up to exponential series where upon the ln's disappear and we are left with power law singularities $\lambda^{\varkappa}$. Where do we find such a " persistent infrared problem" ? Except for special situations we mostly have to consider only two dispersions $\sigma = 1,2$ such that we are led to look at

a) $\sigma = 1$: 1-dimensional quantum systems: massless fermions and bosons,

b) $\sigma = 2$: 2-dimensional classical systems.

Both situations are in fact intimately connected because of the equivalence of a 1-d quantum field theory and a 2-d (classical) Euclidean field theory.

Next I would like to give a survey over some 1-d and 2-d situations which display the same mathematical infrared behaviour, i.e. physical quantities show $\lambda^{\varkappa}$-behaviour, and by proper identification of these parameters we can map one situation on another as far as infrared behaviour is concerned. There are differences, however, not only in the different physical interpretations of physical quantities (temperature in one case may turn out to be a magnetic field in another case), but also differences in related properties. Some of these differences will be pointed out in the following. Let us start with

(A) 2-d classical models: most notably the 8-vertex model (including as special cases the 6-vertex models, Ising-model, etc.) which have been discussed by Dr. Wu. By a suitable transfer matrix representation these models can be related to the

(B) 1-d spin-$\frac{1}{2}$-chain. This model is described by the anisotropic couplings $J_{x,y,z}$, and many properties are known due to the work of Baxter and others. The relevant continuous symmetry shows up for $J_x = J_y > J_z$, the parameter $\lambda = J_x - J_y$ and as $\lambda \to 0$ one has a singular exponent $\varkappa$ in the free energy depending on J_z: $\varkappa(J_z)$. The groundstate for $\lambda \neq 0$ is degenerate (2-fold) giving rise to (Ising-type) ordering. As $\lambda \to 0$ this ordering disappears, but the system stays critical as expressed by the $\lambda^{\varkappa}$-infrared behaviour.

By the Jordan-Wigner transformation we can go over to fermions and consider

(C1) 1-d lattice fermions which model again is related to other (continuum) fermion models, namely

(C2) the massive Thirring model (including the Tomonaga-Luttinger version) and

(C3) the 1-d conductors, discussed by Drs. Gutfreund and Solyom in this panel.

In the (B) and (C) models we have the same infrared behaviour, there are, however, important differences. In contrast to (B) the groundstate for the fermion (C) models is unique, thus no spontaneous ordering, in the situation where the 4-fermion coupling $J_z < J_{x,y}$. Furthermore the (C) models have one-particle (fermion, soliton) excitations which are absent in the spin $\frac{1}{2}$ chain (B). The relationship of (C1) to (C2,3) is not yet clear; the important difference is that charge is conserved in (C2,3) while the (C1) lattice fermions only conserve the quantity $(-1)^Q$ where Q is the charge.

The next step is to go over from Fermions to Bosons. By " Bosonization", discussed before in this panel, one maps the (C) fermion models on the

(D) 1-d Sine-Gordon model. The Euclidean version of this model (on a lattice) is then the

(E) 2-d classical Harmonic Rotator with Hamiltonian

$$\beta H = \sum_{ij} \beta J_{ij} (\varphi_i - \varphi_j)^2 - \sum_i B \cos \varphi_i$$

the angles ranging within $-\infty < \varphi_i < \infty$. This is most easily interpreted as a ferromagnet with spin components $S_i^{x,y} = (\cos \varphi_i , \sin \varphi_i)$ on a 2-d lattice. The continuous symmetry at $B = 0$ (common spin rotation) is lifted by the magnetic field $\lambda = B$. The free enery shows a leading singular term $B^{\varkappa}$ where for $\eta < 4$ ($\eta = \frac{T}{2\pi J}$) the singular exponent $\varkappa = \frac{4}{4-\eta}$ ranges from 1 to ∞, thus displaying a phase transition of continuous order as $B \rightarrow 0$. I like to point out that this harmonic rotator is interesting also in other dimensions. In 1 dimension it is a perfect confinement model (1-d Coulomb plasma, Lenard) and for $d = 3,4,\ldots$ it shows a usual phase transition as $B \rightarrow 0$ with spontaneous magnetization, i.e. the continuous symmetry is broken.

Lastly we can expand the partition function of (E) in powers of B. The resulting expression can then be interpreted as the grand canonical partition function of a

(F) 2-d classical Coulomb plasma. This (neutral) plasma model has been discussed quite frequently in the recent literature as it displays logarithmic infrared behaviour in such an obvious way. In fact the 2-d Coulomb potential is logarithmic

$$V(r) = \pm e^2 \ln \sqrt{\frac{a^2+r^2}{a^2}} \qquad \text{(with soft core cut off)},$$

i.e. the long-range ln-behaviour between relevant degrees of freedom is built into this model already in the model Hamiltonian which makes the 2-d plasma an infrared system par excellence.

The list of related situations which I have discussed may of course be continued. Most notably one can relate the harmonic rotator to the usual rotator or classical XY-model. The discussion of this model, however, is left to the other panelists.

INTERNATIONAL SCIENTIFIC ORGANIZING COMMITTEE

PROF. M. YA. AZBEL
TEL AVIV UNIV.
ISRAEL

PROF. D.D. BETTS
UNIVERSITY OF ALBERTA
CANADA

PROF. JAN DE BOER
UNIV. VAN AMSTERDAM
NETHERLANDS

PROF. M.J. BUCKINGHAM
UNIV. WESTERN AUSTRALIA
AUSTRALIA

PROF. H.B. CALLEN
UNIV. OF PENNSYLVANIA
U.S.A.

PROF. CHARLES P. ENZ
UNIVERSITE DE GENEVE
SWITZERLAND

PROF. MICHAEL E. FISHER
CORNELL UNIVERSITY
U.S.A.

PROF. H. FROHLICH
LIVERPOOL UNIVERSITY
U.K.

PROF. P.G. DE GENNES
COLLEGE DE FRANCE
FRANCE

PROF. DR. H. HAKEN
UNIVERSITAT STUTTGART
GERMANY

PROF. P.C. HEMMER
UNIV. I TRONDHEIM
NORWAY

PROF. G. HORWITZ
HEBREW UNIVERSITY
ISRAEL

PROF. MARK KAC
ROCKEFELLER UNIVERSITY
U.S.A.

PROF. LEO P. KADANOFF
BROWN UNIVERSITY
U.S.A.

PROF. R. KUBO
UNIVERSITY OF TOKYO
JAPAN

PROF. C.G. KUPER
TECHNION
ISRAEL

PROF. P.T. LANDSBERG
SOUTHAMPTON UNIVERSITY
U.K.

PROF. VENIAMIN G. LEVICH
SOVIET ACAD. OF SCIENCES
U.S.S.R.

PROF. SHNEIOR LIFSON
WEIZMANN INST.SCIENCE
ISRAEL

PROF. M. LUBAN
BAR-ILAN UNIVERSITY
ISRAEL

PROF. T. MATSUBARA
UNIVERSITY OF KYOTO
JAPAN

PROF. P. MAZUR
LEIDEN UNIVERSITY
NETHERLANDS

PROF. I. PRIGOGINE
UNIV. LIBRE DE BRUXELLES
BELGIUM

PROF. D. RUELLE
INST. HAUTES ETUDES
FRANCE

PROF. A.J.F. SIEGERT
NORTHWESTERN UNIVERSITY
U.S.A.

PROF. P. SZEPFALUSY
EOTVOS UNIV.
HUNGARY

PROF. R. THIEBERGER
BEN-GURION UNIVERSITY
ISRAEL

PROF. L.E.H. TRAINOR
UNIVERSITY OF TORONTO
CANADA

PROF. ALEXANDER VORONEL
TEL-AVIV UNIVERSITY
ISRAEL

PROF. DR. F.J. WEGNER
UNIVERSITAT HEIDELBERG
GERMANY

LOCAL ORGANIZING COMMITTEE

PROF. AMNON AHARONY
TEL-AVIV UNIVERSITY
ISRAEL

PROF. D.J. AMIT
HEBREW UNIVERSITY
ISRAEL

DR. LUCIEN BENGUIGUI
TECHNION
ISRAEL

DR DARIO CABIB
TECHNION
ISRAEL

DR SHALOM ELIEZER
SOREQ NUCL. RES. CENTRE
ISRAEL

DR MOSHE FIBICH
TECHNION
ISRAEL

PROF. MOSHE GITTERMAN
BAR-ILAN UNIVERSITY
ISRAEL

PROF. JOSEPH IMRY
TEL-AVIV UNIVERSITY
ISRAEL

PROF. C.G. KUPER
TECHNION
ISRAEL

PROF. RAPHAEL D. LEVINE
HEBREW UNIVERSITY
ISRAEL

PROF. ILYA PRIVOROTSKII
TECHNION
ISRAEL

PROF. MICHAEL REVZEN
TECHNION
ISRAEL

DR ILAN RIESS
TECHNION
ISRAEL

PROF. AMIRAM RON
TECHNION
ISRAEL

PROF. LAWRENCE SCHULMAN
TECHNION
ISRAEL

PROF. ISRAEL SENITZKY
TECHNION
ISRAEL

DR BORIS SHAPIRO
TECHNION
ISRAEL

DR CHANOCH WEIL
TECHNION
ISRAEL

CHAIRMEN OF SESSIONS

DR. K.F. BERGGREN
LINKOPING UNIVERSITY
SWEDEN

PROF. D.D. BETTS
UNIVERSITY OF ALBERTA
CANADA

PROF. JAN DE BOER
UNIV. VAN AMSTERDAM
NETHERLANDS

PROF. M.J. BUCKINGHAM
UNIV. WESTERN AUSTRALIA
AUSTRALIA

PROF. CHARLES P. ENZ
UNIVERSITE DE GENEVE
SWITZERLAND

PROF. MELVILLE S. GREEN
TEMPLE UNIVERSITY
U.S.A.

PROF. L. GUNTHER
TUFTS UNIVERSITY
U.S.A.

PROF. P.C. HEMMER
UNIV. I TRONDHEIM
NORWAY

PROF. JOSEPH IMRY
TEL-AVIV UNIVERSITY
ISRAEL

PROF. B. JANCOVICI
UNIVERSITE DE PARIS SUD
FRANCE

DR. P.W. KASTELEYN
LEIDEN UNIVERSITY
NETHERLANDS

DR. R. KLEMM
IOWA STATE UNIV.
U.S.A.

PROF. C.G. KUPER
TECHNION
ISRAEL

PROF. J. LAJZEROWICZ
GRENOBLE
FRANCE

PROF. P.T. LANDSBERG
SOUTHAMPTON UNIVERSITY
U.K.

PROF. J.S. LANGER
CARNEGIE-MELLON UNIV.
U.S.A.

PROF. SHNEIOR LIFSON
WEIZMANN INST. SCIENCE
ISRAEL

PROF. M. LUBAN
BAR-ILAN UNIVERSITY
ISRAEL

PROF. H. MORI
KYUSHU UNIVERSITY
JAPAN

PROF. M. POLLAK
UNIV. CALIF. RIVERSIDE
U.S.A.

DR. E. PRAESTGAARD
UNIVERSITY OF COPENHAGEN
DENMARK

PROF. ILYA PRIVOROTSKII
TECHNION
ISRAEL

PROF. W.F. SAAM
OHIO STATE UNIVERSITY
U.S.A.

PROF. ISRAEL SENITSKY
TECHNION
ISRAEL

DR. G.L. SEWELL
QUEEN MARY COLLEGE
U.K.

PROF. P. SZEPFALUSY
EOTVOS UNIV.
HUNGARY

PROF. A.E.A. THELLUNG
UNIVERSITAT ZURICH
SWITZERLAND

PROF. FLAVIO TOIGO
UNIV. DI PADOVA
ITALY

DR. J. VILLAIN
D.R.F./D.N./C.E.N.G.
FRANCE

DR. F.W. WIEGEL
TWENTE UNIV. OF TECH.
NETHERLANDS

CO-CHAIRMEN OF SESSIONS

[All co-chairmen have as their affiliation: Technion, Israel]

MR. Y. AFEK
MR. D. AGMON
MR. L. BERNY
MR. D. BRAUNSHTEIN
MR. J. CANTOR
MR. Y. DANA
MR. A. ENGLANDER
MR. F. FISHMAN
MR. G. GALECZKI
MR. I. GERTNER
MR. D. GOTTLIEB
MS. C. HERSCOVICI
DR. V. KOGAN
MR. D. MAOR
MR. E. RIBAK
DR. RELI ROSMAN
MS. M. SECHTER
MR. U. SHRETER
MR. M. STERNHEIM